Studies in Logic
Volume 73

Measuring Inconsistency in Information

Volume 63
Argumentation and Reasoned Action. Proceedings of the 1st European Conference on Argumentation, Lisbon 2015. Volume II
Dima Mohammed and Marcin Lewiński, eds

Volume 64
Logic of Questions in the Wild. Inferential Erotetic Logic in Information Seeking Dialogue Modelling
Paweł Łupkowski

Volume 65
Elementary Logic with Applications. A Procedural Perspective for Computer Scientists
D. M. Gabbay and O. T. Rodrigues

Volume 66
Logical Consequences. Theory and Applications: An Introduction.
Luis M. Augusto

Volume 67
Many-Valued Logics: A Mathematical and Computational Introduction
Luis M. Augusto

Volume 68
Argument Technologies: Theory, Analysis and Appplications
Floris Bex, Floriana Grasso, Nancy Green, Fabio Paglieri and Chris Reed, eds

Volume 69
Logic and Conditional Probability. A Synthesis
Philip Calabrese

Volume 70
Proceedings of the International Conference. Philosophy, Mathematics, Linguistics: Aspects of Interaction, 2012 (PhML-2012)
Oleg Prosorov, ed.

Volume 71
Fathoming Formal Logic: Volume I. Theory and Decision Procedures for Propositional Logic
Odysseus Makridis

Volume 72
Fathoming Formal Logic: Volume II. Semantics and Proof Theory for Predicate Logic
Odysseus Makridis

Volume 73
Measuring Inconsistency in Information
John Grant and Maria Vanina Mrtinez, eds.

Studies in Logic Series Editor
Dov Gabbay   dov.gabbay@kcl.ac.uk

Measuring Inconsistency in Information

Edited by

John Grant

and

Maria Vanina Martinez

© Individual author and College Publications 2018
All rights reserved.

ISBN 978-1-84890-244-2

College Publications
Scientific Director: Dov Gabbay
Managing Director: Jane Spurr

http://www.collegepublications.co.uk

Printed by Lightning Source, Milton Keynes, UK

All rights reserved. No part of this publication may be reproduced, stored in a retrieval system or transmitted in any form, or by any means, electronic, mechanical, photocopying, recording or otherwise without prior permission, in writing, from the publisher.

In remembrance of the three people who were with me from the beginning and who will stay with me to the end

My grandmother Margaret 'Mama' (d. 1955)
My father Eugene 'Pop' (d. 1968)
My mother Caroline 'Mom' (d. 2016)

Love cannot be measured

John Grant

Contents

    Preface xiii

    List of Contributors xix
        List of Authors xix
        List of Reviewers xx

1   **From Measuring Infinities to Measuring Inconsistencies** 1
    1.1   Student Days 1
    1.2   From Mathematics to Computer Science via Logic 3
    1.3   Three Classifications for Inconsistent Theories 7
    1.4   Comparing the Inconsistency of Databases 11
    1.5   In Maryland 15
    1.6   Measuring Inconsistencies versus Measuring Infinities 16
    1.7   Book on Measuring Inconsistency 18
            References 18

2   **On the Evaluation of Inconsistency Measures** 19
    2.1   Introduction 19
    2.2   Preliminaries 21
    2.3   Inconsistency Measures 22
    2.4   Rationality Postulates 30
    2.5   Expressivity 37
    2.6   Computational Complexity 40
    2.7   Discussion 42
            References 43
            Appendix: Proofs of Technical Results 47

3   **Measuring Inconsistency in Argument Graphs** 61
    3.1   Introduction 61
    3.2   Review of Abstract Argumentation 63

		3.2.1	Extension-based Semantics	64
		3.2.2	Labelling-based Semantics	66
		3.2.3	Subsidiary Definitions	67
	3.3	Inconsistency Measures for Argument Graphs		70
		3.3.1	Graph Structure Measures	71
		3.3.2	Graph Extension Measures	76
	3.4	Logic-based Instantiations		80
		3.4.1	Review of Deductive Argumentation	80
		3.4.2	Degree of Undercut	84
		3.4.3	Application of Logic-based Measures of Inconsistency .	89
		3.4.4	Related Work	93
	3.5	Resolution through Commitment		94
	3.6	Discussion		99
		References		100
4	**Inconsistency Measures for Disjunctive Logic Programs Under Answer Set Semantics** 105			
	4.1	Introduction		105
	4.2	Preliminaries		107
	4.3	Inconsistency Measures		110
		4.3.1	Measures based on Strong Inconsistency	111
		4.3.2	Distance-based Measures	112
		4.3.3	Modification-based Measures	114
	4.4	Rationality Postulates for Inconsistency Measures		116
		4.4.1	Weakening Monotonicity	117
		4.4.2	Further Postulates	122
	4.5	Compliance with Rationality Postulates		124
	4.6	Computational Complexity		131
		4.6.1	The General Case	134
		4.6.2	Disjunction-free Programs	135
		4.6.3	Stratified Programs	136
		4.6.4	Classical Disjunction-free Programs	138
	4.7	Summary and Discussion		140
		References		142
5	**Inconsistency Measures in General Fuzzy Logic Programming** 147			
	5.1	Introduction		147
	5.2	Preliminary Definitions		149
	5.3	Inconsistency of General Residuated Logic Programs		153
	5.4	Measures of Instability		154
	5.5	Measures of Incoherence		157
		5.5.1	Coherence	157
		5.5.2	Measures of Incoherence on L-interpretations	159

		5.5.3	Measures of Incoherence on Stable General
			Residuated Logic Programs 162
	5.6	Conclusions and Future Work............................ 164
		References... 165

6 Inconsistency Measures in Hybrid Logics 169
  6.1 Introduction 169
    6.1.1 An Overview of Paraconsistency in Logical Systems .. 170
  6.2 Paraconsistency in Hybrid Logic 171
    6.2.1 Quasi-Hybrid Multimodal Logic 172
    6.2.2 Measure of Inconsistency......................... 182
  6.3 Paraconsistency in Other Constituents of Hybrid Logic....... 188
    6.3.1 Quasi-Hybrid Multimodal Logic with Quantifiers..... 188
    6.3.2 The Role of Inconsistency in Nominals.............. 190
    6.3.3 Adding Inconsistency to the Accessibility Relation ... 190
  6.4 Conclusions and Further Work 191
  References... 192

7 Inconsistency Measuring over Multisets of Formulas 195
  7.1 Introduction 195
  7.2 Knowledge Bases as Finite Multisets of Logical Formulas 197
    7.2.1 Inference and Consistency for Multisets of Formulas .. 198
  7.3 Postulates for Inconsistency Measures over Sets of Formulas .. 200
    7.3.1 Form of Postulates 200
    7.3.2 Features about "Set"-postulates 202
  7.4 Inconsistency Measures over Multisets of Formulas 206
    7.4.1 Multiset Postulates do not Need to be Guarded 206
    7.4.2 An Example of an Enduring Failure of an
            Inconsistency Measure 207
    7.4.3 Equivalently Using Multiset Sum or Multiset Difference 208
  7.5 Detailing Inconsistency Postulates for Multisets of Formulas .. 209
    7.5.1 Core Postulates: Consistency Null, Unit Monotony,
            Dominance 209
    7.5.2 Removal Postulates: Free Formula Dilution and
            Penalty 210
    7.5.3 Normalization Postulates 211
    7.5.4 Independence Postulates 211
    7.5.5 Postulates on Conjunction 212
    7.5.6 Additivity 213
    7.5.7 Other Postulates................................ 214
    7.5.8 In General 214
  7.6 Conclusion 215
  References... 216

8 Inconsistency Measures and Paraconsistent Consequence ... 219
- 8.1 Introduction ... 219
- 8.2 Brief Reflections on Consequence and Trivialization ... 221
- 8.3 A Broader View of Consequence ... 223
- 8.4 Logical Explosion and Preservationism ... 224
- 8.5 Motivating Paraconsistent Logics in General ... 226
- 8.6 Examples and Discussion ... 227
- 8.7 Conclusion ... 231
- References ... 231

9 Inconsistency Measurement in Probabilistic Logic ... 235
- 9.1 Introduction ... 235
- 9.2 Preliminaries ... 237
- 9.3 Rationality Postulates ... 238
- 9.4 Approaches ... 243
  - 9.4.1 Classical Approaches ... 243
  - 9.4.2 Distance-based Approaches ... 246
  - 9.4.3 Violation-based Approaches ... 248
  - 9.4.4 Dutch-book Measures ... 252
  - 9.4.5 Fuzzy-logic Measure ... 255
  - 9.4.6 Entropy Measures ... 255
- 9.5 Applications ... 259
- 9.6 Summary ... 264
- References ... 265

10 Measuring Database Inconsistency ... 271
- 10.1 Introduction ... 271
- 10.2 Preliminary Concepts ... 272
  - 10.2.1 Databases, Updates and Integrity Constraints ... 272
  - 10.2.2 Update Checkers ... 275
  - 10.2.3 Syntactic and Semantic Restrictions ... 276
- 10.3 Database Inconsistency Measures ... 276
  - 10.3.1 Definition of Database Inconsistency Measures ... 276
  - 10.3.2 Examples of Database Inconsistency Measures ... 277
- 10.4 Applications of Database Inconsistency Measures ... 279
  - 10.4.1 Inconsistency-tolerant Integrity Checking ... 280
  - 10.4.2 Inconsistency-tolerant Repairs ... 281
  - 10.4.3 Inconsistency-tolerant Repair Checking ... 286
- 10.5 Mathematical and Classical Measures versus Database Inconsistency Measures ... 296
  - 10.5.1 Definitions ... 296
  - 10.5.2 Syntax and Semantics ... 297
  - 10.5.3 Inconsistency ... 298
  - 10.5.4 Postulates ... 299

	10.5.5	Applications 300
10.6		Conclusion 301
		Appendix: Index of Definitions 303
		References 304

11 On Measuring Inconsistency in Spatio-Temporal Databases . 313
- 11.1 Introduction 313
- 11.2 Spatio-Temporal Databases 315
  - 11.2.1 Syntax 315
  - 11.2.2 Semantics 317
- 11.3 Transforming an **ST** Database to a Propositional Knowledge Base 318
- 11.4 Classical Inconsistency Measures for **ST** Databases 321
  - 11.4.1 The Definition of Inconsistency Measure 321
  - 11.4.2 Counting Minimal Inconsistent Subsets 321
  - 11.4.3 Using a Hitting Set 322
  - 11.4.4 Using the Sizes of Inconsistent Subsets 322
  - 11.4.5 Counting the Problematic Formulas 324
  - 11.4.6 Counting the Maximal Consistent Subsets 325
  - 11.4.7 Using a 3-valued Logic 325
- 11.5 New Inconsistency Measures for **ST** Databases 327
  - 11.5.1 Some Rationality Postulates 327
  - 11.5.2 Measuring Inconsistency along the Object and Time Dimensions Individually 329
  - 11.5.3 Measuring Inconsistency along the Object and Time Dimensions Together 331
  - 11.5.4 Measuring Inconsistency along the Space Dimension .. 332
  - 11.5.5 Measuring Inconsistency by Combining the Three Dimensions 335
  - 11.5.6 A Tractable Algorithm to Compute $\mathcal{I}_{OT}$ 337
- 11.6 Conclusion and Future Work 339
- References 340

Preface

The world of information surrounds us in modern society even more perhaps than the world of nature. But the information we can access in many cases contains contradictions. It can be difficult to make good sense out of inconsistent information. In fact, in classical logic from a single inconsistency everything can be concluded; hence it is impossible to use such reasoning in the presence of an inconsistency. However, intuitively we know that not all inconsistencies have the same effect in our view of the world. Once we accept the inevitability of inconsistency and recognize that inconsistency is not good, we may try to at least reduce the amount of inconsistency as much as possible. After all, the less inconsistency there is, the more confidence we may have in our information. But then we need to have a way of measuring inconsistency.

The concept of measuring inconsistency was proposed by one of us (John Grant) in a 1978 paper in the context of first-order logic. Unfortunately, at that time this idea did not fit into any of the areas of interest to researchers. A small number of logicians, primarily philosophers, developed paraconsistent logics that could handle inconsistencies to some extent. But the emphasis was on the rules and semantics of such logics. From the point of view of people working with the storage and retrieval of information, primarily in databases within computer science and information systems, the main issue was the efficient use of the limited storage available at that time as well as the efficient accessing of the data.

By the early 2000s massive amounts of information were stored in large databases. Some AI researchers recognized that dealing in an intelligent way with inconsistencies is unavoidable and developed an interest in finding good ways to measure inconsistency. The initial proposals were primarily in the context of representing information as propositional knowledge bases. The problem turned out to quite complicated and numerous proposals were presented. This led to the study of desirable properties that an inconsistency measure should possess. In the last few years the work was extended to more complex formalisms as well as other considerations such as the expressive range of inconsistency measures and complexity issues.

The aim of this book, the first one on inconsistency measures, is to bring together what has been done so far and point the way to new issues and problems that need to be solved. We hope that it will serve as a starting point for researchers interested in this subject and that it will lead to further research as a result of its content. To start the process of putting such a book together we contacted most of the top researchers in this field as well as some researchers in various aspects of the logical representation of information to probe their interest in contributing to this volume. We are happy that some of them, in several cases with co-authors, were able to write a chapter. In fact every author and co-author is a superb researcher whose contribution we appreciate. We were also fortunate in getting an outstanding group of reviewers, including most of the chapter authors, but also others who were not able to write a chapter but were willing to help. Having read all the chapters and reviews we know that the comments and suggestions of the reviewers substantially improved the book's content. The list of reviewers follows the preface.

The first two chapters differ from the following nine chapters in an important way. The first chapter explains how John Grant originally came up with the idea of comparing inconsistency in information in analogy with the 19th century mathematical discovery of the possibility of comparing the size of infinite sets. The second chapter, by Matthias Thimm, is a masterful survey that covers much of the work on inconsistency measures up to 2017. In fact we are grateful to Matthias not only for writing this critically important chapter but also for being the co-author of two other chapters. The fact of the matter is that we would not have been able to put this book together without his extraordinary effort. The other nine chapters connect inconsistency measures with argumentation, disjunctive logics, fuzzy logics, hybrid logics, multiset representation, paraconsistent logics, probabilistic logics, (relational) databases, and spatio-temporal databases. The rest of the Preface briefly summarizes the main contribution of each chapter individually.

In Chapter 1 John Grant explains how he came to recognize at the beginning of the information age an important difference between mathematical structures and information structures. While mathematical objects such as groups, graphs, and arbitrary relational structures are completely known and consistent, that is not the case for information stored in a database, for instance. So it becomes relevant to consider comparing the inconsistency of two structures such as databases. But in the initial 1978 paper he presented his measures in the framework of first-order logic. Three methods for comparing inconsistency were given there including an absolute measure and a relative measure as well as many mathematical results about them. In this chapter he shows what he had in mind for measuring inconsistency in relational databases. He also gives a possible explanation for the many inconsistency measures devised by researchers using the concept of dimension and proposes its wider use.

In Chapter 2 Matthias Thimm summarizes much of the work on inconsistency measures for propositional knowledge bases. He defines, explains, and illustrates with examples 22 different inconsistency measures proposed by various researchers and discusses compatibilities among them. Given the variety of inconsistency measures and the need for comparing them, some researchers proposed properties that such measures should satisfy. He then defines, explains, and contrasts 18 properties, called rationality postulates, and gives a comprehensive list of which inconsistency measures satisfy which postulates. The expressivity of an inconsistency measure refers to its abilty to distinguish between different knowledge bases and that depends on the size of the range for that measure. Four parametrized measures of expressivity are given and their values are computed for all 22 measures. Finally, a comprehensive listing of the complexity of several computational problems involving inconsistency measures is presented.

In Chapter 3 Anthony Hunter proposes a general framework for measuring inconsistency in abstract argumentation. Argumentation is a natural topic for inconsistency measures; after all, opposing arguments are inconsistent with one another. A common approach in argumentation uses an argument graph where the nodes are arguments and the edges represent opposition between arguments. Seven inconsistency measures as well as seven rationality postulates are proposed for argument graphs. Furthermore, three additional measures are introduced using argument graph extensions and their properties and compatibilities are investigated. Then two approaches are considered in detail for measuring inconsistency in deductive argumentation. The work is illustrated by several illustrative examples. In particular, it is shown how inconsistency measures for argument graphs can be used for inconsistency resolution in argumentation.

In Chapter 4 Markus Ulbricht, Matthias Thimm, and Gerhard Brewka introduce inconsistency measures for disjunctive logic programs under the answer set semantics. This task is challenging because of the non-monotonicity of answer set semantics. Even the basic notion of minimal inconsistent subset no longer applies and must be modified. They present six new inconsistency measures in this framework. The concepts used are minimal strong inconsistent subsets, various distance-based measures, and measuring the effort to repair an inconsistent program. Examples are given for the computation of these measures. They also introduce new rationality postulates that reflect the non-monotonicity of the answer set semantics. Then they study the compliance of all the new measures with respect to the new postulates. Finally, they provide a thorough study of the complexity of the computational problems involving the new inconsistency measures and include special cases where various restrictions are placed on the rules.

In Chapter 5 Nicolás Madrid and Manuel Ojeda-Aciego introduce inconsistency measures for general fuzzy logic programs. The chapter starts with the definition of general residuated logic programs with fuzzy answer set semantics. Inconsistency in this context means that there are no fuzzy answer sets. The two possible reasons for inconsistency are the lack of stable models, called instability, and the incoherence of all the stable models. Then measures of instability are given using the amount of information that has to be added or removed to recover stability. This is followed by measures of incoherence. Both types measure the amount of inconsistency. All these concepts are illustrated by examples.

In Chapter 6 Diana Costa and Manuel Martins introduce inconsistency measures for hybrid logics, a formalism for specifying relational structures with the ability to name worlds (states), including accessibility relations between them, and to assert equalities. This is accomplished by adding to modal logic a new class of atomic formulas called nominals and a new operator called the satisfaction operator. In analogy with quasi-classical logic, a paraconsistent quasi-hybrid multimodal logic is defined. The many subtle issues involving both the syntax and the semantics are carefully explained. Both a measure of inconsistency and a weighted measure of inconsistency are defined for models. These are illustrated on an example involving the path of a patient subject to contradictory diagnoses. The chapter ends with a discussion of inconsistency in several cases: in the presence of quantifiers, inconsistency involving nominals, and inconsistency in the accessibility relation.

In Chapter 7 Philippe Besnard considers the case of inconsistency measures where the knowledge base is a multiset of formulas, that is, where formula duplication is allowed. In that sense this work generalizes the standard concept of an inconsistency measure. For example, the set of minimal inconsistent subsets becomes the multiset of minimal inconsistent subsets. The emphasis in the chapter is on the rationality postulates in the presence of multisets. He presents a careful study of the differences between the postulates in the usual case of sets and the case of multisets. A method is presented that can be used to convert a postulate (for sets) to its multiset version. Several examples are used to illustrate the issues and the process. Altogether a multiset version of each of 18 postulates is given.

In Chapter 8 Bryson Brown explores connections between paraconsistent logics and inconsistency measures. An important aspect of such connections is obtained by generalizing the concept of consequence relation. In classical logic the idea of consequence is that it preserves truth, that is, if all the premises are true the consequence must also be true. One can also go in the opposite direction: if the consequence is false at least some combination of premises must be false also. In order to deal with the presence of inconsistency, the preservationist approach to consequence relations defines some property of the premises that can be preserved without getting into a logical explosion. Using this approach he shows how inconsistency measures relate to three paraconsistent logics.

In Chapter 9 Glauber De Bona, Marcelo Finger, Nico Potyka, and Matthias Thimm primarily survey the research done in recent years on inconsistency measures in probabilistic logics. Many rationality postulates for inconsistency measures in propositional knowledge bases can be simply extended to probabilistic logic. But there is a new property, continuity, that requires smoothness in the behavior of an inconsistency measure. Then the extension of inconsistency measures for the classical case to include probabilities is explored. A different type of inconsistency measure deals with the violations of numerical constraints. Examples are used to illustrate the computation of these measures and their properties in terms of rationality postulates are presented. New results involving Dutch book measures, measures using fuzzy concepts, and entropy are also presented. The chapter ends with several applications of the new inconsistency measures to analyze inconsistent knowledge bases.

In Chapter 10 Hendrik Decker connects inconsistency measures with integrity checking, relaxing repairs, and repair checking for relational and deductive databases. For databases the main significance of inconsistency measures is the comparison of the inconsistency before and after an update. Inconsistency-tolerant integrity checking and repair checking are discussed and their connections are presented in detail. This is followed by a thorough discussion of the differences between classical logic and datalog and how that relates to inconsistency measures. It turns out that in general database inconsistency measures do not have to computed: an integrity checker can be run to obtain the difference in the inconsistency between database states and thereby support a decision to accept or reject a proposed update. All the concepts used in the chapter are carefully explained and the appropriate proofs are given.

In Chapter 11 John Grant, Maria Vanina Martinez, Cristian Molinaro, and Francesco Parisi extend the concept of inconsistency measure to spatio-temporal databases. They use a syntax that allows statements expressing the presence of an object in a region of space at a time value. They then show that a database containing such statements, called an ST database, can be transformed to an equivalent propositional knowledge base. This way all the classical inconsistency measures naturally extend to ST databases. For ST databases they also define new measures that do not have a classical counterpart. These measure inconsistency along the dimensions of objects, space, and time. Their properties in terms of rationality postulates are also investigated. An illustrative example is used throughout to demonstrate the computation of all considered inconsistency measures.

We are happy to acknowledge everyone who helped make this book a reality. We are grateful to Dov Gabbay for accepting our proposal for the book. Jane Spurr's help throughout the process was invaluable. Each chapter author and reviewer made an important contribution. We hope that this book will create more interest and research in the exciting and important topic of measuring inconsistency in information.

John Grant and Maria Vanina Martinez
January 2018

List of Contributors

Authors

Philippe Besnard
Gerhard Brewka
Bryson Brown
Diana Costa
Glauber De Bona
Hendrik Decker
Marcelo Finger
John Grant
Anthony Hunter
Nicolás Madrid
Maria Vanina Martinez
Manuel A. Martins
Cristian Molinaro
Manuel Ojeda-Aciego
Francesco Parisi
Nico Potyka
Matthias Thimm
Markus Ulbricht

Peer Reviewers

Philippe Besnard
Yaxin Bi
Gerhard Brewka
Bryson Brown
Hendrick Decker
Marcelo Finger
Lluis Godo
John Grant
Anthony Hunter
Yue Ma
Nicolás Madrid
Maria Vanina Martinez
David McAreavey
Cristian Molinaro
Manuel Ojeda-Aciego
Francesco Parisi
Nico Potyka
Gerardo I. Simari
Guillermo R. Simari
Matthias Thimm
Guohui Xiao

From Measuring Infinities to Measuring Inconsistencies

John Grant

University of Maryland, College Park, USA
grant@cs.umd.edu

Abstract

In this chapter I write about the beginning of my scholarship and how I came up with the idea of measuring inconsistency. I also give an informal overview of my 1978 paper, [1]. I skip the major mathematical complications that are the result of dealing with infinite models and show how the definitions of the three classifications are simplified when only finite models are considered. Then I present a small database example for illustration. I also briefly discuss how I got back to working on inconsistency measures many years later. Finally, I explore the use of dimensions as they relate to measuring size, show some relationship with measuring inconsistency, including the reason for the existence of many inconsistency measures, and suggest using the dimension concept to measure inconsistency in other contexts.

1 Student Days

Throughout my school years I had many wonderful teachers. By the time I started high school I decided that I wanted to be a teacher also. But I didn't know what I wanted to teach and at what level. As I was going through high school I considered the second issue first. I decided to teach at the highest possible level, at a university. My parents didn't finish high school so they did not have experience at that level, but I found out that to teach at a university requires a Ph.D. degree. The harder part was figuring out what I wanted to teach because I liked so many subjects. I used to go to the Public Library where I took out books on various subjects in order to find out more about them. I recall one time reading a book on mathematics where I was amazed to learn that mathematicians had been able to show that infinity is not one thing but that there are many different infinities.

I liked and did very well in mathematics but it was this discovery, probably more than anything else, that made me study mathematics for my career. I wanted to understand how it is possible to determine that there are many different infinities that can be compared for size. After high school I went to the City College of New York during some of the last years before tuition fees were imposed, which was very fortunate because my family was poor. I was a mathematics major and in addition to taking the required mathematics courses I studied on my own some topics not covered there. I was particularly disappointed that in my courses infinity was treated as one unique concept with the special symbol ∞. But in my extracurricular readings I found out that infinities were studied in a field called mathematical logic. Unfortunately, my college did not offer a course on that topic. However, during my years in college I found out that a major problem about infinite sets, the Continuum Hypothesis, had just been shown to be independent of set theory.

After graduation I entered the Ph.D. program in mathematics at New York University, The Courant Institute of Mathematical Sciences, on an assistantship that was later upgraded to a fellowship. The institute had just moved into a new building on Mercer Street and as a graduate assistant I shared a nice office with another student. Finally I was able to take a course in mathematical logic, but much of the emphasis at the institute was on classical aspects of pure and applied mathematics such as differential equations. In addition to taking courses and studying for comprehensive exams I started reading more about mathematical logic.

One field in particular, model theory, a relatively new area at that time, caught my attention. The basic idea of model theory is the study of properties of very general mathematical structures, called relational structures. A specific relational structure consists of an arbitrary nonempty set, called the domain, and may have relations, functions, and constants. Relational structures are studied using the language of first-order logic. A specific language is obtained by adding symbols for the relations, functions, and constants of the relational structure to the fundamental symbols of first-order logic. A set of sentences, formulas without free variables, closed under the consequence relation of first-order logic is called a theory. A relational structure in which all the sentences of a theory are true is called a model of the theory. For example, an early result states that if a relational structure M is a model for a universal theory T, where a universal theory is the set of logical consequences of a set of sentences where for each sentence all the quantifiers are universal in prenex normal form, then every substructure of M is a model of T. Intuitively, a universal sentence makes a statement about all the elements in the domain; hence for any subset of the domain the universal sentence remains true.

Within model theory I became interested in the work of Abraham Robinson, one of the founders of model theory. He applied concepts of model theory to obtain new results in algebra; he invented non-standard analysis; and around that time introduced the concept of forcing, discovered to show the independence of the Continuum Hypothesis, into model theory. I was very fortunate in graduate school to have the distinguished mathematician, logician, and computer scientist, Martin Davis, as my advisor. He suggested that I contact Professor Robinson who was at Yale University at that time. I did so and he invited me to meet with him; I did so several times. To this day, Abraham Robinson remains an inspiration to me: not only was he a brilliant mathematician and logician but he was also patient and kind to me. His explanations clarified some concepts that I had difficulty learning on my own and gave me great insight into his thinking.

2 From Mathematics to Computer Science via Logic

The 1960s were a great time for getting academic positions. New universities were built; old universities expanded; colleges were turned into universities. This was due to the recognition of the value of higher education for the masses and the consequent large influx of students. There were not enough Ph.D.s to satisfy the demand for faculty. And then, from one year to the next, the academic job market turned sour in 1970, the year I obtained my Ph.D. degree. With great difficulty eventually I got one job offer from the Department of Mathematics at the University of Florida. I took the job but realized after one year that I could not stay there on a long term basis. The emphasis in the department was on writing many papers and working with graduate students primarily in pure mathematics. My interests were in teaching and in expanding my knowledge of topics I had not previously studied, such as mathematics education and modern aspects of applied mathematics. Unfortunately, those activities were of little interest there.

The 1970s were the time when colleges and universities were establishing computer science programs and departments, although some pioneering institutions, including NYU, started earlier. In 1971 the University of Florida created a Department of Computer and Information Sciences but there were not enough faculty to teach even the small number of courses offered. I was asked to teach a course on computability in the Spring of 1972 in this new department. I had never taken any computer science courses: I don't think any were offered when I was in college. My background in computer programming consisted of two short non-credit courses in the mid 1960s. In 1965 during my senior year in college I took a course in MAD and used it to write a program for testing Newton's Method in a Numerical Analysis course. My professor was impressed. Then, in 1966, when I started graduate school, I took a course in FORTRAN but never used it or followed up on it during my graduate studies. Actually, I did more complex programming for Turing machines in my graduate mathematical logic class.

Thus it was not my programming experience, but my knowledge of mathematical logic, namely recursive function theory and computability, that I could apply. During the next several years the university hired a department chairman and several faculty members in computer science. It was a small department and I got to know everyone well. During this time I realized that the future for me was in computer science, not pure mathematics. So I decided to study those aspects of computer science I knew nothing about such as data structures and the theory of programming languages. My many years of studying mathematics on my own, going back to high school, made it possible for me to absorb this material quickly. Soon I was able to help even with the development of the undergraduate curriculum.

In 1974 I transferred from the Department of Mathematics to the Department of Computer and Information Sciences. By that time it wasn't just a matter of learning about computer science; I had to master the concepts well enough to teach the courses. Over the next several years I taught the key courses in the curriculum: introductory programming in PL/I and COBOL, data structures, survey of programming languages, discrete structures, compilers, data base management systems, and computability. I also advised students, was in charge of inviting visitors to campus, and represented the department at the university, state, and regional levels. The passage of time makes me realize what a great privilege it was for me to meet some of the great computer pioneers: Gene Amdahl, John Atanasoff, Grace Hopper, and Daniel McCracken. In retrospect I see the 1970s as the greatest time for teaching computer science. Computers were new and exotic objects; most students in the computer science courses never saw a computer before coming to campus. I remain enormously grateful to the chairman of the department, George Haynam, who gave me such a wonderful opportunity.

In those days I had litle time for research, but I hoped to apply my knowledge of logic to gain insight into aspects of computer science. I had previously seen the use of computers in mathematics using logic: my advisor, Martin Davis, was one of the pioneers in mechanical theorem proving. I also knew about the resolution method in mechanical theorem proving from a class I took with Professor Davis in 1967 but did not do any work in that field.

Because of my knowledge of model theory I started to think about how information, in general, could be represented by relational structures. Unfortunately I didn't know that Ted Codd had recently defined the concept of a relational database, spelled data base in those days, including the relational calculus language using first-order logic. My goal became sketching out how relations could be used to represent information and how representing information might be different from representing mathematical objects.

For a simple example, consider group theory. Groups can be represented mathematically as sets with a binary function (product), a unary function (inverse), and a constant (identity) that satisfy certain rules. Mathematicians typically deal with classes of groups although in some cases individual groups are studied. If the set is finite and fairly small the group can be given by writing out its multiplication table, the table of inverses, and the identity element. Actually, not all of this has to be given, as for example the table of inverses yields immediately the identity element. If the set is large it may not be possible to write out these tables but there has to be a way to find the products and inverses. Such a mathematical object can be represented by a relational structure in several equivalent ways. For example, for the product we can use a binary function symbol or a ternary relation symbol.

My goal was to represent data processing files as relational structures. I realized that a file could be represented as a relation, except that while in mathematics only one set, the domain, is involved, in a file there are different domains, for example for names and dates. But that can be handled by many-sorted logic whose properties are essentially the same as standard first-order logic. After some time I realized that there are some more significant differences between the relational structures representing mathematical objects and my idea of relational structures representing information. First, some information may be unknown. For example, in a file storing data about people, someone's birthday may be missing. I called this incomplete information. Actually, this sort of thing could happen in mathematics as well. Suppose that the product table for a group is written on a page where some of the entries are unreadable or a portion of the page is missing. But mathematicians did not appear to have an interest in exploring the case where information was missing.

The second difference I considered was even more radical than the first. It occurred to me that information may be inconsistent. For example, going back to the file storing data about people, someone might be listed with two different birthdates. I called this inconsistent information. It seemed to me at the time that incomplete information might occur in a mathematical structure, but not inconsistent information. So, for instance, if a "group" had two different results for the product of two elements, it would not be a group. Nevertheless, given a relational structure with the appropriate operations that is not a group, it might be possible to define a distance to measure how "far" it is from a group. Such an inconsistency measure would allow us to compare the closeness of relational structures of the appropriate type to being a group. But, as far as I know, this type of issue is not of interest to mathematicians.

I plan to revisit my work on incomplete information from the 1970s in the future. The focus for this chapter is my work on inconsistency at that time. I should note here that by the time I got to dealing with inconsistency in a serious way I was aware of Codd's work on relational databases. As incompleteness and inconsistency are in a way dual concepts, it is useful to consider the two ways that I expanded the notion of a relational structure. Suppose that R is a relation symbol and a and b are constant symbols in the language under consideration. In a relational structure exactly one of $R(a, b)$ and $\neg R(a, b)$ must be true. I extended the concept of relational structure to an incomplete structure by changing "exactly" to "at most" in the above statement. So in an incomplete structure it is possible for neither $R(a,b)$ nor $\neg R(a,b)$ to be true. Then I extended the concept of a relational structure in the other direction to an inconsistent structure by changing "exactly" to "at least" in that previous statement. Thus in an inconsistent structure it is possible that both $R(a,b)$ and $\neg R(a,b)$ are true.

At this point it is useful to recall what I knew about inconsistency at that time. In my mathematics classes I learnt that a single inconsistency logically implies that every statement is provable. This means that in a proof, once an inconsistency is shown we are finished: getting the contradiction is all that is needed. This is very convenient because it allows proof by contradiction, among other things. In particular, I learnt that showing the irrationality of $\sqrt{2}$ can be done by assuming that it is rational and then using some mathematical results to obtain the negation of the assumption. The critical issue about inconsistency was that it is a single concept that I accepted as obvious. This belief may be a good reason why mathematicians are not interested in investigating "inconsistent" groups, for instance.

But, as I thought about inconsistency in information I realized that it makes sense to consider some information more inconsistent than some other information. Consider the example where the inconsistency is due to two different birthdates for one person and this happens just one time. It seemed to me that such a database should be considered less inconsistent than one where the data contains multiple conflicting birthdays for many people. At some point I remembered how fascinated I was about 15 years earlier when I found out that there are different infinities and their sizes can be compared. I was elated to have discovered something analogous to that case: namely, that there can be different amounts of inconsistency in different chunks of information (databases). My next step was trying to figure out how to do the comparison.

Measuring the size of an infinite set and thereby comparing the sizes of infinite sets uses the concept of an injective function. In particular, if there is an injective (1-1) function $f : A \to B$ then $|A| \leq |B|$ where $|.|$ is used for the size of a set. A bijective function between two sets indicates equal size. This is the way that the sizes of finite sets are compared and the concepts are then extended to infinite sets. Thus for the size of sets there is exactly one way to do the comparison. For this reason I wanted to find the one right way to compare the inconsistency of information. To my surprise I could think of several reasonable ways to compare the inconsistency between two chunks of information. That was when I realized that comparing inconsistencies is a more complex issue than comparing infinities.

Actually, first I needed to figure out exactly what it was about inconsistency in information that I was going to compare. I have already defined the concept of an inconsistent structure; so perhaps I should compare the amount of inconsistency in such structures. That is not what I decided to do for the following reason. All the information I saw in the files that constituted a database was positive information. The data may include a person's birthdate but not days that were not birthdates; there was no negative information. In that sense the concept of an inconsistent structure did not conform to the reality of the storage of information. The reason that we consider two different dates for a person to be inconsistent is additional information: there is a functional dependency $Id \to Birthdate$, where the individual is identified by Id, that does not allow more than a single birthday for each individual.

Therefore I decided that in order to measure inconsistency I had to deal with a set of statements in general; not just tuples in a relation. Actually this contains the previous case because we can consider each tuple as a statement of fact, so a database by itself without integrity constraints is also a set of statements. As I was thinking about all these issues, in 1975, it seemed to me that the people working in databases at that time, mostly data processing people and engineers, would not understand what I was doing and would dismiss my research as irrelevant. I thought that I might have a better chance at acceptance if I presented my work as a logic paper. But the journal rejections and reviewer comments eventually convinced me that researchers in mathematical logic had no interest in my work. I was extremely fortunate that one influential person, Professor Bolesław Sobociński, who was the editor of the Notre Dame Journal of Formal Logic, recognized the value of my pioneering research. I believe that I would not have been able to publish my logic papers, including the one on incomplete models, without his support. To this day I remain very grateful to him.

3 Three Classifications for Inconsistent Theories

In my 1978 paper, [1], written mostly in 1975, I proposed three classifications as measures of inconsistency for theories in first-order logic with equality. Before getting to this matter I need to point out that I changed the definition of "theory" in the process. The reason is that the rules of first-order logic consequence make every first-order logic theory that contains an inconsistency trivial in the sense that it must include all sentences (of the language). Hence for this work a theory consists of a set of sentences only. In this section I review these classifications with a few examples. The main complication in the definitions is due to infinite domains. Since in databases we deal with finite domains, I present here simplified versions of these classifications where only finite domains are considered.

Before getting to the definitions, let me give some insight into how I came up with the three classifications. For finite sets A and B, if $A \subset B$ then $|A| < |B|$. This is not necessarily the case for infinite sets: for example, the set of positive integers is a proper subset of the set of integers, but they have the same size. Nevertheless, if $A \subset B$ then $|A| \leq |B|$: a containing set has size at least as big as any set it contains. The first classification uses this idea where the subset relation is between the models of those theories. But this is just a starting point; it is not a good classification because in most cases of comparison there is no subset relationship between the models; just as in most cases when the sizes of sets are compared one set is not a subset of another set. Another problem is that the set of models may decrease when consistent information is added, just as in case inconsistent information is added, to a theory. The analogy here is with the case where elements are added to an infinite set without increasing its size.

The second and third classifications do not have such a limitation. They differ from each other in the sense that the *level* concept measures absolute inconsistency while the *degree* concept measures relative inconsistency. We can think of a measure for absolute inconsistency as one that answers the question: "How much inconsistency is there?"; while relative inconsistency answers the question: "How inconsistent is it?". Suppose we would like to know how two countries compare in the number of people with higher education. We may obtain the number of college graduates in both countries and compare these numbers. But such a comparison may be unfair if one country has a much larger population than the other one. Instead, we may compare the percentage of the college graduates in the population. The former compares the absolute values; the latter compares the relative values. Both are correct measures in some sense but they compare different things. The relative value is a ratio, often expressed as a percentage.

As an aside let me mention here that even today, as I am writing this chapter, there seems to be some confusion about the distinction between absolute and relative inconsistency. A good example of this is the rationality postulates that have been proposed for inconsistency measures. Most of these are appropriate for abstract measures but at least one proposed postulate is primarily for relative measures. In fact, I have not seen a study of what postulates are appropriate specifically for relative measures.

Going back now to my thinking at that time, for the reasons mentioned above, I introduced three classifications: one uses subsets; another one is an absolute measure; and the third is a relative measure. With the intuition out of the way it is time to give the definitions. The standard definitions for the syntax of first-order logic are used including the equality axioms. We distinguish between atomic formulas, such as $R(x, y)$ where R is a binary relation symbol and equalities, such as $x = y$. The major innovation is the generalization of the concept of a relational structure to a structure where both an atomic formula and its negation (but not an equality and its negation), may hold (be true). We write $dom(A)$ for the domain of a structure A. Hence for $\{a, b\} \subseteq dom(A)$, both literals, $R(a, b)$ and $\neg R(a, b)$ may hold in A. We write

$diag(A)$ for the set of atoms and negations of atoms that hold in A (the diagram of the structure). We call each instance of an atom and its negation holding in a structure an inconsistency and write $Incon(A)$ for the number of inconsistencies in A. The case where either an atom or its negation but not both holds is called a consistency and we write $Con(A)$ for the number of consistencies in A. Thus $|Diag(A)| = 2 \times Incon(A) + Con(A)$.

Γ is a theory, (set of sentences in the given language); $Mod(\Gamma)$ is the class of structures in which every formula of Γ holds; and $Con\ Mod(\Gamma)$ is the class of (classical) relational structures in which every formula of Γ holds.

Example 1 *For a language with the single unary predicate symbol S, let $\Gamma_1 = \{\exists x(S(x) \land \neg S(x))\}$ and $\Gamma_2 = \{\forall x(S(x) \land \neg S(x))\}$. Then $Con\ Mod(\Gamma_1) = \emptyset = Con\ Mod(\Gamma_2)$ but for $dom(A) = \{a_1, a_2\}$ if $Diag(A) = \{S(a_1), \neg S(a_1), S(a_2)\}$ then $A \in Mod(\Gamma_1)$ but $A \notin Mod(\Gamma_2)$.*

The special treatment of equality allows us to write statements that give information about the number of elements allowed in the domain of a structure.

Example 2 $E_3 = \exists x_1 x_2 x_3 \forall y(x_1 \neq x_2 \land x_1 \neq x_3 \land x_2 \neq x_3 \land (y = x_1 \lor y = x_2 \lor y = x_3))$ *restricts the domain to have 3 elements.*
$L_3 = \exists x_1 x_2 x_3 \forall y(y = x_1 \lor y = x_2 \lor y = x_3)$ *restricts the domain to have at most 3 elements.*
$T_3 = \forall y(y = a_1 \lor y = a_2 \lor y = a_3)$ *restricts the domain to the 3 elements a_1, a_2, and a_3.*

The presence of equality means that some theories have no models at all: for example $\Sigma = \{E_3, E_4\}$. We exclude from consideration all theories that have no models. From the logical point of view, all inconsistencies have the same meaning. But when we consider databases, an inconsistency is caused by some information involving a relation, not by asserting, for instance, that a relation contains 3 tuples and also 4 tuples at the same time.

As mentioned earlier, in this presentation the definitions from [1] are simplified by assuming that the domain is finite. In fact, we assume that the domain is restricted in size and cannot be larger than some specified (possibly large) finite integer. We use K for this maximum size. We can think of this as writing L_K for a required formula in every theory under consideration. Furthermore, in the next section we will further restrict the domain size.

The first classification, written simply as $\leq$, is based on model containment. Define $\Gamma \leq \Gamma'$ iff $Mod(\Gamma') \subseteq Mod(\Gamma)$. A weakness of this classification is that it distinguishes even between some consistent theories. If $\Gamma \cup \{\phi\}$ is consistent and $\Gamma \not\vdash \phi$ then $\Gamma \leq \Gamma \cup \{\phi\}$. And if, as usual, we define $\Gamma < \Gamma'$ iff $\Gamma \leq \Gamma'$ and $\Gamma' \not\leq \Gamma$, then $\Gamma < \Gamma \cup \{\phi\}$.

Example 3 *Continuing with Example 1 we obtain $\Gamma_1 \leq \Gamma_2$, but as shown there, $\Gamma_2 \not\leq \Gamma_1$. Hence $\Gamma_1 < \Gamma_2$. Next, let*

$\Gamma_3 = \{\neg E_3, \exists x(S(x) \wedge \neg S(x))\}$ and
$\Gamma_4 = \{\neg E_2, \exists x(S(x) \wedge \neg S(x))\}$.
Here also $Con\ Mod(\Gamma_3) = \emptyset = Con\ Mod(\Gamma_4)$.
Let B be the structure with $dom(B) = \{b_1, b_2, b_3\}$ and
$Diag(B) = \{S(b_1), \neg S(b_1), S(b_2), S(b_3)\}$.
Then $A \in Mod(\Gamma_3)$, $A \notin Mod(\Gamma_4)$, $B \in Mod(\Gamma_4)$, $B \notin Mod(\Gamma_3)$. This shows that Γ_3 and Γ_4 are incomparable using $\leq$.

The second classification is called the level of a theory. This defines an absolute value for inconsistency in the form of a set C and a function f. C is the set of cardinalities for which the theory has a model. Then f assigns to each cardinal number in C a value which is the minimal number of inconsistencies for models whose domain size is that cardinal. Recall that we are dealing for this simplification only with cardinalities up to some finite K. We then define the level of Γ to be less than or equal to the level of Γ' iff Γ' does not have models in any cardinality for which Γ doesn't have a model and for all the common cardinalities the value of f is less than or equal to the value of f'. The formal definition is as follows.

Definition 1 $level(\Gamma) = <C, f>$ where
$C = \{n \in \{1, \ldots, K\} | \exists A \in Mod(\Gamma)\ and\ |dom(A)| = n\}$
and $f(n) = min\{Incon(A) | A \in Mod(\Gamma)\ and\ |dom(A)| = n\}$.
Then $level(\Gamma) \leq level(\Gamma')$ iff $C' \subseteq C$ and for every $n \in C'$, $f(n) \leq f'(n)$.
Again, $level(\Gamma) < level(\Gamma')$ iff $level(\Gamma) \leq level(\Gamma')$ and $level(\Gamma') \not\leq level(\Gamma)$.

Example 4 Continuing with Example 3 for Γ_3 (resp. Γ_4) there are no models with 3 (resp. 4) elements; hence $level(\Gamma_i) = <C_i, f_i>$ for $i = 1, 2, 3, 4$ as follows:
$C_1 = C_2 = \{1, \ldots, K\}$, $C_3 = \{1, \ldots, K\} \setminus \{3\}$, $C_4 = \{1, \ldots, K\} \setminus \{2\}$.
In the case of Γ_i, $i = 1, 3, 4$, for each allowed domain we can obtain a model with exactly one inconsistency, which is the minimal number of inconsistencies. However, for Γ_2 a model with size n must have n inconsistencies. Thus we obtain
$\forall n \in C_i$, $f_1(n) = f_3(n) = f_4(n) = 1$, $f_2(n) = n$.
Hence $level(\Gamma_1) \leq level(\Gamma_2), level(\Gamma_3), level(\Gamma_4)$, but no level relationship exists among Γ_2, Γ_3, and Γ_4.

Example 5 Suppose the language has a unary predicate symbol S and a binary predicate symbol P. Γ_i for $i = 1, 2, 3, 4$ are still theories in this language. Let $\Gamma_5 = \{\exists x \forall y (P(x, y) \wedge \neg P(x, y))\}$.
Then $C_5 = \{1, \ldots, K\}$, $f_5(n) = n$ for all $n \in C_5$. Hence, $level(\Gamma_2) = level(\Gamma_5)$.

The third definition is the degree of a theory which is also greatly simplified by the assumption that there is a finite limit on the size of the domains and is presented here in a slightly modified form. It is the smallest possible ratio of the number of inconsistencies over the sum of the number of consistencies and inconsistencies taken over all models of the theory. This is a relative inconsistency measure with values between 0 and 1 that assigns 0 to all consistent theories.

Definition 2 $deg(\Gamma) = min_{A \in Mod(\Gamma)}\{\frac{Incon(A)}{Con(A)+Incon(A)}\}$

Example 6 *Continuing with the previous examples using the language with a single unary predicate symbol S, we obtain*
$deg(\Gamma_1) = \frac{1}{K} = deg(\Gamma_3) = deg(\Gamma_4)$, $deg(\Gamma_2) = 1$. *Consider now the language given in Example 5 where a binary predicate symbol P was added to S. Using this language we obtain*
$deg(\Gamma_1) = \frac{1}{k^2+k} = deg(\Gamma_3) = deg(\Gamma_4)$, $deg(\Gamma_2) = \frac{k}{k^2+k} = \frac{1}{k+1} = deg(\Gamma_5)$.

From this example we see that the degree measure, unlike the level classification, is language dependent. Hence if we wish to compare the degree of inconsistency of two theories we should use the same language for them. It turns out that these three classifications are partially compatible in the sense that if Γ is less than Γ' in one of them then Γ' cannot be less than Γ in another one. However, they are not totally compatible in the sense that if Γ is less than Γ' in one classification then Γ need not be less than Γ' in another one.

The bulk of [1] gives numerous mathematical results about the three classifications. These results do not appear to be useful for (finite) information structures; the interested reader is referred to the original paper for them.

4 Comparing the Inconsistency of Databases

In retrospect I now realize that I should also have written a paper about comparing the inconsistency of databases when I was working on the 1978 paper. Although it may not have been possible to publish it at that time, by the late 1970s a sophisticated theory of relational databases was being developed by several prominent researchers. Certainly I could have at least presented this work at the 1981 XP2 Workshop that brought together many of the people working on theoretical aspects of database systems at that time and perhaps gotten some of them interested in working on inconsistency in databases. Incidentally, this workshop was the precursor to the very successful Principles of Database Systems (PODS) conferences still going on at this time. I attended the workshop but just gave a talk on numerical dependencies there.

It is impossible to know now exactly how I would have presented this material about inconsistency in databases. Nevertheless I remember the type of database examples I was working with in those days. I was thinking mainly of supplier-part-order and employee-department type data. For simplicity I use

Eno	Ename	Age	Salary	Dno
111	Jones	36	22000	1
222	Smith	42	21000	2
111	Jones	37	22000	1
333	Roberts	22	15000	2

Dno	Dname	Loc
1	Sales	A1
2	Service	B2

Table 1: DB1: the EMPLOYEE and DEPARTMENT relations

Eno	Ename	Age	Salary	Dno
111	Jones	36	22000	1
222	Smith	42	21000	2
111	Jones	37	22000	1
333	Roberts	22	15000	2
444	Adams	51	26000	3

Dno	Dname	Loc
1	Sales	A1
2	Service	B2

Table 2: DB2: the EMPLOYEE and DEPARTMENT relations

here a database with 2 relations and 3 integrity constraints:

EMPLOYEE(Eno,Ename,Age,Sal,Dno)
DEPARTMENT(Dno,Dname,Loc)
Eno → Ename,Age,Sal,Dno
Dno → Dname,Loc
EMPLOYEE[Dno] ⊆ DEPARTMENT[Dno]

That is, Eno is the key for EMPLOYEE; Dno is the key for DEPARTMENT; and every Dno in the EMPLOYEE relation must already exist in the DEPARTMENT relation.

In order to write these databases in logic we need to define the syntax first. Let's start with a sorted logic with equality where each attribute, such as Eno, is a sort and there is a domain for each sort, such as dom(Eno). There are two relations, EMPLOYEE and DEPARTMENT, with the appropriate sorts for the attributes. The elements of all the domains are constant symbols in the language. We need all the inequalities for different elements of each domain,

Eno	Ename	Age	Salary	Dno
222	Smith	42	21000	2
111	Jones	37	22000	1
333	Roberts	22	15000	2
333	Roberts	21	14000	2

Dno	Dname	Loc
1	Sales	A1
2	Service	B2

Table 3: DB3: the EMPLOYEE and DEPARTMENT relations

such as, $Jones \neq Smith$. The equality axioms are assumed. Additionally, we also need a statement for each attribute stating that the only elements allowed for an attribute are the constants for that attribute. As all the domains are assumed finite, such statements can be written in first-order logic but I am skipping these long statements. So the above statements are part of each DB.

Only for DB1, as the others are similar, I write the actual theory, that is, the additional statements pertaining to that database:

EMPLOYEE(111,Jones,36,22000,1)
EMPLOYEE(222,Smith,42,21000,2)
EMPLOYEE(111,Jones,37,22000,1)
EMPLOYEE(333,Roberts,22,15000,2)
DEPARTMENT(1,Sales,A1)
DEPARTMENT(2,Service,B2)

$\forall Eno \forall Ename_1 \forall Ename_2 \forall Age_1 \forall Age_2 \forall Sal_1 \forall Sal_2 \forall Dno_1 \forall Dno_2$
$((EMPLOYEE(Eno, Ename_1, Age_1, Sal_1, Dno_1) \wedge$
$EMPLOYEE(Eno, Ename_2, Age_2, Sal_2, Dno_2)) \rightarrow$
$(Ename_1 = Ename_2 \wedge Age_1 = Age_2 \wedge Sal_1 = Sal_2 \wedge Dno_1 = Dno_2))$

$\forall Dno \forall Dname_1 \forall Dname_2 \forall Loc_1 \forall Loc_2$
$((DEPARTMENT(Dno, Dname_1, Loc_1) \wedge$
$DEPARTMENT(Dno, Dname_2, Loc_2)) \rightarrow$
$(Dname_1 = Dname_2 \wedge Loc_1 = Loc_2))$

$\forall Eno \forall Ename \forall Age \forall Sal \forall Dno \exists Dname \exists Loc$
$(EMPLOYEE(Eno, Ename, Age, Sal, Dno) \rightarrow$
$DEPARTMENT(Dno, Dname, Loc))$

We have two choices at this point. If we use the Open World Assumption (OWA), then, indeed, this is the theory for DB1. But for databases usually the Closed World Assumption (CWA), all atoms not given are assumed false, is what is meant. So if we adopt the CWA then we would add a statement for each relation stating that there are no additional tuples for the relation. For example, for the DEPARTMENT relation, instead of writing the two atoms: $DEPARTMENT(1, Sales, A1)$ and $DEPARTMENT(2, Service, B2)$ we would write the statement:
$\forall Dno \forall Dname \forall Loc (DEPARTMENT(Dno, Dname, Loc) \rightarrow$
$((Dno = 1 \wedge Dname = Sales \wedge Loc = A1) \vee (Dno = 2 \wedge Dname = Service \wedge Loc = B2)))$

It turns out that for comparing the amount of inconsistency, which approach we choose makes no difference.

Next we consider structures for these theories. Each structure will have exactly the given constants in the domain for each attribute. Hence the number of possible atoms is the same for each database:

$K = (|dom(Eno)| \times |dom(Ename)| \times |dom(Age)| \times |dom(Sal)| \times |dom(Dno)|) + (|dom(Dno)| \times |dom(Dname)| \times |dom(Loc)|)$.

The following structure is part of a model for the theory DB1:
$EMPLOYEE(111, Jones, 36, 22000, 1)$
$\neg EMPLOYEE(111, Jones, 36, 22000, 1)$
$EMPLOYEE(222, Smith, 42, 21000, 2)$
$EMPLOYEE(111, Jones, 37, 22000, 1)$
$EMPLOYEE(333, Roberts, 22, 15000, 2)$
$DEPARTMENT(1, Sales, A1)$
$DEPARTMENT(2, Service, B2)$

If we assume the CWA then the structure must also include the negations of all other possible atoms, such as $\neg EMPLOYEE(111, Smith, 22, 21000, 1)$. In the case of the OWA we can substitute the positive version for any of these negated atoms.

Next we consider the 3 classifications and consider how DB1, DB2, and DB3 are related. For the first classification we obtain $DB1 < DB2$ but there is no $\leq$ relationship between DB1 and DB3 nor between DB2 and DB3.

Next we consider the second classification: levels. As shown above, the number of possible atoms K is the same for all of these structures, hence there is only one possible value for C, that is, $C = \{K\}$. This is different from the way we defined the number of elements in a structure where there were no sorts, but it is equivalent to it. Then $lev(DB1) = <\{K\}, f_1>$ where $f_1(K) = 1$. Similarly, $lev(DB2) = <\{K\}, f_2>$ where $f_2(K) = 2$. Then $lev(DB3) = lev(DB1) < lev(DB2)$. The third classification, degrees, follows the pattern of levels. We obtain $deg(DB1) = deg(DB3) = \frac{1}{K} < deg(DB2) = \frac{2}{K}$.

These databases, DB1, DB2, and DB3, have the same schema. So we may wonder how their inconsistency would compare to a different database. Without getting into details, suppose that DB4 has relations SUPPLIER, PART, and ORDER and there is some inconsistency present. For the first classification, the different schemas prevent a comparison. As far as the level classification is concerned, the problem is that for DB4, the number of all possible atoms, K' is almost certainly different from K. Hence the level of DB4 cannot be compared to the levels of DB1-DB3. For the third classification, $deg(DB4) = \frac{n}{K'}$, where n is the minimal number of inconsistencies in a model for DB4. For example, if $n = 1$ then we obtain $deg(DB4) = \frac{1}{K'}$. So the issue of which database has higher degree depends in an important way on the domains for the attributes in the schemas. This does not appear to be a good intuitive way of determining which database is more inconsistent. A more sensible ratio for a relative measure would be obtained by taking the minimal number of inconsistencies over the number of tuples in the database instead of the number K obtained from the domain sizes.

5 In Maryland

After I got tenure and promotion I was very happy at the University of Florida. But in 1976 George Haynam left and in 1977 the university hired a new department chairman. By the beginning of 1978 I was looking for another job. I got one job offer: to be the computer science coordinator at a university near Baltimore, Maryland. I discussed the offer with two senior colleagues: both of them strongly urged me not to accept: they warned me that the heavy teaching load and administrative duties would end my career in research. However, even though with the help of my faculty union, the United Faculty of Florida (UFF), I won my grievance case against the new chairman, the stress he created was overwhelming and I accepted the job. But I also had serious reservations and reached out to my advisor, Martin Davis, for advice. He had been extremely helpful to me not only throughout my graduate studies but also had helped me obtain my first job at the University of Florida; furthermore, he had given me some excellent advice when I moved to the Department of Computer and Information Sciences from the Department of Mathematics; and then he had helped me in securing tenure and promotion. Once again, when I needed him at a critical time he was there to help. He told me that one of his students, who started at NYU after I had left, Barry Jacobs, just got a faculty position in the Department of Computer Science at the University of Maryland. He suggested that I contact him and thought that this would open up some research opportunities for me.

So in the Fall of 1978 I contacted Barry and then arranged to meet him at the University of Maryland. He explained to me his project of developing a logic for databases and was interested in working with me. Furthermore, he introduced me to Jack Minker, who was in his last year as department chairman, and was setting up a research group in logic and databases. I joined his group right away. I learnt a lot from both of them quickly: how to collaborate on research and how to write up and present research in the best possible way. Barry left the university in 1984 to work full time for NASA but I continued working with Jack and some of his students and visitors for an amazing 25 years. None of this work touched on inconsistencies.

In 1989 V. S. Subrahmanian joined the department. I met with him and was surprised to find that he had done work on inconsistencies and knew about my 1978 paper. We then started a collaboration that had some gaps but lasted more than 20 years. Our joint work included reasoning in and semantics for inconsistent knowledge bases but not inconsistency measurement.

In the early 2000s I became aware of research in measuring inconsistency in knowledge bases. I became most interested in the work of Anthony Hunter and contacted him about it. This led to a very fruitful and ongoing collaboration with Tony for nearly 15 years. Finally, working with him I was able to get back to study inconsistency measurement, but in a different framework than the one

I devised 30 years earlier. This is not the place to review what we have done but I can mention that one of our papers, [2], can be thought of as a greatly improved and expanded version of [1].

6 Measuring Inconsistencies versus Measuring Infinities

Back more than 40 years ago when I started thinking about inconsistencies, one of the first things I realized was that there appeared to be several different reasonable ways to compare them. This contrasts sharply with comparing infinities, which was my inspiration, where there is a unique way. As I explained earlier, for [1] I then investigated three methods. In recent years a highly sophisticated analysis by multiple researchers has come up with many inconsistency measures for knowledge bases. I would like to end this paper by presenting my thoughts about why there is such a difference between measuring infinities and inconsistencies.

Consider two rectangles (using the same units): A has length 5 and width 2; B has length 8 and width 4. Which is bigger? The only reasonable answer is that B is bigger. But now consider a third rectangle C with length 10 and width 3. It is easy to see that C is bigger than A. But what about B and C: which is bigger? If by bigger we mean longer then C is bigger. If by bigger we mean wider then B is bigger. Perhaps we should consider both length and width. If by bigger we mean larger area then B is bigger. But if by bigger we mean larger perimeter then C is bigger. We may even consider the sum of 3 sides, the 2 widths and a length, perhaps to build a fence around 3 sides. In that case B and C have the same value.

So for a very simple 2-dimensional structure the question of which is bigger can be answered reasonably in several different ways that give different answers. The situation becomes even more complicated if we deal with more complex figures and go into higher dimensions. On the other hand, if we consider only one dimensional objects, line segments, then there is only one way to compare their size. I think that essentially this is what happens when we compare the size of sets. Although we may obtain infinite sets in different ways, a set is simply composed of a collection of elements. Each element has an independent status and even if the elements have some structure, for example they might be sets themselves, such structures are not used to determine the size of the set. In that sense a set is a 1-dimensional object and so there is only one way to compare and measure the size of sets.

When we consider an inconsistent knowledgebase we deal with a set of formulas. Each formula may play a different role in the formation of inconsistencies. We are no longer dealing with a 1-dimensional object. I think that this is the reason for the existence of many different inconsistency measures. What is not clear is if it is possible to distinguish among orthogonal dimensions whose combinations account somehow for such a multitude of incompatible inconsistency measures. I doubt that this will be the case; however, in some special kinds of knowledge bases dimensions can be formulated.

In particular, in [3], where measuring inconsistency is studied for spatio-temporal databases; space, time, and objects can be used as such dimensions for some inconsistency measures. But even for the kind of databases considered in Section 4 we may think of 3 ways of measuring the inconsistency by considering the minimal number of tuples to add, the minimal number of tuples to delete, and the minimal number of integrity constraints to delete in order to restore the consistency of (repair) the database and these three ways can be used as dimensions. In fact in many other cases also dimensions can be found. In the case of the inconsistent groups we may consider changes in the multiplication table and changes in the inverse table as two dimensions. For graphs that are inconsistent with respect to some graph property we may consider making changes to vertices versus edges as 2 dimensions. For that matter 2 other dimensions may be doing insertions and deletions. In fact, we may combine these two and have 4 dimensions: inserting vertices, deleting vertices, inserting edges, and deleting edges. For example, if the graph property is connectivity, then only the dimensions deleting vertices and inserting edges are relevant, but for other properties other combinations of dimensions may be useful.

Going back to logic now, if we want to limit the various roles that different formulas can take in obtaining inconsistencies, we can try to put limits on what formulas are allowed. We need negation to get an inconsistency. So the simplest situation is one where the only formulas are literals, that is, atoms and negated atoms. Consider how we can measure inconsistency in such a case. Suppose we are given an infinite number of atoms: $a_1, a_2, \ldots$ and each knowledgebase is a finite set of literals. There is not much choice in how we measure the inconsistency of knowledgebases: count the number of atoms for which both the atom and its negation appear. This is about as close as we can get to a 1-dimensional situation. This is what I called earlier an absolute measure and so it is somewhat analogous to the level concept. Actually, there is another measure, a relative measure that may also be reasonable. For this one we take the ratio of the number of atoms for which both the atom and its negation appear with the total number of atoms that appear in the knowledgebase. This measure is somewhat analogous to the degree concept and is in line with the suggestion of using the number of tuples in the denominator for databases. In any case we have reduced the dimensionality involved in measuring inconsistency.

7 Book on Measuring Inconsistency

For several years I have been thinking about what I could do in getting a book out on measuring inconsistency. I am very happy that this book [4] is now a reality and I hope that it will give more publicity to what I believe is an important research topic. I am very grateful to my wonderful co-editor and all the outstanding researchers who made such important contributions to this endeavor.

Acknowledgements I wish to thank the reviewers and my sons for helpful comments.

References

[1] J. Grant. Classifications for Inconsistent Theories. *The Notre Dame Journal of Formal Logic*, XIX (3): 435–444, July 1978.

[2] J. Grant and A. Hunter. Analysing Inconsistent First-order Knowledgebases. *Artificial Intelligence*, 172: 1064–1093, 2008.

[3] J. Grant, M.V. Martinez, C. Molinaro, and F. Parisi. On Measuring Inconsistency in Spatio-Temporal Databases. In: [4].

[4] J. Grant and M.V. Martinez, Editors. Measuring Inconsistency in Information. College Publications, 2018.

On the Evaluation of Inconsistency Measures

Matthias Thimm

Institute for Web Science and Technologies
University of Koblenz-Landau, Germany
thimm@uni-koblenz.de

Abstract

We discuss the issue of evaluating inconsistency measures along the three dimensions of rationality postulates, expressivity, and computational complexity. We survey a broad selection on inconsistency measures and evaluate their performance on these three dimensions.

1 Introduction

In classical logic, the notion of *inconsistency* is a binary concept. Either a formula (or a set of formulas) is inconsistent or not. However, quantifying inconsistency is an important challenge for logical accounts to knowledge representation as differences in the severity of inconsistency may indeed be recognised for certain types of applications. Consider the following two classical logic knowledge bases

$$\mathcal{K}_1 = \{a, \neg a, b, \neg b\} \qquad \mathcal{K}_2 = \{\neg b, a, a \rightarrow b, c\}$$

Both $\mathcal{K}_1$ and $\mathcal{K}_2$ are classically inconsistent, i.e., there is no interpretation satisfying any of them. But looking closer into the structure of the knowledge bases one can identify differences in the severity of the inconsistency. In $\mathcal{K}_1$ there are two apparent contradictions, i.e., $\{a, \neg a\}$ and $\{b, \neg b\}$ are directly conflicting formulas. In $\mathcal{K}_2$, the conflict is a bit more hidden. Here, three formulas are necessary to produce a contradiction ($\{\neg b, a, a \rightarrow b\}$). Moreover, there is one formula in $\mathcal{K}_2$ (c), which is not participating in any conflict and one could still infer meaningful information from this by relying on e.g. paraconsistent reasoning techniques [5]. In conclusion, one should regard $\mathcal{K}_1$ as *more inconsistent* than $\mathcal{K}_2$.

The analysis of the severity of inconsistency in the knowledge bases $\mathcal{K}_1$ and $\mathcal{K}_2$ above was informal. Formal accounts to the problem of assessing the severity of inconsistency are given by *inconsistency measures* and there have been a lot of proposals of those in recent years (see the other chapters in this volume and the measures reviewed in Section 3). Up to today, the concept of *severity of inconsistency* has not been axiomatized in a satisfactory manner and the series of different inconsistency measures approach this challenge from different points of view and focus on different aspects on what constitutes *severity*. Consider the next two knowledge bases

$$\mathcal{K}_3 = \{a, \neg a, b\} \qquad \mathcal{K}_4 = \{a \vee b, \neg a \vee b, a \vee \neg b, \neg a \vee \neg b\}$$

Again both $\mathcal{K}_3$ and $\mathcal{K}_4$ are inconsistent, but which one is more inconsistent than the other? Our reasoning from above cannot be applied here in the same fashion. The knowledge base $\mathcal{K}_3$ contains an apparent contradiction ($\{a, \neg a\}$) but also a formula not participating in the inconsistency ($\{b\}$). The knowledge base $\mathcal{K}_4$ contains a "hidden" conflict as four formulas are necessary to produce a contradiction, but all formulas of $\mathcal{K}_4$ are participating in this. In this case, it is not clear how to quantitatively assess the inconsistency of these knowledge bases and different measures may order these knowledge bases differently. More generally speaking, it is not universally agreed upon which so-called *rationality postulates* should be satisfied by a reasonable account of inconsistency measurement, see [3] for a discussion. Besides concrete approaches to inconsistency measurement the community has also proposed a series of those rationality postulates in order to describe general desirable behaviour and the classification of inconsistency measures by the postulates they satisfy is still one the most important ways to evaluate the quality of a measure, even if the set of desirable postulates is not universally accepted. For example, one of the most popular rationality postulates is *monotonicity* which states that for any $\mathcal{K} \subseteq \mathcal{K}'$, the knowledge base $\mathcal{K}$ cannot be regarded as more inconsistent as $\mathcal{K}'$. The justification for this demand is that inconsistency cannot be resolved when adding new information but only increased. While this is usually regarded as a reasonable demand there are also situations where *monotonicity* may be seen as counterintuitive. Consider the next two knowledge bases

$$\mathcal{K}_5 = \{a, \neg a\} \qquad \mathcal{K}_6 = \{a, \neg a, b_1, \ldots, b_{998}\}$$

We have $\mathcal{K}_5 \subseteq \mathcal{K}_6$ and following *monotonicity*, $\mathcal{K}_6$ should be regarded as least as inconsistent as $\mathcal{K}_5$. However, when judging the content of the knowledge bases in a "relative" manner, $\mathcal{K}_5$ may seem more inconsistent. More precisely, $\mathcal{K}_5$ contains no useful information and all formulas of $\mathcal{K}_5$ are in conflict with another formula. In $\mathcal{K}_6$, however, only 2 out of 1000 formulas are participating in the contradiction. So it may also be a reasonable point of view to judge $\mathcal{K}_5$ more inconsistent than $\mathcal{K}_6$.

In this chapter, we will not give a final answer to the discussion on which rationality postulate is desirable or not. We will, however, provide a comprehensive overview of the existing rationality postulates and the compliance of different measures wrt. those, continuing work from [41]. It is up to the reader and future work to conclude said discussion. Besides satisfaction of rationality postulates we will address two more "objective" evaluation metrics, namely *expressivity* [38] and *computational complexity* [44, 43]. The former refers to the ability of an inconsistency measure to differentiate many levels of the severity of inconsistency. Consider the following family of knowledge bases

$$\mathcal{K}_7^1 = \{a_1, \neg a_1\} \quad \mathcal{K}_7^2 = \{a_1, \neg a_1, a_2, \neg a_2\} \quad \ldots \quad \mathcal{K}_7^n = \{a_1, \neg a_1, \ldots, a_n, \neg a_n\}$$

Each knowledge base $\mathcal{K}_7^{i+1}$ contains one more apparent contradiction than $\mathcal{K}_7^i$, so it is reasonable to assess $\mathcal{K}_7^{i+1}$ as *strictly* more inconsistent than $\mathcal{K}_7^i$. Following [38] we will present a formal framework for assessing the expressivity of inconsistency measures and provide a comprehensive overview of the expressivity of different measures. Finally, we will consider the computational complexity involved in computing the value of an inconsistency measure, following [44, 43]. As detecting inconsistency alone is an intractable problem, we cannot hope to determine inconsistency values in an efficient manner. However, inconsistency measures can be classified into different levels of the polynomial hierarchy and thus algorithms determining them may exhibit significant differences in performance. As before, we provide a comprehensive overview of the computational complexity landscape of different measures as well.

This chapter summarises the works [38, 41, 44, 43] and complements their results by additionally considering more recent approaches to inconsistency measurement. The rest of this chapter is organised as follows. In Section 2 we present necessary preliminaries on propositional logic and we review our selection of inconsistency measures to be studied in Section 3. In Sections 4, 5, and 6 presents the evaluation measures of rationality postulates, expressivity, and computational complexity, respectively. We conclude in Section 7. Appendix 7 contains proofs of new technical results.

2 Preliminaries

Let At be some fixed propositional signature, i.e., a (possibly infinite) set of propositions, and let $\mathcal{L}(\mathsf{At})$ be the corresponding propositional language constructed using the usual connectives $\wedge$ (*conjunction*), $\vee$ (*disjunction*), $\rightarrow$ (*implication*), and $\neg$ (*negation*). A literal is a proposition p or a negated proposition $\neg p$.

Definition 1. A knowledge base $\mathcal{K}$ is a finite set of formulas $\mathcal{K} \subseteq \mathcal{L}(\mathsf{At})$. Let $\mathbb{K}$ be the set of all knowledge bases.

A clause is a disjunction of literals. A formula is in conjunctive normal form (CNF) if the formula is a conjunction of clauses. If X is a formula or a set of formulas we write $\mathsf{At}(X)$ to denote the set of propositions appearing in X. For a set $S = \{\phi_1, \ldots, \phi_n\}$ let $\bigwedge S = \phi_1 \wedge \ldots \wedge \phi_n$.

Semantics for a propositional language is given by *interpretations* where an *interpretation* ω on At is a function $\omega : \mathsf{At} \to \{\mathsf{true}, \mathsf{false}\}$. Let $\Omega(\mathsf{At})$ denote the set of all interpretations for At. An interpretation ω *satisfies* (or is a *model* of) an atom $a \in \mathsf{At}$, denoted by $\omega \models a$, if and only if $\omega(a) = \mathsf{true}$. The satisfaction relation $\models$ is extended to formulas in the usual way.

For $\Phi \subseteq \mathcal{L}(\mathsf{At})$ we also define $\omega \models \Phi$ if and only if $\omega \models \phi$ for every $\phi \in \Phi$. Define furthermore the set of models $\mathsf{Mod}(X) = \{\omega \in \Omega(\mathsf{At}) \mid \omega \models X\}$ for every formula or set of formulas X. Two formulas or sets of formulas X_1, X_2 are *equivalent*, denoted by $X_1 \equiv X_2$, if and only if $\mathsf{Mod}(X_1) = \mathsf{Mod}(X_2)$. Furthermore, two sets of formulas X_1, X_2 are *semi-extensionally equivalent* [36, 37]—or *bijection equivalent* [10]—if and only if there is a bijection $s : X_1 \to X_2$ such that for all $\alpha \in X_1$ we have $\alpha \equiv s(\alpha)$. We denote this by $X_1 \equiv_b X_2$. If $\mathsf{Mod}(X) = \emptyset$ we also write $X \models \bot$ and say that X is *inconsistent*.

3 Inconsistency Measures

Let $\mathbb{R}^\infty_{\geq 0}$ be the set of non-negative real values including ∞. Inconsistency measures are functions $\mathcal{I} : \mathbb{K} \to \mathbb{R}^\infty_{\geq 0}$ that aim at assessing the severity of inconsistency in a knowledge base $\mathcal{K}$, cf. [10]. The basic idea is that the larger the inconsistency in $\mathcal{K}$ the larger the value $\mathcal{I}(\mathcal{K})$. Formally, we define inconsistency measures as follows, cf. e.g. [15].

Definition 2. An inconsistency measure $\mathcal{I}$ is any function $\mathcal{I} : \mathbb{K} \to \mathbb{R}^\infty_{\geq 0}$.

There is a wide variety of inconsistency measures in the literature. In this work, we select 22 inconsistency measures in order to discuss issues pertaining to their evaluation. We briefly introduce these measures in this section for the sake of completeness, but we refer for a detailed explanation to the corresponding original papers.[1]

The measure $\mathcal{I}_d(\mathcal{K})$ is usually referred to as a baseline for inconsistency measures as it only distinguishes between consistent and inconsistent knowledge bases.

Definition 3 ([15]). The *drastic inconsistency measure* $\mathcal{I}_d : \mathbb{K} \to \mathbb{R}^\infty_{\geq 0}$ is defined as

$$\mathcal{I}_d(\mathcal{K}) = \begin{cases} 1 & \text{if } \mathcal{K} \models \bot \\ 0 & \text{otherwise} \end{cases}$$

for $\mathcal{K} \in \mathbb{K}$.

---

[1] Implementations of these measures can also be found in the *Tweety Libraries for Artificial Intelligence* [42] and an online interface is available at http://tweetyproject.org/w/incmes

A set $M \subseteq \mathcal{K}$ is called a *minimal inconsistent subset* (MI) of $\mathcal{K}$ if $M \models \perp$ and there is no $M' \subset M$ with $M' \models \perp$. Let $\mathsf{MI}(\mathcal{K})$ be the set of all MIs of $\mathcal{K}$.

Definition 4 ([15]). The *MI-inconsistency measure* $\mathcal{I}_{\mathsf{MI}} : \mathbb{K} \to \mathbb{R}^\infty_{\geq 0}$ is defined as

$$\mathcal{I}_{\mathsf{MI}}(\mathcal{K}) = |\mathsf{MI}(\mathcal{K})|$$

for $\mathcal{K} \in \mathbb{K}$.

Definition 5 ([15]). The *MI^c-inconsistency measure* $\mathcal{I}_{\mathsf{MI}^c} : \mathbb{K} \to \mathbb{R}^\infty_{\geq 0}$ is defined as

$$\mathcal{I}_{\mathsf{MI}^c}(\mathcal{K}) = \sum_{M \in \mathsf{MI}(\mathcal{K})} \frac{1}{|M|}$$

for $\mathcal{K} \in \mathbb{K}$.

A *probability function* P on $\mathcal{L}(\mathsf{At})$ is a function $P : \Omega(\mathsf{At}) \to [0,1]$ with $\sum_{\omega \in \Omega(\mathsf{At})} P(\omega) = 1$. We extend P to assign a probability to any formula $\phi \in \mathcal{L}(\mathsf{At})$ by defining

$$P(\phi) = \sum_{\omega \models \phi} P(\omega)$$

Let $\mathcal{P}(\mathsf{At})$ be the set of all those probability functions.

Definition 6 ([22]). The *η-inconsistency measure* $\mathcal{I}_\eta : \mathbb{K} \to \mathbb{R}^\infty_{\geq 0}$ is defined as

$$\mathcal{I}_\eta(\mathcal{K}) = 1 - \max\{\xi \mid \exists P \in \mathcal{P}(\mathsf{At}) : \forall \alpha \in \mathcal{K} : P(\alpha) \geq \xi\}$$

for $\mathcal{K} \in \mathbb{K}$.

A *three-valued interpretation* v on At is a function $v : \mathsf{At} \to \{T, F, B\}$ where the values T and F correspond to the classical **true** and **false**, respectively. The additional truth value B stands for *both* and is meant to represent a conflicting truth value for a proposition. The function v is extended to arbitrary formulas as shown in Table 1. An interpretation v satisfies a formula α, denoted by $v \models^3 \alpha$ if either $v(\alpha) = T$ or $v(\alpha) = B$. Inconsistency can be measured by seeking an interpretation v that assigns B to a minimal number of propositions.

Definition 7 ([10]). The *contension inconsistency measure* $\mathcal{I}_c : \mathbb{K} \to \mathbb{R}^\infty_{\geq 0}$ is defined as

$$\mathcal{I}_c(\mathcal{K}) = \min\{|v^{-1}(B)| \mid v \models^3 \mathcal{K}\}$$

for $\mathcal{K} \in \mathbb{K}$.

α	β	$v(\alpha \wedge \beta)$	$v(\alpha \vee \beta)$
T	T	T	T
T	B	B	T
T	F	F	T
B	T	B	T
B	B	B	B
B	F	F	B
F	T	F	T
F	B	F	B
F	F	F	F

α	$v(\neg \alpha)$
T	F
B	B
F	T

Table 1: Truth tables for propositional three-valued logic.

Let $\mathsf{MC}(\mathcal{K})$ be the set of maximal consistent subsets of $\mathcal{K}$, i.e.

$$\mathsf{MC}(\mathcal{K}) = \{\mathcal{K}' \subseteq \mathcal{K} \mid \mathcal{K}' \not\models \bot \wedge \forall \mathcal{K}'' \supsetneq \mathcal{K}' : \mathcal{K}'' \models \bot\}$$

Furthermore, let $\mathsf{SC}(\mathcal{K})$ be the set of self-contradictory formulas of $\mathcal{K}$, i.e.

$$\mathsf{SC}(\mathcal{K}) = \{\phi \in \mathcal{K} \mid \phi \models \bot\}$$

Definition 8 ([10]). *The* MC-*inconsistency measure* $\mathcal{I}_{mc} : \mathbb{K} \to \mathbb{R}_{\geq 0}^{\infty}$ *is defined as*

$$\mathcal{I}_{mc}(\mathcal{K}) = |\mathsf{MC}(\mathcal{K})| + |\mathsf{SC}(\mathcal{K})| - 1$$

for $\mathcal{K} \in \mathbb{K}$.

Definition 9 ([10]). *The problematic inconsistency measure* $\mathcal{I}_p : \mathbb{K} \to \mathbb{R}_{\geq 0}^{\infty}$ *is defined as*

$$\mathcal{I}_p(\mathcal{K}) = |\bigcup_{M \in \mathsf{MI}(\mathcal{K})} M|$$

for $\mathcal{K} \in \mathbb{K}$.

A subset $H \subseteq \Omega(\mathsf{At})$ is called a *hitting set* of $\mathcal{K}$ if for every $\phi \in \mathcal{K}$ there is $\omega \in H$ with $\omega \models \phi$.

Definition 10 ([39]). *The hitting-set inconsistency measure* $\mathcal{I}_{hs} : \mathbb{K} \to \mathbb{R}_{\geq 0}^{\infty}$ *is defined as*

$$\mathcal{I}_{hs}(\mathcal{K}) = \min\{|H| \mid H \text{ is a hitting set of } \mathcal{K}\} - 1$$

for $\mathcal{K} \in \mathbb{K}$ *with* $\min \emptyset = \infty$.

An *interpretation distance* d is a function $d : \Omega(\mathsf{At}) \times \Omega(\mathsf{At}) \to [0, \infty)$ that satisfies (let $\omega, \omega', \omega'' \in \Omega(\mathsf{At})$)

1. $d(\omega, \omega') = 0$ if and only if $\omega = \omega'$ (*reflexivity*),
2. $d(\omega, \omega') = d(\omega', \omega)$ (*symmetry*), and
3. $d(\omega, \omega'') \leq d(\omega, \omega') + d(\omega', \omega'')$ (*triangle inequality*).

One prominent example of such a distance is the *Dalal distance* d_{d} defined via

$$d_{\mathrm{d}}(\omega, \omega') = |\{a \in \mathsf{At} \mid \omega(a) \neq \omega'(a)\}|$$

for all $\omega, \omega' \in \Omega(\mathsf{At})$. If $X \subseteq \Omega(\mathsf{At})$ is a set of interpretations we define $d_{\mathrm{d}}(X, \omega) = \min_{\omega' \in X} d_{\mathrm{d}}(\omega', \omega)$ (if $X = \emptyset$ we define $d_{\mathrm{d}}(X, \omega) = \infty$). For definitions 11, 12, and 13 below we assume d_{d} fixed but the measures could be defined using arbitrary distances.

Definition 11 ([12]). The Σ-*distance inconsistency measure* $\mathcal{I}_{\mathrm{dalal}}^{\Sigma} : \mathbb{K} \to \mathbb{R}_{\geq 0}^{\infty}$ is defined as

$$\mathcal{I}_{\mathrm{dalal}}^{\Sigma}(\mathcal{K}) = \min\left\{ \sum_{\alpha \in \mathcal{K}} d_{\mathrm{d}}(\mathsf{Mod}(\alpha), \omega) \mid \omega \in \Omega(\mathsf{At}) \right\}$$

for $\mathcal{K} \in \mathbb{K}$.

Definition 12 ([12]). The max-*distance inconsistency measure* $\mathcal{I}_{\mathrm{dalal}}^{\max} : \mathbb{K} \to \mathbb{R}_{\geq 0}^{\infty}$ is defined as

$$\mathcal{I}_{\mathrm{dalal}}^{\max}(\mathcal{K}) = \min\left\{ \max_{\alpha \in \mathcal{K}} d_{\mathrm{d}}(\mathsf{Mod}(\alpha), \omega) \mid \omega \in \Omega(\mathsf{At}) \right\}$$

for $\mathcal{K} \in \mathbb{K}$.

Definition 13 ([12]). The hit-*distance inconsistency measure* $\mathcal{I}_{\mathrm{dalal}}^{\mathrm{hit}} : \mathbb{K} \to \mathbb{R}_{\geq 0}^{\infty}$ is defined as

$$\mathcal{I}_{\mathrm{dalal}}^{\mathrm{hit}}(\mathcal{K}) = \min \{ |\{\alpha \in \mathcal{K} \mid d_{\mathrm{d}}(\mathsf{Mod}(\alpha), \omega) > 0\}| \mid \omega \in \Omega(\mathsf{At})\}$$

for $\mathcal{K} \in \mathbb{K}$.

For $\mathcal{K} \in \mathbb{K}$ define

$\mathsf{MI}^{(i)}(\mathcal{K}) = \{M \in \mathsf{MI}(\mathcal{K}) \mid |M| = i\}$

$\mathsf{CN}^{(i)}(\mathcal{K}) = \{C \subseteq \mathcal{K} \mid |C| = i \wedge C \not\models \bot\}$

$$R_i(\mathcal{K}) = \begin{cases} 0 & \text{if } |\mathsf{MI}^{(i)}(\mathcal{K})| + |\mathsf{CN}^{(i)}(\mathcal{K})| = 0 \\ |\mathsf{MI}^{(i)}(\mathcal{K})|/(|\mathsf{MI}^{(i)}(\mathcal{K})| + |\mathsf{CN}^{(i)}(\mathcal{K})|) & \text{otherwise} \end{cases}$$

for $i = 1, \ldots, |\mathcal{K}|$.

Definition 14 ([31]). The D_f-inconsistency measure $\mathcal{I}_{D_f} : \mathbb{K} \to \mathbb{R}_{\geq 0}^\infty$ is defined as

$$\mathcal{I}_{D_f}(\mathcal{K}) = 1 - \Pi_{i=1}^{|\mathcal{K}|}(1 - R_i(\mathcal{K})/i)$$

for $\mathcal{K} \in \mathbb{K}$.

A *minimal proof* for $\alpha \in \{x, \neg x \mid x \in \mathsf{At}\}$ in $\mathcal{K}$ is a set $\pi \subseteq \mathcal{K}$ such that

1. α appears as a literal in π
2. $\pi \models \alpha$, and
3. π is minimal wrt. set inclusion.

Let $P_m^\mathcal{K}(x)$ be the set of all minimal proofs of x in $\mathcal{K}$.

Definition 15 ([21]). The *proof-based inconsistency measure* $\mathcal{I}_{P_m} : \mathbb{K} \to \mathbb{R}_{\geq 0}^\infty$ is defined as

$$\mathcal{I}_{P_m}(\mathcal{K}) = \sum_{a \in \mathsf{At}} |P_m^\mathcal{K}(a)| \cdot |P_m^\mathcal{K}(\neg a)|$$

for $\mathcal{K} \in \mathbb{K}$.

Definition 16 ([46]). The *mv inconsistency measure* $\mathcal{I}_{mv} : \mathbb{K} \to \mathbb{R}_{\geq 0}^\infty$ is defined as

$$\mathcal{I}_{mv}(\mathcal{K}) = \frac{|\bigcup_{M \in \mathsf{MI}(\mathcal{K})} \mathsf{At}(M)|}{|\mathsf{At}(\mathcal{K})|}$$

for $\mathcal{K} \in \mathbb{K}$.

Definition 17 ([7]). The *nc-inconsistency measure* $\mathcal{I}_{nc} : \mathbb{K} \to \mathbb{R}_{\geq 0}^\infty$ is defined as

$$\mathcal{I}_{nc}(\mathcal{K}) = |\mathcal{K}| - \max\{n \mid \forall \mathcal{K}' \subseteq \mathcal{K} : |\mathcal{K}'| = n \Rightarrow \mathcal{K}' \not\models \bot\}$$

for $\mathcal{K} \in \mathbb{K}$.

The work [40] considers different families of inconsistency measures based on many-valued logics. We focus here on the three instantiations $\mathcal{I}_{t_\mathrm{prod}}^\mathrm{fuz}$, $\mathcal{I}_{t_\mathrm{min}}^{\mathrm{fuz},\Sigma}$, $\mathcal{I}_{t_\mathrm{prod}}^{\mathrm{fuz},\Sigma}$ based on Fuzzy logic.

A *fuzzy product interpretation* ω is a function $\omega : \mathcal{L}(\mathsf{At}) \to [0,1]$ satisfying $\omega(\neg \alpha) = 1 - \omega(\alpha)$, $\omega(\alpha \wedge \beta) = \omega(\alpha)\omega(\beta)$, and $\omega(\alpha \vee \beta) = \omega(\alpha) + \omega(\beta) - \omega(\alpha)\omega(\beta)$. A *fuzzy minimum interpretation* ω is a function $\omega : \mathcal{L}(\mathsf{At}) \to [0,1]$ satisfying $\omega(\neg \alpha) = 1 - \omega(\alpha)$, $\omega(\alpha \wedge \beta) = \min(\omega(\alpha), \omega(\beta))$, and $\omega(\alpha \vee \beta) = \max(\omega(\alpha), \omega(\beta))$. Let Ω_prod and Ω_min be the sets of all fuzzy product interpretations and fuzzy minimum interpretations, respectively.

Definition 18 ([40]). The product fuzzy inconsistency measure $\mathcal{I}^{\text{fuz}}_{t_{\text{prod}}}$ is defined as

$$\mathcal{I}^{\text{fuz}}_{t_{\text{prod}}}(\mathcal{K}) = \min\{1 - \omega(\bigwedge \mathcal{K}) \mid \omega \in \Omega_{\text{prod}}\}$$

for $\mathcal{K} \in \mathbb{K}$.

Definition 19 ([40]). The minimum-sum fuzzy inconsistency measure $\mathcal{I}^{\text{fuz},\Sigma}_{t_{\min}}$ is defined as

$$\mathcal{I}^{\text{fuz},\Sigma}_{t_{\min}}(\mathcal{K}) = \min\{\sum_{\alpha \in \mathcal{K}}(1 - \omega(\alpha)) \mid \omega \in \Omega_{\min}\}$$

for $\mathcal{K} \in \mathbb{K}$.

Definition 20 ([40]). The product-sum fuzzy inconsistency measure $\mathcal{I}^{\text{fuz},\Sigma}_{t_{\text{prod}}}$ is defined as

$$\mathcal{I}^{\text{fuz},\Sigma}_{t_{\text{prod}}}(\mathcal{K}) = \min\{\sum_{\alpha \in \mathcal{K}}(1 - \omega(\alpha)) \mid \omega \in \Omega_{\text{prod}}\}$$

for $\mathcal{K} \in \mathbb{K}$.

Note that we do not consider the *minimum fuzzy inconsistency measure*, i.e., the variant of Definition 18 with minimum product interpretation, as this is equivalent to $\mathcal{I}_d$ [40].

A set of maximal consistent subsets $\mathcal{C} \subseteq \text{MC}(\mathcal{K})$ is called an **MC**-*cover* if

$$\bigcup_{C \in \mathcal{C}} C = K$$

An **MC**-cover $\mathcal{C}$ is *normal* if no proper subset of $\mathcal{C}$ is an **MC**-cover. A normal **MC**-cover is maximal if

$$\lambda(\mathcal{C}) = |\bigcap_{C \in \mathcal{C}} C|$$

is maximal for all normal **MC**-covers.

Definition 21 ([1]). The *MCSC inconsistency measure* $\mathcal{I}_{mcsc} : \mathbb{K} \to \mathbb{R}^\infty_{\geq 0}$ is defined as

$$\mathcal{I}_{mcsc}(\mathcal{K}) = |\mathcal{K}| - \lambda(\mathcal{C})$$

for all $\mathcal{K} \in \mathbb{K}$ and any maximal **MC**-cover $\mathcal{C}$. If there is no maximal **MC**-cover we define $\mathcal{I}_{mcsc}(\mathcal{K}) = |\mathcal{K}|$.

Note that the case of the non-existence of a (maximal) MC-cover happens when $\mathcal{K}$ contains an inconsistent formula such as $a \wedge \neg a$. This special case was only implicit in [1].

For a formula ϕ let $\phi[a_1, i_1 \to \psi_1; \ldots, a_k, i_k \to \psi_k]$ denote the formula ϕ where the i_jth occurrence of the proposition a_j is replaced by the formula ψ_j, for all $j = 1, \ldots, k$. For example,

$$(a \wedge b \vee (\neg a \wedge b))[a, 2 \to \top; b, 1 \to \bot] = (a \wedge \bot \vee (\neg \top \wedge b))$$

Definition 22 ([4][2]). The *forgetting-based inconsistency measure* $\mathcal{I}_{\text{forget}} : \mathbb{K} \to \mathbb{R}_{\geq 0}^{\infty}$ is defined as

$$\mathcal{I}_{\text{forget}}(\mathcal{K}) = \min\{k \mid (\bigwedge \mathcal{K})[a_1, i_1 \to \phi_1; \ldots; a_k, i_k \to \phi_k] \not\models \bot, \phi_j \in \{\bot, \top\}\}$$

for all $\mathcal{K} \in \mathbb{K}$.

A set $\{K_1, \ldots, K_n\}$ of pairwise disjoint subsets of $\mathcal{K}$ is called a *conditional independent MUS partition* of $\mathcal{K}$, iff each K_i is inconsistent and $\mathsf{MI}(K_1 \cup \ldots \cup K_n)$ is the disjoint union of all $\mathsf{MI}(K_i)$.

Definition 23 ([18]). The *CC inconsistency measure* $\mathcal{I}_{CC} : \mathbb{K} \to \mathbb{R}_{\geq 0}^{\infty}$ is defined as

$$\mathcal{I}_{CC}(\mathcal{K}) = \max\{n \mid \{K_1, \ldots, K_n\} \text{ is a conditional independent MUS partition of } \mathcal{K}\}$$

for all $\mathcal{K} \in \mathbb{K}$.

Definition 24 ([17]). The *independent set-based inconsistency measure* $\mathcal{I}_{\text{is}} : \mathbb{K} \to \mathbb{R}_{\geq 0}^{\infty}$ is defined as

$$\mathcal{I}_{\text{is}}(\mathcal{K}) = \ln |\{K \subseteq \mathsf{MI}(\mathcal{K}) \mid K \text{ consists of pairwise disjoint subsets}\}|$$

for all $\mathcal{K} \in \mathbb{K}$.

Note that [17] did not explicitly define the basis of the logarithm used in the previous definition. As the exact choice only changes the scaling behaviour of the measure, we make it explicit and use the natural logarithm.

The formal definitions of the considered inconsistency measures are summarised in Figure 1.

We conclude this section with a small example illustrating the behavior of the considered inconsistency measures.

Example 1. Let $\mathcal{K}_8$ and $\mathcal{K}_9$ be given as

$$\mathcal{K}_8 = \{a, b \vee c, \neg a \wedge \neg b, d\} \qquad \mathcal{K}_9 = \{a, \neg a, b, \neg b\}$$

---

[2]Note that we give a slightly different but equivalent formalization.

$$\mathcal{I}_d(\mathcal{K}) = \begin{cases} 1 & \text{if } \mathcal{K} \models \bot \\ 0 & \text{otherwise} \end{cases}$$

$$\mathcal{I}_{\mathsf{MI}}(\mathcal{K}) = |\mathsf{MI}(\mathcal{K})|$$

$$\mathcal{I}_{\mathsf{MI}^C}(\mathcal{K}) = \sum_{M \in \mathsf{MI}(\mathcal{K})} \frac{1}{|M|}$$

$$\mathcal{I}_\eta(\mathcal{K}) = 1 - \max\{\xi \mid \exists P \in \mathcal{P}(\mathsf{At}) : \forall \alpha \in \mathcal{K} : P(\alpha) \geq \xi\}$$

$$\mathcal{I}_c(\mathcal{K}) = \min\{|v^{-1}(B)| \mid v \models^3 \mathcal{K}\}$$

$$\mathcal{I}_{mc}(\mathcal{K}) = |\mathsf{MC}(\mathcal{K})| + |\mathsf{SC}(\mathcal{K})| - 1$$

$$\mathcal{I}_p(\mathcal{K}) = |\bigcup_{M \in \mathsf{MI}(\mathcal{K})} M|$$

$$\mathcal{I}_{hs}(\mathcal{K}) = \min\{|H| \mid H \text{ is a hitting set of } \mathcal{K}\} - 1$$

$$\mathcal{I}_{\mathrm{dalal}}^\Sigma(\mathcal{K}) = \min\{\sum_{\alpha \in \mathcal{K}} d_{\mathsf{d}}(\mathsf{Mod}(\alpha), \omega) \mid \omega \in \Omega(\mathsf{At})\}$$

$$\mathcal{I}_{\mathrm{dalal}}^{\max}(\mathcal{K}) = \min\{\max_{\alpha \in \mathcal{K}} d_{\mathsf{d}}(\mathsf{Mod}(\alpha), \omega) \mid \omega \in \Omega(\mathsf{At})\}$$

$$\mathcal{I}_{\mathrm{dalal}}^{\mathrm{hit}}(\mathcal{K}) = \min\{|\{\alpha \in \mathcal{K} \mid d_{\mathsf{d}}(\mathsf{Mod}(\alpha), \omega) > 0\}| \mid \omega \in \Omega(\mathsf{At})\}$$

$$\mathcal{I}_{D_f}(\mathcal{K}) = 1 - \Pi_{i=1}^{|\mathcal{K}|}(1 - R_i(\mathcal{K})/i)$$

$$\mathcal{I}_{P_m}(\mathcal{K}) = \sum_{a \in \mathsf{At}} |P_m^{\mathcal{K}}(a)| \cdot |P_m^{\mathcal{K}}(\neg a)|$$

$$\mathcal{I}_{mv}(\mathcal{K}) = \frac{|\bigcup_{M \in \mathsf{MI}(\mathcal{K})} \mathsf{At}(M)|}{|\mathsf{At}(\mathcal{K})|}$$

$$\mathcal{I}_{nc}(\mathcal{K}) = |\mathcal{K}| - \max\{n \mid \forall \mathcal{K}' \subseteq \mathcal{K} : |\mathcal{K}'| = n \Rightarrow \mathcal{K}' \not\models \bot\}$$

$$\mathcal{I}_{t_{\mathrm{prod}}}^{\mathrm{fuz}}(\mathcal{K}) = \min\{1 - \omega(\bigwedge \mathcal{K}) \mid \omega \in \Omega_{\mathrm{prod}}\}$$

$$\mathcal{I}_{t_{\min}}^{\mathrm{fuz},\Sigma}(\mathcal{K}) = \min\{\sum_{\alpha \in \mathcal{K}}(1 - \omega(\alpha)) \mid \omega \in \Omega_{\min}\}$$

$$\mathcal{I}_{t_{\mathrm{prod}}}^{\mathrm{fuz},\Sigma}(\mathcal{K}) = \min\{\sum_{\alpha \in \mathcal{K}}(1 - \omega(\alpha)) \mid \omega \in \Omega_{\mathrm{prod}}\}$$

$$\mathcal{I}_{mcsc}(\mathcal{K}) = |\mathcal{K}| - \lambda(\mathcal{C})$$

$$\mathcal{I}_{\mathrm{forget}}(\mathcal{K}) = \min\{k \mid (\bigwedge \mathcal{K})[a_1, i_1 \to \phi_1; \ldots; a_k, i_k \to \phi_k] \not\models \bot, \phi_j \in \{\bot, \top\}\}$$

$$\mathcal{I}_{CC}(\mathcal{K}) = \max\{n \mid \{K_1, \ldots, K_n\} \text{ is a CI partition of } \mathcal{K}\}$$

$$\mathcal{I}_{\mathrm{is}}(\mathcal{K}) = \ln|\{K \subseteq \mathsf{MI}(\mathcal{K}) \mid K \text{ consists of pairwise disjoint subsets}\}|$$

Figure 1: Definitions of the considered inconsistency measures

Then

$$\mathcal{I}_d(\mathcal{K}_8) = 1 \qquad \mathcal{I}_d(\mathcal{K}_9) = 1$$
$$\mathcal{I}_{\mathsf{MI}}(\mathcal{K}_8) = 1 \qquad \mathcal{I}_{\mathsf{MI}}(\mathcal{K}_9) = 2$$
$$\mathcal{I}_{\mathsf{MI}^{\mathsf{C}}}(\mathcal{K}_8) = 1/2 \qquad \mathcal{I}_{\mathsf{MI}^{\mathsf{C}}}(\mathcal{K}_9) = 1$$
$$\mathcal{I}_\eta(\mathcal{K}_8) = 1/2 \qquad \mathcal{I}_\eta(\mathcal{K}_9) = 1/2$$
$$\mathcal{I}_c(\mathcal{K}_8) = 1 \qquad \mathcal{I}_c(\mathcal{K}_9) = 2$$
$$\mathcal{I}_{mc}(\mathcal{K}_8) = 1 \qquad \mathcal{I}_{mc}(\mathcal{K}_9) = 3$$
$$\mathcal{I}_p(\mathcal{K}_8) = 2 \qquad \mathcal{I}_p(\mathcal{K}_9) = 4$$
$$\mathcal{I}_{hs}(\mathcal{K}_8) = 1 \qquad \mathcal{I}_{hs}(\mathcal{K}_9) = 1$$
$$\mathcal{I}_{\mathrm{dalal}}^{\Sigma}(\mathcal{K}_8) = 1 \qquad \mathcal{I}_{\mathrm{dalal}}^{\Sigma}(\mathcal{K}_9) = 2$$
$$\mathcal{I}_{\mathrm{dalal}}^{\max}(\mathcal{K}_8) = 1 \qquad \mathcal{I}_{\mathrm{dalal}}^{\max}(\mathcal{K}_9) = 1$$
$$\mathcal{I}_{\mathrm{dalal}}^{\mathrm{hit}}(\mathcal{K}_8) = 1 \qquad \mathcal{I}_{\mathrm{dalal}}^{\mathrm{hit}}(\mathcal{K}_9) = 2$$
$$\mathcal{I}_{D_f}(\mathcal{K}_8) \approx 0.083 \qquad \mathcal{I}_{D_f}(\mathcal{K}_9) \approx 0.167$$
$$\mathcal{I}_{P_m}(\mathcal{K}_8) = 1 \qquad \mathcal{I}_{P_m}(\mathcal{K}_9) = 2$$
$$\mathcal{I}_{mv}(\mathcal{K}_8) = 1/2 \qquad \mathcal{I}_{mv}(\mathcal{K}_9) = 1$$
$$\mathcal{I}_{nc}(\mathcal{K}_8) = 3 \qquad \mathcal{I}_{nc}(\mathcal{K}_9) = 3$$
$$\mathcal{I}_{t_{\mathrm{prod}}}^{\mathrm{fuz}}(\mathcal{K}_8) = 0.75 \qquad \mathcal{I}_{t_{\mathrm{prod}}}^{\mathrm{fuz}}(\mathcal{K}_9) = 0.9375$$
$$\mathcal{I}_{t_{\min}}^{\mathrm{fuz},\Sigma}(\mathcal{K}_8) = 1 \qquad \mathcal{I}_{t_{\min}}^{\mathrm{fuz},\Sigma}(\mathcal{K}_9) = 2$$
$$\mathcal{I}_{t_{\mathrm{prod}}}^{\mathrm{fuz},\Sigma}(\mathcal{K}_8) = 1 \qquad \mathcal{I}_{t_{\mathrm{prod}}}^{\mathrm{fuz},\Sigma}(\mathcal{K}_9) = 2$$
$$\mathcal{I}_{mcsc}(\mathcal{K}_8) = 2 \qquad \mathcal{I}_{mcsc}(\mathcal{K}_9) = 4$$
$$\mathcal{I}_{\mathrm{forget}}(\mathcal{K}_8) = 1 \qquad \mathcal{I}_{\mathrm{forget}}(\mathcal{K}_9) = 2$$
$$\mathcal{I}_{CC}(\mathcal{K}_8) = 1 \qquad \mathcal{I}_{CC}(\mathcal{K}_9) = 2$$
$$\mathcal{I}_{\mathrm{is}}(\mathcal{K}_8) \approx 0.693 \qquad \mathcal{I}_{\mathrm{is}}(\mathcal{K}_9) \approx 1.386$$

4 Rationality Postulates

In the following, we recall 18 rationality postulates that have been proposed in the literature [14, 35, 16, 31, 30, 37, 3]. A previous survey of rationality postulates can be found in [41].

The first set of rationality postulates has been proposed in [14] in order to provide a definition of a *basic inconsistency measure*. In order to state these postulates we need one further definition.

Definition 25. A formula $\alpha \in \mathcal{K}$ is called a *free formula* if $\alpha \notin \bigcup \mathsf{MI}(\mathcal{K})$. Let $\mathsf{Free}(\mathcal{K})$ be the set of all free formulas of $\mathcal{K}$.

In other words, a free formula is basically a formula that is not directly participating in any derivation of a contradiction. Using this definition and the concepts already introduced before, the first five rationality postulates of [14] can be stated as follows. For the remainder of this section, let $\mathcal{I}$ be any function $\mathcal{I} : \mathbb{K} \to \mathbb{R}_{\geq 0}^\infty$, $\mathcal{K}, \mathcal{K}' \in \mathbb{K}$, and $\alpha, \beta \in \mathcal{L}(\mathsf{At})$.

Consistency (**CO**) $\mathcal{I}(\mathcal{K}) = 0$ if and only if $\mathcal{K}$ is consistent

Normalization (**NO**) $0 \leq \mathcal{I}(\mathcal{K}) \leq 1$

Monotony (**MO**) If $\mathcal{K} \subseteq \mathcal{K}'$ then $\mathcal{I}(\mathcal{K}) \leq \mathcal{I}(\mathcal{K}')$

Free-formula independence (**IN**) If $\alpha \in \mathsf{Free}(\mathcal{K})$ then
$\mathcal{I}(\mathcal{K}) = \mathcal{I}(\mathcal{K} \setminus \{\alpha\})$

Dominance (**DO**) If $\alpha \not\models \bot$ and $\alpha \models \beta$ then $\mathcal{I}(\mathcal{K} \cup \{\alpha\}) \geq \mathcal{I}(\mathcal{K} \cup \{\beta\})$

The first postulate, **CO**, requires that consistent knowledge bases receive the minimal inconsistency value zero and every inconsistent knowledge base have a strictly positive inconsistency value. This postulate is actually the only generally accepted postulate and describes the minimal requirement for an inconsistency measure. An inconsistency measure $\mathcal{I}$ that satisfies **CO** does not distinguish between consistent knowledge bases and can, at least, distinguish inconsistent knowledge bases from consistent ones.

The postulate **NO** states that the inconsistency value is always in the unit interval, thus allowing inconsistency values to be comparable even if knowledge bases are of different sizes. In later works, this postulate is usually regarded as an optional feature, because many measures tend to assess inconsistency *absolutely* and not *relatively*. The distinction between these two points of view was already made in [9], but a thorough investigation of the implications for taking either view on the validity of other postulates has still to be made.

MO requires that adding formulas to the knowledge base cannot decrease the inconsistency value. Besides **CO** this is the least disputed postulate and most inconsistency measures do satisfy it (see below).

IN states that removing free formulas from the knowledge base should not change the inconsistency value. The motivation here is that free formulas do not participate in inconsistencies and should not contribute to having a certain inconsistency value.

DO says that substituting a consistent formula α by a weaker formula β should not increase the inconsistency value. Here, as β carries less information than α there should be less opportunities for inconsistencies to occur.[3]

---

[3] A weaker version of **DO** has also been discussed in [2, 6]. In this version the additional condition $\alpha \notin \mathcal{K}$ is added to the postulate. The special case $\alpha \in \mathcal{K}$ is usually the reason that measures do not satisfy the original version of **DO**; we leave a thorough study of this weaker version for future work.

The set of postulates was extended in [35] for the case of inconsistency measurement in probabilistic logics. However, we can state these postulates also for propositional logic.

Definition 26. A formula $\alpha \in \mathcal{K}$ is called a *safe formula* if it is consistent and $\mathsf{At}(\alpha) \cap \mathsf{At}(\mathcal{K} \setminus \{\alpha\}) = \emptyset$. Let $\mathsf{Safe}(\mathcal{K})$ be the set of all safe formulas of $\mathcal{K}$.

A formula is safe if its signature is disjoint from the signature of the rest of the knowledge base, cf. the concept of language splitting in belief revision [34, 24]. Every safe formula is also a free formula [35].

Safe-formula independence **(SI)** If $\alpha \in \mathsf{Safe}(\mathcal{K})$ then
$\mathcal{I}(\mathcal{K}) = \mathcal{I}(\mathcal{K} \setminus \{\alpha\})$

Super-Additivity **(SA)** If $\mathcal{K} \cap \mathcal{K}' = \emptyset$ then $\mathcal{I}(\mathcal{K} \cup \mathcal{K}') \geq \mathcal{I}(\mathcal{K}) + \mathcal{I}(\mathcal{K}')$

Penalty **(PY)** If $\alpha \notin \mathsf{Free}(\mathcal{K})$ then $\mathcal{I}(\mathcal{K}) > \mathcal{I}(\mathcal{K} \setminus \{\alpha\})$

The postulate SI requires that removing isolated formulas from a knowledge base cannot change the inconsistency value. This postulate is a weakening of IN, i.e., if a measure $\mathcal{I}$ satisfies IN it also satisfies SI, cf. [35, 41] and Theorem 1.

SA is a strengthening of MO [35] and requires that the sum of the inconsistency values of two disjoint knowledge bases not be larger than the inconsistency value of the joint knowledge base.

PY is the complementary postulate to IN and states that adding a formula participating in an inconsistency must have a positive impact on the inconsistency value.

The following two postulates have been first used in [16]:

MI-separability **(MI)** If $\mathsf{MI}(\mathcal{K} \cup \mathcal{K}') = \mathsf{MI}(\mathcal{K}) \cup \mathsf{MI}(\mathcal{K}')$ and $\mathsf{MI}(\mathcal{K}) \cap \mathsf{MI}(\mathcal{K}') = \emptyset$ then $\mathcal{I}(\mathcal{K} \cup \mathcal{K}') = \mathcal{I}(\mathcal{K}) + \mathcal{I}(\mathcal{K}')$

MI-normalization **(MN)** If $M \in \mathsf{MI}(\mathcal{K})$ then $\mathcal{I}(M) = 1$

MI focuses particularly on the role of minimal inconsistent subsets in the determination of the inconsistency value. It states that the sum of the inconsistency values of two knowledge bases that have "non-interfering" sets of minimal inconsistent subsets should be the same as the inconsistency value of their union.

MN demands that a minimal inconsistent subset is the atomic unit for measuring inconsistency by requiring that the inconsistency value of any minimal inconsistent subset be one.

The following postulates have been proposed in [30] to further define the role of minimal inconsistent subsets in measuring inconsistency[4]:

Attenuation **(AT)** $M, M' \in \mathsf{MI}(\mathcal{K})$ and $|M| > |M'|$ implies $\mathcal{I}(M) < \mathcal{I}(M')$

---

[4]Note that in the previous study on compliance of rationality postulates [41] the postulates AT and EC were stated in a slightly different way, we give here the original definitions.

***Equal Conflict* (EC)** $M, M' \in \mathsf{MI}(\mathcal{K})$ and $|M| = |M'|$ implies $\mathcal{I}(M) = \mathcal{I}(M')$

***Almost Consistency* (AC)** Let $M_1, M_2, \ldots$ be a sequence of minimal inconsistent sets M_i with $\lim_{i \to \infty} |M_i| = \infty$, then $\lim_{i \to \infty} \mathcal{I}(M_i) = 0$

The postulate **AT** states that minimal inconsistent sets of smaller size should have a larger inconsistency value. The motivation of this postulate stems from the *lottery paradox*[5] [25].

The postulate **EC** is the counterpart of **AT** and requires minimal inconsistent subsets of the same size to have the same inconsistency value.

AC considers the inconsistency values on arbitrarily large minimal inconsistent subsets in the limit and requires this to be zero.

The following postulates are from [31].

***Contradiction* (CD)** $\mathcal{I}(\mathcal{K}) = 1$ if and only if for all $\emptyset \neq \mathcal{K}' \subseteq \mathcal{K}$, $\mathcal{K}' \models \bot$

***Free Formula Dilution* (FD)** If $\alpha \in \mathsf{Free}(\mathcal{K})$ then $\mathcal{I}(\mathcal{K}) \geq \mathcal{I}(\mathcal{K} \setminus \{\alpha\})$

CD is meant as an extension of **NO** and states that a knowledge base is maximally inconsistent if all non-empty subsets are inconsistent. Note that **CD** only makes sense if **NO** is satisfied as well. **FD** has been introduced to serve as a weaker version of IN for normalised measures, i.e., measures satisfying **NO**. For those, it may be the case that adding free formulas decreases the inconsistency value as they measure a "relative" amount of inconsistency. We do not consider here the property *Monotony w.r.t. Conflict Ratio* from [31] as it is too specifically tailored for the measure $\mathcal{I}_{D_f}$.

The following property has been mentioned independently in [36] and [10]:

***Irrelevance of Syntax* (SY)** If $\mathcal{K} \equiv_b \mathcal{K}'$ then $\mathcal{I}(\mathcal{K}) = \mathcal{I}(\mathcal{K}')$

SY states that knowledge bases with pairwise equivalent formulas should receive the same inconsistency value.

In [3] a series of further postulates have been discussed. For our current study, we only consider the following two:

***Exchange* (EX)** If $\mathcal{K}' \not\models \bot$ and $\mathcal{K}' \equiv \mathcal{K}''$ then $\mathcal{I}(\mathcal{K} \cup \mathcal{K}') = \mathcal{I}(\mathcal{K} \cup \mathcal{K}'')$

***Adjunction Invariance* (AI)** $\mathcal{I}(\mathcal{K} \cup \{\alpha, \beta\}) = \mathcal{I}(\mathcal{K} \cup \{\alpha \wedge \beta\})$

---

[5]Consider a lottery of n tickets and let a_i be the proposition that ticket i, $i = 1, \ldots, n$ will win. It is known that exactly one ticket will win ($a_1 \vee \ldots \vee a_n$) but each ticket owner assumes that his ticket will not win ($\neg a_i$, $i = 1, \ldots, n$). For $n = 1000$ it is reasonable for each ticket owner to believe that he will not win but for e.g., $n = 2$ it is not. Therefore larger minimal inconsistent subsets can be regarded less inconsistent than smaller ones.

EX is similar in spirit to SY and demands that exchanging consistent parts of the knowledge base with equivalent ones should not change the inconsistency value.

AI demands that the set notation of knowledge bases should be equivalent to the conjunction of its formulas in terms of inconsistency values. In difference to EX note that AI has no precondition on the consistency of the considered formulas.

Note that not all postulates are independent and that some are incompatible. Some relationships are summarised in the following theorem, see [41] for proofs of items 1–8, [2] for proofs of items 9 and 10, proofs of items 11 and 12 are trivial and omitted. In the theorem, a statement "A implies B" is meant to be read as "if a measure satisfies A then it satisfies B"; a statement "A_1, ..., A_n are incompatible" means "there is no measure satisfying A_1, ..., A_n at the same time".

Theorem 1.

1. IN implies SI

2. IN implies FD

3. SA implies MO

4. MN and AC are incompatible

5. MN and CD are incompatible

6. MO implies FD

7. MN, MI, and NO are incompatible

8. MN, SA, and NO are incompatible

9. CO, DO, and SA are incompatible

10. CO, DO, and MI are incompatible

11. MN implies EC

12. MN and AT are incompatible

See also [3, 2] for some more detailed discussions.

Tables 2 and 3 give the complete picture on which inconsistency measure satisfies (✓) or violates (✗) the previously discussed rationality postulates. Some of these results have been shown before in [23, 15, 16, 31, 10, 46, 37, 12, 21, 18, 39, 4, 41, 40][6], marked correspondingly in Tables 2 and 3. The

---

[6]Note that proofs of [37] are for propositional probabilistic logic. As this is a generalization of propositional logic, the results apply here as well.

$\mathcal{I}$	CO	NO	MO	IN	DO	SI	SA	PY	MI	MN
$\mathcal{I}_d$	✓[16]	✓[41]	✓[16]	✓[16]	✓[16]	✓[37]	✗[37]	✗[37]	✗[37]	✓[41]
$\mathcal{I}_{\mathsf{MI}}$	✓[15]	✗[41]	✓[15]	✓[15]	✗[31]	✓[37]	✓[37]	✓[37]	✓[16]	✓[16]
$\mathcal{I}_{\mathsf{MI}^C}$	✓[10]	✗[37]	✓[10]	✓[10]	✗[41]	✓[37]	✓[37]	✓[37]	✓[37]	✗[41]
$\mathcal{I}_\eta$	✓[23]	✓[23]	✓[37]	✓[37]	✓[41]	✓[37]	✗[37]	✗[37]	✗[37]	✗[41]
$\mathcal{I}_c$	✓[10]	✗[41]	✓[10]	✓[10]	✓[41]	✓[41]	✗[41]	✗[41]	✗[41]	✗[41]
$\mathcal{I}_{mc}$	✓[10]	✗[41]	✓[10]	✓[10]	✗[41]	✓[41]	✗[41]	✗[41]	✗[18]	✗[41]
$\mathcal{I}_p$	✓[10]	✗[41]	✓[10]	✓[10]	✗[41]	✓[41]	✗[41]	✓[41]	✗[41]	✗[41]
$\mathcal{I}_{hs}$	✓[39]	✗[41]	✓[39]	✓[39]	✓[39]	✗[39]	✗[41]	✗[39]	✗[41]	✗[41]
$\mathcal{I}_{\mathrm{dalal}}^{\Sigma}$	✓[12]	✗[41]	✓[12]	✓[12]	✗	✓[41]	✗[41]	✗[41]	✗[41]	✗[41]
$\mathcal{I}_{\mathrm{dalal}}^{\max}$	✓[12]	✗[41]	✓[12]	✓[12]	✓[12]	✓[41]	✗[41]	✗[41]	✗[41]	✗[41]
$\mathcal{I}_{\mathrm{dalal}}^{\mathrm{hit}}$	✓[12]	✗[41]	✓[12]	✓[12]	✗	✓[41]	✗[41]	✗[41]	✗[41]	✓[41]
$\mathcal{I}_{D_f}$	✓[31]	✓[31]	✗[41]	✗[41]	✗[41]	✗[41]	✗[41]	✗[41]	✗[41]	✗[41]
$\mathcal{I}_{P_m}$	✗	✗[41]	✓[21]	✗[21]	✗[21]	✓[41]	✓[41]	✓[41]	✗[41]	✗[41]
$\mathcal{I}_{mv}$	✓[46]	✓[41]	✗[41]	✗[41]	✗[41]	✗[41]	✗[41]	✗[41]	✗[41]	✓[41]
$\mathcal{I}_{nc}$	✓[41]	✗[41]	✓[41]	✗	✗	✗	✓[41]	✓[41]	✗[41]	✓[41]
$\mathcal{I}^{\mathrm{fuz}}_{t_{\mathrm{prod}}}$	✓[40]	✓[40]	✓[40]	✗[40]	✗[40]	✓[40]	✗[40]	✗[40]	✗[40]	✗[40]
$\mathcal{I}^{\mathrm{fuz},\Sigma}_{t_{\min}}$	✓[40]	✗[40]	✓[40]	✗[40]	✓[40]	✓[40]	✗[40]	✗[40]	✗[40]	✗[40]
$\mathcal{I}^{\mathrm{fuz},\Sigma}_{t_{\mathrm{prod}}}$	✓[40]	✗[40]	✓[40]	✗[40]	✓[40]	✓[40]	✗[40]	✗[40]	✗[40]	✗[40]
$\mathcal{I}_{mcsc}$	✓[1]	✗	✓[1]	✓[1]	✗[1]	✓	✓[1]	✗	✗[1]	✗
$\mathcal{I}_{\mathrm{forget}}$	✓[4]	✗	✓[4]	✓[4]	✗[4]	✓	✓	✗	✗[18]	✗
$\mathcal{I}_{CC}$	✓	✗	✓[18]	✓	✗	✓	✓	✗[20]	✓[18]	✓
$\mathcal{I}_{\mathrm{is}}$	✓[17]	✗	✓[17]	✓[17]	✗	✓	✓	✓	✓[17]	✓[17]

Table 2: Compliance of inconsistency measures with rationality postulates CO, NO, MO, IN, DO, SI, SA, PY, MI, and MN; previous results are indicated by a super-scripted reference of the original work (some of the results have been shown in multiple publications, only the first occurrence is cited)

proofs and counterexamples of the remaining statements are given in the appendix. Note that in [46] it has been shown that $\mathcal{I}_{mv}$ satisfies restricted versions of MO and IN where only formulas are considered that do not use fresh propositions. Some results reported here correct previous statements. In particular, $\mathcal{I}_{P_m}$ does not satisfy CO as claimed in [21], $\mathcal{I}_{nc}$ does not satisfy IN, SI, and DO as claimed in [41], and both $\mathcal{I}_{\mathrm{dalal}}^{\Sigma}$ and $\mathcal{I}_{\mathrm{dalal}}^{\mathrm{hit}}$ do not satisfy DO as claimed in [12]. Due to a different phrasing of the postulates AT and EC in [41] compared to their original definitions in [30], we also corrected some results pertaining to these. See Appendix 7 for the corresponding proofs and counterexamples.

The only rationality postulate that almost all considered measures agree upon is CO, which is not surprising as it captures the minimal requirement for any inconsistency measure.[7] Most measures also satisfy MO, which is also the least disputed in the literature. The only cases where MO fails is usually

---
[7]The fact that $\mathcal{I}_{P_m}$ violates CO is also unintentional as the original paper [21] falsely claimed that CO is satisfied

$\mathcal{I}$	AT	EC	AC	CD	FD	SY	EX	AI
$\mathcal{I}_d$	✗	✓	✗[41]	✗[41]	✓[41]	✓[37]	✓[41]	✓[41]
$\mathcal{I}_{MI}$	✗	✗	✗[41]	✗[41]	✓[41]	✓[10]	✗[41]	✗[41]
$\mathcal{I}_{MI^C}$	✓[41]	✓[41]	✓[41]	✗[41]	✓[41]	✓[10]	✗[41]	✗[41]
$\mathcal{I}_\eta$	✓[41]	✓[41]	✓[41]	✗[41]	✓[41]	✓[37]	✗[41]	✗[41]
$\mathcal{I}_c$	✗[41]	✗[41]	✗[41]	✗[41]	✓[41]	✗[10]	✓[41]	✓[41]
$\mathcal{I}_{mc}$	✗[41]	✓	✗[41]	✗[41]	✓[41]	✓[10]	✗[41]	✗[41]
$\mathcal{I}_p$	✗[41]	✓[41]	✗[41]	✗[41]	✓[41]	✓[10]	✗[41]	✗[41]
$\mathcal{I}_{hs}$	✗	✓	✗[41]	✗[41]	✓[41]	✓[39]	✗[41]	✗[41]
$\mathcal{I}_{dalal}^\Sigma$	✗[41]	✗[41]	✗[41]	✗[41]	✓[41]	✓[41]	✗[41]	✗[41]
$\mathcal{I}_{dalal}^{max}$	✗[41]	✗[41]	✗[41]	✗[41]	✓[41]	✓[41]	✗[41]	✗[41]
$\mathcal{I}_{dalal}^{hit}$	✗	✓	✗[41]	✗[41]	✓[41]	✓[41]	✗[41]	✗[41]
$\mathcal{I}_{D_f}$	✓[31]	✓[41]	✓[31]	✓[31]	✓[31]	✓[41]	✗[41]	✗[41]
$\mathcal{I}_{P_m}$	✗[41]	✗[41]	✗[41]	✗[41]	✓[41]	✗[41]	✗[41]	✗[41]
$\mathcal{I}_{mv}$	✗	✓	✗[41]	✗[41]	✗[41]	✗[41]	✗[41]	✗[41]
$\mathcal{I}_{nc}$	✗	✓	✗[41]	✗[41]	✓[41]	✓[41]	✗[41]	✗[41]
$\mathcal{I}_{t_{prod}}^{fuz}$	✗[40]	✗[40]	✗[40]	✗[40]	✓[40]	✗[40]	✗[40]	✓[40]
$\mathcal{I}_{t_{min}}^{fuz,\Sigma}$	✗[40]	✓	✗[40]	✗[40]	✓[40]	✓[40]	✗[40]	✗[40]
$\mathcal{I}_{t_{prod}}^{fuz,\Sigma}$	✗[40]	✗[40]	✗[40]	✗[40]	✓[40]	✗[40]	✗[40]	✗[40]
$\mathcal{I}_{mcsc}$	✗	✓	✗	✗	✓	✓	✗	✗
$\mathcal{I}_{forget}$	✗	✗	✗	✗	✓	✗	✗[4]	✓[4]
$\mathcal{I}_{CC}$	✗	✓	✗	✗	✓	✓	✗	✗
$\mathcal{I}_{is}$	✗	✓	✗	✗	✓	✓	✗	✗

Table 3: Compliance of inconsistency measures with rationality postulates AT, EC, AC, CD, FD, SY, EX, and AI; previous results are indicated by a superscripted reference of the original work (some of the results have been shown in multiple publications, only the first occurrence is cited)

when NO is satisfied, cf. $\mathcal{I}_{D_f}$ and $\mathcal{I}_{mv}$. However, note that MO and NO are not generally incompatible as e.g. $\mathcal{I}_\eta$ satisfies both. Some other postulates are violated by most of the considered inconsistency measures, in particular if they address a very specific feature. For example, CD is motivated by the measure $\mathcal{I}_{D_f}$—which is also the only one satisfying it—and can be seen as the counterpart to CO as it describes a concept of *maximal inconsistency*. Of course, requiring that a *maximally inconsistent* knowledge base receive the maximal possible inconsistency value is a desirable property. The specific instance of this requirement in CD, i.e., that *maximal inconsistency* is defined by not having non-empty consistent subsets and that the maximal value is 1, is very specific to $\mathcal{I}_{D_f}$. The value 1 only makes sense when the measure is normalized, so that 1 is indeed the maximal possible value. Moreover, the definition of *maximal inconsistency* requires some more investigation.

One important thing to note from the results shown in Tables 2 and 3, is that all investigated inconsistency measures satisfy different sets of postulates. More precisely, there are no two inconsistency measures $\mathcal{I}$ and $\mathcal{I}'$ that satisfy and violate the exact same set of postulates. This also means that we can find knowledge bases $\mathcal{K}, \mathcal{K}'$ such that $\mathcal{I}(\mathcal{K}) < \mathcal{I}(\mathcal{K}')$ and $\mathcal{I}'(\mathcal{K}) \geq \mathcal{I}'(\mathcal{K}')$, meaning that all considered inconsistency measures are essentially different.[8]

5 Expressivity

The drastic inconsistency measure $\mathcal{I}_d$ (see Figure 1) is usually considered as a very naive baseline approach for inconsistency measurement. Surprisingly, this measure already satisfies many rationality postulates, cf. Tables 2 and 3. What sets it apart from other more sophisticated inconsistency measures is that it cannot differentiate between different inconsistent knowledge bases. However, this demand is exactly what inconsistency measures are supposed to satisfy. To address this issue, the work [38] initiated the analysis of the *expressivity* of inconsistency measures. With expressivity of inconsistency measures we here mean the number of different values an inconsistency measure can attain.

Example 2. Consider the knowledge bases $\mathcal{K}_{10}$ and $\mathcal{K}_{11}$ defined via

$$\mathcal{K}_{10} = \{a, b, \neg a \vee \neg b, c, d, \neg c \vee \neg d\}$$
$$\mathcal{K}_{11} = \{a, \neg a, b, \neg b\}$$

Both knowledge bases contain two minimal inconsistent subsets and, thus, $\mathcal{I}_{\mathsf{MI}}$ is not able to differentiate their severity of inconsistency (recall that $\mathcal{I}_{\mathsf{MI}}$ takes the number of minimal inconsistent subsets as the inconsistency values)

$$\mathcal{I}_{\mathsf{MI}}(\mathcal{K}_{10}) = \mathcal{I}_{\mathsf{MI}}(\mathcal{K}_{11}) = 2$$

On the other hand, $\mathcal{I}_{\mathsf{MI^c}}$ does distinguish $\mathcal{K}_{10}$ and $\mathcal{K}_{11}$ (recall that $\mathcal{I}_{\mathsf{MI^c}}$ sums the reciprocal sizes of all minimal inconsistent subsets)

$$\mathcal{I}_{\mathsf{MI^c}}(\mathcal{K}_{10}) = 2/3 \qquad \mathcal{I}_{\mathsf{MI^c}}(\mathcal{K}_{11}) = 1$$

Therefore, $\mathcal{I}_{\mathsf{MI^c}}$ can be regarded as more *expressive* than $\mathcal{I}_{\mathsf{MI}}$ wrt. $\mathcal{K}_{10}$ and $\mathcal{K}_{11}$

Example 3. Consider the family of knowledge bases $\mathcal{K}_{12}^i$ (for $i \in \mathbb{N}$)

$$\mathcal{K}_{12}^i = \{a_1 \wedge \ldots \wedge a_i, \neg a_1 \wedge \ldots \wedge \neg a_i\}$$

Observe that $\mathcal{K}_{12}^i$ contains one minimal inconsistent subset (independently of i) and therefore

$$\mathcal{I}_{\mathsf{MI}}(\mathcal{K}_{12}^i) = 1$$

---

[8]An earlier observation regarding a subset of the investigated measures has been made in [10].

for all $i \in \mathbb{N}$. However, $\mathcal{I}_c$ is able to distinguish every single member of the family (recall that, roughly, $\mathcal{I}_c$ counts the number of propositions which are involved in conflicts)

$$\mathcal{I}_c(\mathcal{K}_{12}^i) = i$$

for $i \in \mathbb{N}$. Therefore, $\mathcal{I}_c$ can be regarded as more *expressive* than $\mathcal{I}_{\mathsf{MI}}$ wrt. $\mathcal{K}_{12}^i$.

In the following, we recall the framework of [38] and investigate the expressivity of inconsistency measures along four different dimensions of subclasses of knowledge bases.

Definition 27. Let ϕ be a formula. The *length* $\texttt{len}(\phi)$ of ϕ is recursively defined as

$$\texttt{len}(\phi) = \begin{cases} 1 & \text{if } \phi \in \mathsf{At} \\ 1 + \texttt{len}(\phi') & \text{if } \phi = \neg \phi' \\ 1 + \texttt{len}(\phi_1) + \texttt{len}(\phi_2) & \text{if } \phi = \phi_1 \wedge \phi_2 \\ 1 + \texttt{len}(\phi_1) + \texttt{len}(\phi_2) & \text{if } \phi = \phi_1 \vee \phi_2 \end{cases}$$

In other words $\texttt{len}(\phi)$ is the number of connectives plus the number of occurrences of atom in ϕ. Furthermore, we treat $\phi_1 \to \phi_2$ as an abbreviation of $\neg \phi_1 \vee \phi_2$ and therefore $\texttt{len}(\phi_1 \to \phi_2) = 2 + \texttt{len}(\phi_1) + \texttt{len}(\phi_2)$.

Definition 28. Define the following subclasses of the set of all knowledge bases $\mathbb{K}$:

$$\mathbb{K}^v(n) = \{\mathcal{K} \in \mathbb{K} \mid |\mathsf{At}(\mathcal{K})| \leq n\}$$
$$\mathbb{K}^f(n) = \{\mathcal{K} \in \mathbb{K} \mid |\mathcal{K}| \leq n\}$$
$$\mathbb{K}^l(n) = \{\mathcal{K} \in \mathbb{K} \mid \forall \phi \in \mathcal{K} : \texttt{len}(\phi) \leq n\}$$
$$\mathbb{K}^p(n) = \{\mathcal{K} \in \mathbb{K} \mid \forall \phi \in \mathcal{K} : |\mathsf{At}(\phi)| \leq n\}$$

In other words, $\mathbb{K}^v(n)$ is the set of all knowledge bases that mention at most n different propositions; $\mathbb{K}^f(n)$ is the set of all knowledge bases that contain at most n formulas; $\mathbb{K}^l(n)$ is the set of all knowledge bases that contain only formulas with maximal length n; and $\mathbb{K}^p(n)$ is the set of all knowledge bases that contain only formulas that mention at most n different propositions each. The motivation for considering these particular subclasses of knowledge bases is that each of them considers a different aspect of the size of a knowledge base. As a syntactical object, a knowledge base is a set of formulas, and both the number of formulas (considered by the class $\mathbb{K}^f(n)$) and the length of each formula ($\mathbb{K}^l(n)$) are the essential parameters that define its size. From a semantical point of view, the number of propositions appearing in each formula ($\mathbb{K}^p(n)$) and in the complete knowledge base ($\mathbb{K}^v(n)$) define the scope of the knowledge. Larger numbers for both of them also indicate larger scope and thus greater size. Inconsistency measures should adhere to the size of the knowledge base in terms of their expressivity. For example, the number of possible inconsistency values of a particular measure should not decrease when moving from a set $\mathbb{K}^v(n)$ to a set $\mathbb{K}^v(n')$ with $n' > n$, as knowledge bases with n' formulas should provide a larger variety in terms of inconsistency than knowledge bases of size n. Indeed, this property is true for all considered measures as $\mathbb{K}^v(n) \subseteq \mathbb{K}^v(n')$ (the same holds for all classes above).

The aim of an expressivity analysis is to investigate the number of different values that a specific inconsistency measure can attain on different subclasses of knowledge bases. This idea can be formalised by *expressivity characteristics* [38].

Definition 29. Let $\mathcal{I}$ be an inconsistency measure and $n > 0$. Let $\alpha \in \{v, f, l, p\}$. The α-*characteristic* $\mathcal{C}^\alpha(\mathcal{I}, n)$ of $\mathcal{I}$ wrt. n is defined as

$$\mathcal{C}^\alpha(\mathcal{I}, n) = |\{\mathcal{I}(\mathcal{K}) \mid \mathcal{K} \in \mathbb{K}^\alpha(n)\}|$$

In other words, $\mathcal{C}^\alpha(\mathcal{I}, n)$ is the number of different inconsistency values $\mathcal{I}$ assigns to knowledge bases from $\mathbb{K}^\alpha(n)$.

Table 4 shows the expressivity characteristics for all measures considered in this paper. Proofs pertaining to measures $\mathcal{I}_d$, $\mathcal{I}_{\mathsf{MI}}$, $\mathcal{I}_{\mathsf{MI}^c}$, $\mathcal{I}_\eta$, $\mathcal{I}_c$, $\mathcal{I}_{mc}$, $\mathcal{I}_p$, $\mathcal{I}_{hs}$, $\mathcal{I}_{\text{dalal}}^{\Sigma}$, $\mathcal{I}_{\text{dalal}}^{\max}$, $\mathcal{I}_{\text{dalal}}^{\text{hit}}$, $\mathcal{I}_{D_f}$, $\mathcal{I}_{P_m}$, $\mathcal{I}_{mv}$, and $\mathcal{I}_{nc}$ can be found in [38]. Proofs pertaining to measures $\mathcal{I}_{t_{\text{prod}}}^{\text{fuz}}$, $\mathcal{I}_{t_{\min}}^{\text{fuz},\Sigma}$, and $\mathcal{I}_{t_{\text{prod}}}^{\text{fuz},\Sigma}$ can be found in [40]. The remaining proofs are given in the appendix.

The evaluation shows that inconsistency measures behave quite differently wrt. expressivity. First, the analysis clearly shows that the inconsistency measure $\mathcal{I}_d$ is indeed a poor inconsistency measure as it has a constant expressivity value in all four considered dimensions. Second, one can categorise measures into different clusters pertaining to each expressivity characteristic and with significant differences between the values in multiple order of magnitudes. For example, for $\mathcal{C}^v$ we have one measure with constant expressivity value ($\mathcal{I}_d$) and several with an expressivity value of linear size ($\mathcal{I}_c$, $\mathcal{I}_{\text{dalal}}^{\max}$, and $\mathcal{I}_{mv}$). Next, there are two measures with an expressivity value of exponential size ($\mathcal{I}_\eta$ and $\mathcal{I}_{hs}$) and, finally, several measures with infinite expressivity values. This gives us a clear superiority relation wrt. each concrete expressivity characteristic. Third, evaluating expressivity depends highly on the characteristic. As one can see from Table 4, the rankings on expressivity induced by the characteristics $\mathcal{C}^v$ and $\mathcal{C}^f$ are reversed in some places. Consider e.g., the measures $\mathcal{I}_p$ and $\mathcal{I}_c$. The measure $\mathcal{I}_p$ has a rather low expressivity value wrt. $\mathcal{C}^f$ but a high value wrt. $\mathcal{C}^v$. Conversely, $\mathcal{I}_c$ has a high expressivity value wrt. $\mathcal{C}^f$ but a rather low value wrt. $\mathcal{C}^v$. Similar observations can be made for other measures. The reason for this is that the expressivity characteristics $\mathcal{C}^f$ and $\mathcal{C}^v$ provide a means to differentiate between so-called syntactic measures and semantical measures, cf. [13]. This categorisation aims at classifying measures on whether they operate on the formula level (syntactic measures) or on the proposition level (semantic measures). While the original definition of syntactical and semantical measure is rather informal, the expressivity characteristics $\mathcal{C}^f$ and $\mathcal{C}^v$ make this distinction more precise. In particular, $\mathcal{C}^v$ measures how susceptible a measure is when the vocabulary, i.e. the semantical side of the knowledge base, is restricted. Semantical measures such as $\mathcal{I}_c$ have a low expressivity when the vocabulary is restricted. On the other hand, $\mathcal{C}^f$ measures how susceptible a measure is when the number of formulas, i.e., the syntactical part, is restricted. Syntactical measures such as $\mathcal{I}_p$ have a low expressivity in this case.

There are some measures ($\mathcal{I}_{\text{dalal}}^{\Sigma}$, $\mathcal{I}_{P_m}$, $\mathcal{I}_{t_{\text{prod}}}^{\text{fuz}}$, $\mathcal{I}_{t_{\text{prod}}}^{\text{fuz},\Sigma}$, $\mathcal{I}_{\text{forget}}$) that have infinite expressivity values in all considered dimensions. Just from the point of view of expressivity, these measures seem to be good candidates for "good" measures. However, as the previous section already discussed, the satisfaction of certain rationality postulates may be of more importance than expressivity.

6 Computational Complexity

The final evaluation criterion we consider is *computational complexity* [32]. Following [44, 43], we consider the following three decision problems one can consider for inconsistency measures. Let $\mathcal{I}$ be some inconsistency measure.

EXACT$_\mathcal{I}$ **Input:** $\mathcal{K} \in \mathbb{K}$, $x \in \mathbb{R}_{\geq 0}^{\infty}$
           Output: TRUE iff $\mathcal{I}(\mathcal{K}) = x$

UPPER$_\mathcal{I}$ **Input:** $\mathcal{K} \in \mathbb{K}$, $x \in \mathbb{R}_{\geq 0}^{\infty}$
           Output: TRUE iff $\mathcal{I}(\mathcal{K}) \leq x$

LOWER$_\mathcal{I}$ **Input:** $\mathcal{K} \in \mathbb{K}$, $x \in \mathbb{R}_{\geq 0}^{\infty} \setminus \{0\}$
           Output: TRUE iff $\mathcal{I}(\mathcal{K}) \geq x$

In other words, EXACT$_\mathcal{I}$ is the problem of deciding whether a given value x is the inconsistency value of a given knowledge base. The problems UPPER$_\mathcal{I}$ and LOWER$_\mathcal{I}$ are about deciding whether a given value x is an upper/lower bound of the inconsistency value of a given knowledge base, respectively.

Furthermore, we consider the following natural function problem:

VALUE$_\mathcal{I}$ **Input:** $\mathcal{K} \in \mathbb{K}$
           Output: The value of $\mathcal{I}(\mathcal{K})$

Table 5 gives an overview of the computational complexity landscape of the considered measures. Proofs of the results pertaining to $\mathcal{I}_d, \mathcal{I}_{\text{MI}}, \mathcal{I}_{\text{MI}^c}, \mathcal{I}_\eta, \mathcal{I}_c, \mathcal{I}_{mc}$, $\mathcal{I}_p, \mathcal{I}_{hs}, \mathcal{I}_{\text{dalal}}^{\text{hit}}, \mathcal{I}_{\text{dalal}}^{\Sigma}, \mathcal{I}_{\text{dalal}}^{\max}, \mathcal{I}_{nc}, \mathcal{I}_{mcsc}, \mathcal{I}_{\text{forget}}, \mathcal{I}_{CC}, \mathcal{I}_{\text{is}}$ can be found in [43], see also [28, 45] for proofs pertaining to some generalisations of $\mathcal{I}_c$. Proofs pertaining to the measure $\mathcal{I}_{mv}$ can be found in [46]. Proofs pertaining to the measures

$\mathcal{I}^{\text{fuz}}_{t_{\text{prod}}}$, $\mathcal{I}^{\text{fuz},\Sigma}_{t_{\text{min}}}$, and $\mathcal{I}^{\text{fuz},\Sigma}_{t_{\text{prod}}}$ for the problems EXACT$_\mathcal{I}$, UPPER$_\mathcal{I}$, and LOWER$_\mathcal{I}$ can be found in [40]. Proofs pertaining to the measures $\mathcal{I}_{D_f}$ and $\mathcal{I}_{P_m}$ can be found in the appendix. We refer to [32] and [44, 43] for the exact definitions of the mentioned complexity classes, which will only informally be discussed below.

The analysis of the computational complexity of different measures shows that measuring inconsistency can be significantly more or less complex depending on the actual measure. In general, measures can be categorised into four different classes [43]. The first class contains measures on the first level of the polynomial hierarchy, i.e., those where the problem UPPER$_\mathcal{I}$ is NP-complete. Under standard complexity-theoretic assumptions (such as assuming that P $\neq$ NP) these measures are significantly easier to deal with than the other measures. In particular, the decision problem UPPER$_\mathcal{I}$ itself is not harder than a satisfiability test in propositional logic and implementations for these measures may benefit from the use of SAT solvers. The next class contains measures on the second level of the polynomial hierarchy, i.e., those where the problem UPPER$_\mathcal{I}$ is Π_2^p-complete. The increase in complexity here is similar (roughly) to the increase in complexity when going from the satisfiability problem in propositional logic to the satisfiability problem in e.g. disjunctive logic programs under the answer set semantics [8]. Solvers for the latter could also be used for the development of implementations for those measures. The measure $\mathcal{I}_{CC}$ is presumably not contained in this second class (although a formal proof is still missing) but Table 5 shows that it is at most on the third level of the polynomial hierarchy, thus again presumably significantly more complex than the previous measures. The final class contains measures beyond the polynomial hierarchy (under standard complexity-theoretic assumptions), i.e., the remaining measures. These measures are inherently more complex as they need to count structures of exponential number (therefore, most of them can be shown to be complete for some "counting" complexity class, those with prefix C). For example, the measure $\mathcal{I}_{\text{MI}}$ is defined to be the number of minimal inconsistent subsets of the knowledge base. This task is hard for two reasons: first, the number of minimal inconsistent subsets may be exponential in the size of the knowledge base, and second, verifying whether some set is indeed a minimal inconsistent subset is hard itself, in fact it is D_1^p-complete [33]. However, for minimal inconsistent sets there are systems available such as [29, 26] that allow the enumeration of those in an effective manner for reasonable problem sizes.

In general, inconsistency measurement is an inherently intractable problem. As the problem of recognising inconsistency is already on the first level of the polynomial hierarchy (it is coNP-complete), we cannot hope for efficient algorithms to *measure* inconsistency (unless P = NP). But this also means that measures in the first category from above are *optimal* wrt. computational complexity (there are minor complexity-theoretic differences in problems other than UPPER$_\mathcal{I}$ for these measures, but this is, arguably, negligible compared to the increase in complexity when moving to the second category of measures).

Implementations of the inconsistency measures discussed in this chapter can also be found in the *Tweety Libraries for Artificial Intelligence* [42] and an online interface is available as well[9].

7 Discussion

Inconsistency measurement is a problem that is not easily defined in a formal manner. Many approaches have been proposed, in particular in recent years, each taking a different perspective on this issue. In this chapter, we addressed the issue of evaluating the appropriateness of these different approaches by considering three different evaluation metrics. First, we discussed rationality postulates. Those aim at prescribing general desirable behaviour of an inconsistency measure and there have also been a lot of proposals in the recent past. Many of them are mutually exclusive, describe orthogonal requirements, and are not generally accepted in the community. Second, we discussed the expressivity of inconsistency measures, i.e., the capability of an inconsistency measure to discriminate between many inconsistent knowledge bases. In general, we expect inconsistency measures to be sensitive towards the addition and deletion of inconsistent parts, so a high expressivity can be regarded as a favourable argument for an inconsistency measure. Finally, we discussed the computational complexity of determining inconsistency values. As deciding inconsistency is (presumably) an intractable problem itself, the task of measuring inconsistency cannot be easier than that. Still, there are differences in the computational complexity of different approaches and it usually better to focus on approaches which are on e.g. the lower levels of the polynomial hierarchy. In order to illustrate the behaviour of these three evaluation metrics, we evaluated a selection of 22 inconsistency measures from the recent literature wrt. those.

This chapter is not intended to identify the best inconsistency measure currently available, but only to highlight their advantages and disadvantages. Different measures behave differently wrt. to the evaluation metrics. The measure $\mathcal{I}_c$ is computationally attractive but its expressivity is limited by the size of the vocabulary. The measure $\mathcal{I}_{\text{dalal}}^{\Sigma}$ has maximal expressivity but fails to satisfy the **DO** rationality postulate. From these observations, only few general assessments on the quality of each measure can be given. In particular, expressivity and computational complexity are objective evaluation metrics. If given two inconsistency measures $\mathcal{I}_1$ and $\mathcal{I}_2$ with identical behaviour, except that $\mathcal{I}_1$ has strictly higher expressivity or strictly lower computational complexity than $\mathcal{I}_2$, then $\mathcal{I}_1$ should be preferred to $\mathcal{I}_2$ (abstractly speaking). On the other hand, rationality postulates are a subjective means to evaluate inconsistency measures as the appropriateness of many of those is not generally agreed upon,

---

[9] http://tweetyproject.org/w/incmes

see e. g. [3]. Tables 2 and 3 showed that the behaviour of the evaluated measures differs significantly in light of the available rationality postulates. But in contrast to expressivity and computational complexity, rationality postulates actually address the underlying issue of formally defining "severity of inconsistency". As a generally agreed upon definition is still an open question, the rationality postulates discussed in this chapter can still serve as a guideline to select an appropriate inconsistency measure wrt. some application. If an application demands the satisfaction of one or more given rationality postulates, among all measures that satisfy those postulates one can select a measure that behaves well wrt. the other evaluation criteria expressivity and computational complexity.

This survey points to a series of open research questions that may be interesting to pursue. For example, the discussion on the "right" set of postulates is not over. The analysis on the compliance of rationality postulates showed that for all postulates we can find an inconsistency measures that satisfies it and another one that violates it. Of course, this situation will only worsen the more measures and postulates are being proposed. What is needed is a characterising definition of an inconsistency measure using few postulates, as the *entropy* is characterised by few simple properties as an information measure. However, we are currently far away from a complete understanding of what an inconsistency constitutes.

Furthermore, our analysis of computational complexity showed that inconsistency measurement may be significantly harder than inconsistency detection (under the usual complexity theoretical assumptions). So far, the *algorithmic study* of inconsistency measurement has (almost) not been investigated at all. Although straightforward prototype implementations of most measures are available (see the remark at the end of the previous section), those implementations do not necessarily optimise runtime performance. Only a few papers [27, 28, 29, 39] have addressed this challenge previously, mainly by developing approximation algorithms. Besides more work on approximation algorithms, another venue for future work is also to develop algorithms that work effectively on certain language fragments—such as certain description logics—and thus may work well in practical applications.

Although we surveyed a rather large selection of inconsistency measures, the analysis is, of course, not complete. Incorporating recent works such as [6, 2, 19] may shed some new light on the issues discussed in this chapter.

References

[1] M. Ammoura, B. Raddaoui, Y. Salhi, and B. Oukacha. On Measuring Inconsistency using Maximal Consistent Sets. In *Proceedings of the 13th*

European Conference on Symbolic and Quantitative Approaches to Reasoning with Uncertainty (ECSQARU'15), pages 267–276. Springer, 2015.

[2] M. Ammoura, Y. Salhi, B. Oukacha, and B. Raddaoui. On an MCS-based Inconsistency Measure. *International Journal of Approximate Reasoning*, 80:443–459, 2017.

[3] Ph. Besnard. Revisiting Postulates for Inconsistency Measures. In *Proceedings of the 14th European Conference on Logics in Artificial Intelligence (JELIA'14)*, pages 383–396, 2014.

[4] Ph. Besnard. Forgetting-based Inconsistency Measure. In *Proceedings of the 10th International Conference on Scalable Uncertainty Management (SUM'16)*, pages 331–337, 2016.

[5] J. Béziau, W. Carnielli, and D. Gabbay, editors. *Handbook of Paraconsistency*. College Publications, London, 2007.

[6] G. De Bona and A. Hunter. Localising Iceberg Inconsistencies. *Artificial Intelligence*, 246:118–151, 2017.

[7] D. Doder, M. Raskovic, Z. Markovic, and Z. Ognjanovic. Measures of Inconsistency and Defaults. *International Journal of Approximate Reasoning*, 51:832–845, 2010.

[8] T. Eiter and G. Gottlob. On the Computational Cost of Disjunctive Logic Programming: Propositional case. *Annals of Mathematics and Artificial Intelligence*, 15(3-4):289–323, 1995.

[9] J. Grant. Classifications for Inconsistent Theories. *Notre Dame Journal of Formal Logic*, 19(3):435–444, 1978.

[10] J. Grant and A. Hunter. Measuring Consistency Gain and Information Loss in Stepwise Inconsistency Resolution. In *Proceedings of the 11th European Conference on Symbolic and Quantitative Approaches to Reasoning with Uncertainty (ECSQARU 2011)*, number 6717 in Lecture Notes in Artificial Intelligence, pages 362–373. Springer-Verlag, 2011.

[11] J. Grant and A. Hunter. Distance-based Measures of Inconsistency. In *Proceedings of the 12th Europen Conference on Symbolic and Quantitative Approaches to Reasoning with Uncertainty (ECSQARU'13)*, volume 7958 of *Lecture Notes in Computer Science*, pages 230–241. Springer, 2013.

[12] J. Grant and A. Hunter. Analysing Inconsistent Information using Distance-based Measures. *International Journal of Approximate Reasoning*, 89:3–26, 2017.

[13] A. Hunter and S. Konieczny. Approaches to Measuring Inconsistent Information. In *Inconsistency Tolerance*, volume 3300 of *Lecture Notes in Computer Science*, pages 189–234. Springer, 2004.

[14] A. Hunter and S. Konieczny. Shapley Inconsistency Values. In *Proceedings of the 10th International Conference on Knowledge Representation (KR'06)*, pages 249–259. AAAI Press, 2006.

[15] A. Hunter and S. Konieczny. Measuring Inconsistency through Minimal Inconsistent Sets. In *Proceedings of the Eleventh International Conference on Principles of Knowledge Representation and Reasoning (KR'2008)*, pages 358–366, Sydney, Australia, September 2008. AAAI Press, Menlo Park, California.

[16] A. Hunter and S. Konieczny. On the Measure of Conflicts: Shapley Inconsistency Values. *Artificial Intelligence*, 174(14):1007–1026, July 2010.

[17] S. Jabbour. On Inconsistency Measuring and Resolving. In *Proceedings of the 22nd European Conference on Artificial Intelligence (ECAI'16)*, pages 1676–1677, 2016.

[18] S. Jabbour, Y. Ma, and B. Raddaoui. Inconsistency Measurement thanks to MUS Decomposition. In *Proceedings of the 13th International Conference on Autonomous Agents and Multiagent Systems (AAMAS 2014)*, pages 877–884, 2014.

[19] S. Jabbour, Y. Ma, B. Raddaoui, and L. Sais. Quantifying Conflicts in Propositional Logic through Prime Implicates. *International Journal of Approximate Reasoning*, 2017.

[20] S. Jabbour, Y. Ma, B. Raddaoui, L. Sais, and Y. Salhi. A MIS Partition based Framework for Measuring Inconsistency. In *Proceedings of the 15th International Conference on Principles of Knowledge Representation and Reasoning (KR'16)*, 2016.

[21] S. Jabbour and B. Raddaoui. Measuring Inconsistency through Minimal Proofs. In *Proceedings of the 12th European Conference on Symbolic and Quantitative Approaches to Reasoning with Uncertainty*, ECSQARU'13, pages 290–301, Berlin, Heidelberg, 2013.

[22] K.M. Knight. Measuring Inconsistency. *Journal of Philosophical Logic*, 31:77–98, 2001.

[23] K.M. Knight. A Theory of Inconsistency. *PhD thesis, University Of Manchester*, July 2002.

[24] G. Kourousias and D.C. Makinson. Parallel Interpolation, Splitting, and Relevance in Belief Change. *Journal of Symbolic Logic*, 72:994–1002, 2007.

[25] H.E. Kyburg. Probability and the Logic of Rational Belief. *Wesleyan University Press*, 1961.

[26] M.H. Liffiton, A. Previti, A. Malik, , and J. Marques-Silva. Fast, Flexible MUS Enumeration. *Constraints*, 21(2):223–250, 2016.

[27] Y. Ma, G. Qi, G. Xiao, P. Hitzler, and Z. Lin. An anytime Algorithm for Computing Inconsistency Measurement. In *Knowledge Science, Engineering and Management*, number 5914 in Lecture Notes in Computer Science, pages 29–40. Springer Berlin/Heidelberg, 2009.

[28] Y. Ma, G. Qi, G. Xiao, P.Hitzler, and Z. Lin. Computational Complexity and Anytime Algorithm for Inconsistency Measurement. *International Journal of Software and Informatics*, 4(1):3–21, 2010.

[29] K. McAreavey, W. Liu, and P. Miller. Computational Approaches to Finding and Measuring Inconsistency in Arbitrary Knowledge bases. *International Journal of Approximate Reasoning*, 55:1659–1693, 2014.

[30] K. Mu, W. Liu, and Z. Jin. A General Framework for Measuring Inconsistency through Minimal Inconsistent Sets. *Knowledge and Information Systems*, 27:85–114, 2011.

[31] K. Mu, W. Liu, Z. Jin, and D. Bell. A Syntax-based Approach to Measuring the Degree of Inconsistency for Belief Bases. *International Journal of Approximate Reasoning*, 52(7), 2011.

[32] C.H. Papadimitriou. Computational Complexity. Addison-Wesley, 1994.

[33] C.H. Papadimitriou and D. Wolfe. The Complexity of Facets Resolved. *Journal of Computer and System Sciences*, 37(1):2–13, 1988.

[34] R. Parikh. Beliefs, Belief revision, and Splitting languages. In *Logic, Language and Computation, Vol. 2*, pages 266–278. CSLI Publications, 1999.

[35] M. Thimm. Measuring Inconsistency in Probabilistic Knowledge bases. In *Proceedings of the Twenty-Fifth Conference on Uncertainty in Artificial Intelligence (UAI'09)*, pages 530–537. AUAI Press, June 2009.

[36] M. Thimm. Analyzing Inconsistencies in Probabilistic Conditional Knowledge Bases using Continuous Inconsistency Measures. In *Proceedings of the Third Workshop on Dynamics of Knowledge and Belief (DKB'11)*, pages 31–45, October 2011.

[37] M. Thimm. Inconsistency Measures for Probabilistic Logics. *Artificial Intelligence*, 197:1–24, April 2013.

[38] M. Thimm. On the Expressivity of Inconsistency Measures. *Artificial Intelligence*, 234:120–151, February 2016.

[39] M. Thimm. Stream-based Inconsistency Measurement. *International Journal of Approximate Reasoning*, 68:68–87, January 2016.

[40] M. Thimm. Measuring Inconsistency with Many-valued Logics. *International Journal of Approximate Reasoning*, 86:1–23, July 2017.

[41] M. Thimm. On the Compliance of Rationality Postulates for Inconsistency Measures: A more or less Complete Picture. *Künstliche Intelligenz*, 31(1):37–39, March 2017.

[42] M. Thimm. The Tweety Library Collection for Logical Aspects of Artificial Intelligence and Knowledge Representation. *Künstliche Intelligenz*, 31(1):93–97, March 2017.

[43] M. Thimm and J.P. Wallner. On the Complexity of Inconsistency Measurement, 2017. Submitted for publication. http://mthimm.de/misc/tw_compim.pdf.

[44] M. Thimm and J.P. Wallner. Some Complexity Results on Inconsistency Measurement. In *Proceedings of the 15th International Conference on Principles of Knowledge Representation and Reasoning (KR'16)*, pages 114–123, April 2016.

[45] G. Xiao, Z. Lin, Y. Ma, and G. Qi. Computing Inconsistency Measurements under Multi-valued Semantics by Partial Max-sat Solvers. In *Proceedings of the Twelfth International Conference on the Principles of Knowledge Representation and Reasoning*, pages 340–349, 2010.

[46] G. Xiao and Y. Ma. Inconsistency Measurement based on Variables in Minimal Unsatisfiable Subsets. In *Proceedings of the 20th European Conference on Artificial Intelligence (ECAI'12)*, 2012.

Appendix: Proofs of Technical Results

Examples 4–6 give counterexamples for some false claims given in the literature, see Tables 2 and 3. General corrections regarding the postulates AT and EC can be found in Theorem 6 below.

Example 4. $\mathcal{I}_{P_m}$ does not satisfy CO as falsely claimed in [21]. Consider the knowledge base $\mathcal{K} = \{\neg(a \land a), a\}$ where $\{a\}$ is the only minimal proof of a and there is no minimal proof for $\neg a$ (as $\{\neg(a \land a), a\}$ does not contain $\neg a$ as a literal). It follows $|P_m^{\mathcal{K}}(a))| = 1$, $|P_m^{\mathcal{K}}(\neg a)| = 0$ and therefore $\mathcal{I}_{P_m}(\mathcal{K}) = 0$, despite the fact that $\mathcal{K}$ is inconsistent.

Example 5. $\mathcal{I}_{nc}$ does not satisfy IN, SI, and DO as falsely claimed in [41]. For SI, consider the knowledge base $\mathcal{K} = \{a \land \neg a\}$ and observe that b is a safe formula in $\mathcal{K} \cup \{b\}$. However, we have $\mathcal{I}_{nc}(\mathcal{K}) = 1$ and $\mathcal{I}_{nc}(\mathcal{K} \cup \{b\}) = 2$ contradicting SI. Due to Theorem 1 $\mathcal{I}_{nc}$ cannot satisfy IN as well. For DO, consider the knowledge base $\mathcal{K}' = \{a, a \land a, a \land a \land a, \neg a\}$, the formula $\alpha = \neg a$, and the formula $\beta = \neg a \land \neg a$. Observe that $\alpha \not\models \bot$ and $\alpha \models \beta$. We have

$$\mathcal{I}_{nc}(\mathcal{K}' \cup \{\alpha\}) = \mathcal{I}_{nc}(\{a, a \land a, a \land a \land a, \neg a\}) = 3$$
$$\mathcal{I}_{nc}(\mathcal{K}' \cup \{\beta\}) = \mathcal{I}_{nc}(\{a, a \land a, a \land a \land a, \neg a, \neg a \land \neg a\}) = 4$$

contradicting DO for $\mathcal{I}_{nc}$.

Example 6. $\mathcal{I}_{\text{dalal}}^{\text{hit}}$ and $\mathcal{I}_{\text{dalal}}^{\Sigma}$ do not satisfy DO as falsely claimed in [11, 12]. Consider the knowledge base $\mathcal{K} = \{a, a \land a, a \land a \land a, \neg a\}$, the formula $\alpha = \neg a$, and the formula $\beta = \neg a \land \neg a$. Observe that $\alpha \not\models \bot$ and $\alpha \models \beta$. We have

$$\mathcal{I}_{\text{dalal}}^{\text{hit}}(\mathcal{K} \cup \{\alpha\}) = \mathcal{I}_{\text{dalal}}^{\text{hit}}(\{a, a \land a, a \land a \land a, \neg a\}) = 1$$
$$\mathcal{I}_{\text{dalal}}^{\text{hit}}(\mathcal{K} \cup \{\beta\}) = \mathcal{I}_{\text{dalal}}^{\text{hit}}(\{a, a \land a, a \land a \land a, \neg a, \neg a \land \neg a\}) = 2$$
$$\mathcal{I}_{\text{dalal}}^{\Sigma}(\mathcal{K} \cup \{\alpha\}) = \mathcal{I}_{\text{dalal}}^{\Sigma}(\{a, a \land a, a \land a \land a, \neg a\}) = 1$$
$$\mathcal{I}_{\text{dalal}}^{\Sigma}(\mathcal{K} \cup \{\beta\}) = \mathcal{I}_{\text{dalal}}^{\Sigma}(\{a, a \land a, a \land a \land a, \neg a, \neg a \land \neg a\}) = 2$$

contradicting DO for both $\mathcal{I}_{\text{dalal}}^{\text{hit}}$ and $\mathcal{I}_{\text{dalal}}^{\Sigma}$.

We now provide proofs for the missing statements regarding the compliance of the rationality postulates of the measures $\mathcal{I}_{mcsc}$, $\mathcal{I}_{\text{forget}}$, $\mathcal{I}_{CC}$, and $\mathcal{I}_{\text{is}}$; see the Tables 2 and 3. For all proofs in the Appendix we denote by $+X$ a proof that shows that property X is satisfied and by $-X$ a proof that shows that property X is violated.

Theorem 2. $\mathcal{I}_{mcsc}$ satisfies SI, EC, FD, and SY. $\mathcal{I}_{mcsc}$ does not satisfy NO, PY, MN, AT, AC, CD, EX, and AI.

Proof.

−NO We have $\mathcal{I}_{mcsc}(\{a, \neg a\}) = 2$, so $\mathcal{I}_{mcsc}$ violates NO.

+SI This follows from IN due to Theorem 1.

−PY Consider $\mathcal{K} = \{a, b, \neg a, \neg b, a \lor b\}$ and observe that $a \lor b \notin \text{Free}(\mathcal{K})$. However, we have $\mathcal{I}_{mcsc}(\mathcal{K}) = \mathcal{I}_{mcsc}(\mathcal{K} \setminus \{a \lor b\}) = 4$.

−**MN** Proposition 2 in [1] showed that $\mathcal{I}_{mcsc}(M) = 2$ for minimal inconsistent sets M with $|M| > 1$.

−**AT** Consider the minimal inconsistent sets $M = \{a, \neg a\}$ and $M' = \{a, b, \neg a \vee \neg b\}$. We have $|M| < |M'|$ but $\mathcal{I}_{mcsc}(M) = \mathcal{I}_{mcsc}(M') = 2$.

+**EC** For any minimal inconsistent set M with $|M| > 1$ we have $\mathcal{I}_{mcsc}(M) = 2$ due to Proposition 6 in [1]. If $|M| = 1$ we have $\mathcal{I}_{mcsc}(M) = 1$.

−**AC** Consider $M_i = \{a_1, \ldots, a_i, \neg a_1 \vee \ldots \vee \neg a_i\}$ for $i \in \mathbb{N}$. Then $\lim_{i \to \infty} |M_i| = \infty$ but $\lim_{i \to \infty} \mathcal{I}_{mcsc}(M_i) = 2$.

−**CD** We have $\mathcal{I}_{mcsc}(\{a \wedge \neg a, b \wedge \neg b\}) = 2$ but every non-empty subset of $\{a \wedge \neg a, b \wedge \neg b\}$ is inconsistent.

+**FD** This follows from **MO** due to Theorem 1.

+**SY** Let $\mathcal{K}, \mathcal{K}'$ be knowledge bases with $\mathcal{K} \equiv_b \mathcal{K}'$ and let s be a bijection $s : \mathcal{K} \to \mathcal{K}'$ such that $\alpha \equiv s(\alpha)$ for all $\alpha \in \mathcal{K}$. Then $|\mathcal{K}| = |\mathcal{K}'|$ and observe that $\{\alpha_1, \ldots, \alpha_k\} \subseteq \mathcal{K}$ is a consistent set if and only if $\{s(\alpha_1), \ldots, s(\alpha_k)\} \subseteq \mathcal{K}'$ is a consistent set. It follows that $\mathcal{I}_{mcsc}(\mathcal{K}) = \mathcal{I}_{mcsc}(\mathcal{K}')$.

−**EX** Consider $\mathcal{K} = \{a \wedge \neg a\}$, $\mathcal{K}' = \{b \wedge c\}$, and $\mathcal{K}'' = \{b, c\}$ and observe that $\mathcal{K}' \equiv \mathcal{K}''$. However, we have $\mathcal{I}_{mcsc}(\mathcal{K} \cup \mathcal{K}') = 2 \neq 3 = \mathcal{I}_{mcsc}(\mathcal{K} \cup \mathcal{K}'')$.

−**AI** We have $\mathcal{I}_{mcsc}(\{a \wedge \neg a\}) = 1 \neq 2 = \mathcal{I}_{mcsc}(\{a, \neg a\})$.

□

Theorem 3. $\mathcal{I}_{\text{forget}}$ satisfies **SI**, **SA**, and **FD**. $\mathcal{I}_{\text{forget}}$ does not satisfy **NO**, **PY**, **MI**, **MN**, **AT**, **EC**, **AC**, **CD**, and **SY**.

Proof.

−**NO** We have $\mathcal{I}_{\text{forget}}(\{a \wedge \neg a, b \wedge \neg b\}) = 2$, so $\mathcal{I}_{\text{forget}}$ violates **NO**.

+**SI** This follows from **IN** due to Theorem 1.

+**SA** Let $\mathcal{K} = \mathcal{K}_1 \cup \mathcal{K}_2$ with $\mathcal{K}_1 \cap \mathcal{K}_2 = \emptyset$ and $\mathcal{I}_{\text{forget}}(\mathcal{K}) = k$. Let $a_1, \ldots, a_k \in \text{At}$, $i_1, \ldots, i_k \in \mathbb{N}$, and $\phi_1, \ldots, \phi_k \in \{\bot, \top\}$ be such that

$$(\bigwedge \mathcal{K})[a_1, i_1 \to \phi_1; \ldots; a_k, i_k \to \phi_k] \not\models \bot$$

As each triple (a_j, i_j, ϕ_j) for $j = 1, \ldots, k$ identifies a replacement in either $\mathcal{K}_1$ or $\mathcal{K}_2$ we can write the above as

$$(\bigwedge \mathcal{K}_1)[a_1, i_1 \to \phi_1; \ldots; a_{k'}, i_{k'} \to \phi_{k'}] \wedge$$
$$(\bigwedge \mathcal{K}_2)[a_{k'+1}, i_{k'+1} \to \phi_{k'+1}; \ldots; a_k, i_k \to \phi_k] \not\models \bot$$

assuming that the $a_1, \ldots, a_k$ are numbered adequately and $1 \le k' \le k$. It follows that

$$(\bigwedge \mathcal{K}_1)[a_1, i_1 \to \phi_1; \ldots; a_{k'}, i_{k'} \to \phi_{k'}] \not\models \bot \quad \text{and}$$

$$(\bigwedge \mathcal{K}_2)[a_{k'+1}, i_{k'+1} \to \phi_{k'+1}; \ldots; a_k, i_k \to \phi_k] \not\models \bot$$

and therefore $\mathcal{I}_{\text{forget}}(\mathcal{K}_1) \le k'$ and $\mathcal{I}_{\text{forget}}(\mathcal{K}_2) \le k - k'$ and therefore $\mathcal{I}_{\text{forget}}(\mathcal{K}) \ge \mathcal{I}_{\text{forget}}(\mathcal{K}_1) + \mathcal{I}_{\text{forget}}(\mathcal{K}_2)$.

−**PY** We have $\mathcal{I}_{\text{forget}}(\{a, \neg a\}) = \mathcal{I}_{\text{forget}}(\{a, a \wedge a, \neg a\}) = 1$ but $a \wedge a \notin \mathsf{Free}(\{a, a \wedge a, \neg a\})$.

−**MI** Consider $\mathcal{K} = \{a \wedge \neg a \wedge c\}$ and $\mathcal{K}' = \{b \wedge \neg b \wedge \neg c\}$ and observe that $\mathsf{MI}(\mathcal{K} \cup \mathcal{K}') = \{\mathcal{K}, \mathcal{K}'\}$, $\mathsf{MI}(\mathcal{K}) = \{\mathcal{K}\}$, $\mathsf{MI}(\mathcal{K}') = \{\mathcal{K}'\}$ but $\mathcal{I}_{\text{forget}}(\mathcal{K} \cup \mathcal{K}') = 3 \ne 2 = 1 + 1 = \mathcal{I}_{\text{forget}}(\mathcal{K}) + \mathcal{I}_c(\mathcal{K}')$.

−**MN** We have $\mathcal{I}_{\text{forget}}(\{a \wedge b, \neg a \wedge \neg b\}) = 2$ but $\{a \wedge b, \neg a \wedge \neg b\}$ is minimally inconsistent.

−**AT** Consider the minimal inconsistent sets $M = \{a \wedge \neg a\}$ and $M' = \{a, \neg a\}$. We have $|M| < |M'|$ but $\mathcal{I}_{\text{forget}}(M) = \mathcal{I}_{\text{forget}}(M') = 1$.

−**EC** Consider the minimal inconsistent sets $M = \{a \wedge \neg a\}$ and $M' = \{a \wedge b \wedge \neg a \wedge \neg b\}$. We have $|M| = |M'|$ but $\mathcal{I}_{\text{forget}}(M) = 1 < 2 = \mathcal{I}_{\text{forget}}(M')$.

−**AC** Consider $M_i = \{a_1, \ldots, a_i, \neg a_1 \vee \ldots \vee \neg a_i\}$ for $i \in \mathbb{N}$. Then $\lim_{i \to \infty} |M_i| = \infty$ but $\lim_{i \to \infty} \mathcal{I}_{\text{forget}}(M_i) = 1$.

−**CD** We have $\mathcal{I}_{\text{forget}}(\{a, \neg a\}) = 1$ but $\{a\} \subseteq \{a, \neg a\}$ is consistent.

+**FD** This follows from **MO** due to Theorem 1.

−**SY** Consider $\mathcal{K} = \{a \wedge \neg a\}$ and $\mathcal{K}' = \{a \wedge \neg a \wedge b \wedge \neg b\}$. Then $\mathcal{K} \equiv_b \mathcal{K}'$ but $\mathcal{I}_{\text{forget}}(\mathcal{K}) = 1 \ne 2 = \mathcal{I}_{\text{forget}}(\mathcal{K}')$.

□

Theorem 4. $\mathcal{I}_{CC}$ satisfies CO, IN, SI, MN, EC, FD, and SY. $\mathcal{I}_{CC}$ does not satisfy NO, DO, PY, AT, AC, CD, EX, and AI.

Proof.

+**CO** For consistent $\mathcal{K}$ the set $\emptyset$ is the only conditional independent MUS partition of $\mathcal{K}$ and therefore $\mathcal{I}_{CC}(\mathcal{K}) = 0$. For inconsistent $\mathcal{K}$, any set $\{M\}$ with $M \in \mathsf{MI}(\mathcal{K})$ is a conditional independent MU partition of $\mathcal{K}$ and therefore $\mathcal{I}_{CC}(\mathcal{K}) \ge 1$.

−**NO** We have $\mathcal{I}_{CC}(\{a, \neg a, b, \neg b\}) = 2$, so $\mathcal{I}_{CC}$ violates NO.

+IN This follows from the fact that S is a conditional independent MUS partition of $\mathcal{K}$ if S is a conditional independent MUS partition of $\mathcal{K} \setminus \{\alpha\}$ for $\alpha \in \mathsf{Free}(\mathcal{K} \setminus \{\alpha\})$.

−DO Consider $\mathcal{K} = \{a, \neg a \wedge b, \neg c\}$, $\alpha = \neg b \wedge c$ and $\beta = c$. Observe $\alpha \not\models \bot$ and $\alpha \models \beta$. However, we have $\mathcal{I}_{CC}(\mathcal{K} \cup \{\alpha\}) = 1 < 2 = \mathcal{I}_{CC}(\mathcal{K} \cup \{\beta\})$.

+SI This follows from IN due to Theorem 1.

−PY We have $\mathcal{I}_{CC}(\{a, \neg a, a \wedge a\}) = \mathcal{I}_{CC}(\{a, \neg a\}) = 1$ but $a \wedge a \notin \mathsf{Free}(\{a, \neg a, a \wedge a\})$.

+MN For a minimal inconsistent M the set $\{M\}$ is the maximal conditional independent MUS partition of M and therefore $\mathcal{I}_{CC}(M) = 1$.

−AT This follows from MN due to Theorem 1.

+EC This follows from MN due to Theorem 1.

−AC This follows from MN due to Theorem 1.

−CD This follows from MN due to Theorem 1.

+FD This follows from MO due to Theorem 1.

+SY Let $\mathcal{K}, \mathcal{K}'$ be knowledge bases with $\mathcal{K} \equiv_b \mathcal{K}'$ and let s be a bijection $s: \mathcal{K} \to \mathcal{K}'$ such that $\alpha \equiv s(\alpha)$ for all $\alpha \in \mathcal{K}$. Then $|\mathcal{K}| = |\mathcal{K}'|$ and observe that $\{M_1, \ldots, M_k\}$ is a conditionally independent MUS partition of $\mathcal{K}$ if and only if

$$\{\{s(\alpha) \mid \alpha \in M_i\} \mid i = 1, \ldots, k\}$$

is a conditionally independent MUS partition of $\mathcal{K}'$. It follows that $\mathcal{I}_{mcsc}(\mathcal{K}) = \mathcal{I}_{mcsc}(\mathcal{K}')$.

−EX Consider $\mathcal{K} = \{\neg a, \neg b\}$, $\mathcal{K}' = \{a, b\}$, and $\mathcal{K}'' = \{a \wedge b\}$. Observe $\mathcal{K}' \equiv \mathcal{K}''$ but $\mathcal{I}_{CC}(\mathcal{K} \cup \mathcal{K}') = 2 \neq 1 = \mathcal{I}_{CC}(\mathcal{K} \cup \mathcal{K}'')$.

−AI The counterexample for EX above also serves as a counterexample for AI. □

Theorem 5. $\mathcal{I}_{is}$ satisfies SI, SA, PY, EC, FD, and SY. $\mathcal{I}_{is}$ does not satisfy NO, DO, AT, AC, CD, EX, and AI.

Proof. **−NO** We have $\mathcal{I}_{is}(\{a, \neg a, b, \neg b\}) = \ln 4 \approx 1.39$, so $\mathcal{I}_{is}$ violates NO.

−DO Consider $\mathcal{K} = \{a, \neg a\}$, $\alpha = a$ and $\beta = a \wedge a$. Observe $\alpha \not\models \bot$ and $\alpha \models \beta$. However, we have $\mathcal{I}_{is}(\mathcal{K} \cup \{\alpha\}) = 1 < \ln 3 = \mathcal{I}_{is}(\mathcal{K} \cup \{\beta\})$.

+**SI** This follows from **IN** due to Theorem 1.

+**SA** Let $\mathcal{K}, \mathcal{K}'$ with $\mathcal{K} \cap \mathcal{K}' = \emptyset$. Then $\mathsf{MI}(\mathcal{K}) \cap \mathsf{MI}(\mathcal{K}') = \emptyset$ and $\mathsf{MI}(\mathcal{K}) \cup \mathsf{MI}(\mathcal{K}') \subseteq \mathsf{MI}(\mathcal{K} \cup \mathcal{K}')$. Then by taking the union of any set of pairwise disjoint subsets of $\mathsf{MI}(\mathcal{K})$ and any set of pairwise disjoint subsets of $\mathsf{MI}(\mathcal{K}')$ one obtains a set of pairwise disjoint subsets of $\mathsf{MI}(\mathcal{K} \cup \mathcal{K}')$ (note that the empty set is a set of pairwise disjoint subsets of both $\mathsf{MI}(\mathcal{K})$ and $\mathsf{MI}(\mathcal{K}')$). If i_S is the number of sets of pairwise disjoint subsets of a set S then $i_{\mathsf{MI}(\mathcal{K} \cup \mathcal{K}')} \geq i_{\mathsf{MI}(\mathcal{K})} i_{\mathsf{MI}(\mathcal{K}')}$. Therefore

$$\begin{aligned}\mathcal{I}_{\mathrm{is}}(\mathcal{K} \cup \mathcal{K}') &= \ln i_{\mathsf{MI}(\mathcal{K} \cup \mathcal{K}')} \\ &\geq \ln i_{\mathsf{MI}(\mathcal{K})} i_{\mathsf{MI}(\mathcal{K}')} \\ &= \ln i_{\mathsf{MI}(\mathcal{K})} + \ln i_{\mathsf{MI}(\mathcal{K}')} \\ &= \mathcal{I}_{\mathrm{is}}(\mathcal{K}) + \mathcal{I}_{\mathrm{is}}(\mathcal{K}')\end{aligned}$$

+**PY** If $\alpha \notin \mathsf{Free}(\mathcal{K})$ then $\mathsf{MI}(\mathcal{K} \setminus \{\alpha\}) \subsetneq \mathsf{MI}(\mathcal{K})$. Then every set of pairwise disjoint subsets S of $\mathsf{MI}(\mathcal{K} \setminus \{\alpha\})$ is also a set of pairwise disjoint subsets of $\mathsf{MI}(\mathcal{K})$. Let $M \in \mathsf{MI}(\mathcal{K})$ with $\alpha \in M$. Then $M \notin \mathsf{MI}(\mathcal{K} \setminus \{\alpha\})$ and $\{M\}$ is a set of pairwise disjoint subsets of $\mathsf{MI}(\mathcal{K})$. Therefore, the set of sets of pairwise disjoint subsets of $\mathsf{MI}(\mathcal{K})$ is a strict superset of the set of sets of pairwise disjoint subsets of $\mathsf{MI}(\mathcal{K} \setminus \{\alpha\})$. The claim follows from the monotonicity of the logarithm.

−**AT** This follows from **MN** due to Theorem 1.

+**EC** This follows from **MN** due to Theorem 1.

−**AC** This follows from **MN** due to Theorem 1.

−**CD** This follows from **MN** due to Theorem 1.

+**FD** This follows from **MO** due to Theorem 1.

+**SY** Let $\mathcal{K}, \mathcal{K}'$ be knowledge bases with $\mathcal{K} \equiv_b \mathcal{K}'$ and let s be a bijection $s : \mathcal{K} \to \mathcal{K}'$ such that $\alpha \equiv s(\alpha)$ for all $\alpha \in \mathcal{K}$. Observe that $M = \{\alpha_1, \ldots, \alpha_k\} \in \mathsf{MI}(\mathcal{K})$ iff $s(M) = \{s(\alpha_1,) \ldots, s(\alpha_k)\} \in \mathsf{MI}(\mathcal{K}')$. It follows that $\{M_1, \ldots, M_l\}$ is a set of pairwise disjoint subsets of $\mathsf{MI}(\mathcal{K})$ iff $\{s(M_1), \ldots, s(M_l)\}$ is a set of pairwise disjoint subsets of $\mathsf{MI}(\mathcal{K}')$ and therefore the claim.

−**EX** Consider $\mathcal{K} = \{\neg a, \neg b\}$, $\mathcal{K}' = \{a, b\}$, and $\mathcal{K}'' = \{a \wedge b\}$. Observe $\mathcal{K}' \equiv \mathcal{K}''$ but $\mathcal{I}_{\mathrm{is}}(\mathcal{K} \cup \mathcal{K}') = \ln 4 \neq \ln 3 = \mathcal{I}_{\mathrm{is}}(\mathcal{K} \cup \mathcal{K}'')$.

−**AI** The counterexample for **EX** above also serves as a counterexample for **AI**. □

The following theorem corrects some previous statements from [41, 40] where the postulates **AT** and **EC** have been stated in different way compared to the original definition from [30]. More precisely, the following results shows the compliance of all considered measures (except $\mathcal{I}_{mcsc}, \mathcal{I}_{\text{forget}}, \mathcal{I}_{CC}$, and $\mathcal{I}_{is}$ which have been dealt with above) with the postulates **AT** and **EC**. It corrects previous results by showing that 1.) $\mathcal{I}_d, \mathcal{I}_{\text{MI}}, \mathcal{I}_{hs}, \mathcal{I}_{mv}, \mathcal{I}_{\text{dalal}}^{\text{hit}}$, and $\mathcal{I}_{nc}$ do not satisfy **AT**, 2.) $\mathcal{I}_{\text{MI}}$ does not satisfy **EC**, and 3.) $\mathcal{I}_d, \mathcal{I}_{mc}, \mathcal{I}_{hs}, \mathcal{I}_{\text{dalal}}^{\text{hit}}, \mathcal{I}_{mv}, \mathcal{I}_{nc}$, and $\mathcal{I}_{t_{\min}}^{\text{fuz},\Sigma}$ satisfy **EC**. All other statements remain unchanged.

Theorem 6. For $\mathcal{I} \in \{\mathcal{I}_{\text{MI}^c}, \mathcal{I}_\eta, \mathcal{I}_{D_f}\}$, $\mathcal{I}$ satisfies **AT**. For $\mathcal{I} \in \{\mathcal{I}_d, \mathcal{I}_{\text{MI}}, \mathcal{I}_c, \mathcal{I}_{mc}, \mathcal{I}_p, \mathcal{I}_{hs}, \mathcal{I}_{\text{dalal}}^{\Sigma}, \mathcal{I}_{\text{dalal}}^{\max}, \mathcal{I}_{\text{dalal}}^{\text{hit}}, \mathcal{I}_{P_m}, \mathcal{I}_{mv}, \mathcal{I}_{nc}, \mathcal{I}_{t_{\text{prod}}}^{\text{fuz}}, \mathcal{I}_{t_{\min}}^{\text{fuz},\Sigma}, \mathcal{I}_{t_{\text{prod}}}^{\text{fuz},\Sigma}\}$, $\mathcal{I}$ violates **AT**. For $\mathcal{I} \in \{\mathcal{I}_d, \mathcal{I}_{\text{MI}}, \mathcal{I}_{\text{MI}^c}, \mathcal{I}_\eta, \mathcal{I}_{mc}, \mathcal{I}_p, \mathcal{I}_{hs}, \mathcal{I}_{\text{dalal}}^{\text{hit}}, \mathcal{I}_{D_f}, \mathcal{I}_{mv}, \mathcal{I}_{nc}, \mathcal{I}_{t_{\min}}^{\text{fuz},\Sigma}\}$, $\mathcal{I}$ satisfies **EC**. For $\mathcal{I} \in \{\mathcal{I}_c, \mathcal{I}_{\text{dalal}}^{\Sigma}, \mathcal{I}_{\text{dalal}}^{\max}, \mathcal{I}_{P_m}, \mathcal{I}_{t_{\text{prod}}}^{\text{fuz}}, \mathcal{I}_{t_{\text{prod}}}^{\text{fuz},\Sigma}\}$, $\mathcal{I}$ violates **EC**.

Proof.

$\mathcal{I}_d$ −**AT** Consider $M = \{a, b, \neg a \vee \neg b\}$ and $M' = \{\neg a, a\}$. We have $|M| > |M'|$ but $\mathcal{I}_d(M) = 1 = \mathcal{I}_d(M')$.

$\mathcal{I}_d$ +**EC** For any pair of minimal inconsistent sets M, M' (independently of whether they have the same cardinality) we always have $\mathcal{I}_d(M) = 1 = \mathcal{I}_d(M')$.

$\mathcal{I}_{\text{MI}}$ −**AT** Consider $M = \{a, b, \neg a \vee \neg b\}$ and $M' = \{\neg a, a\}$. We have $|M| > |M'|$ but $\mathcal{I}_{\text{MI}}(M) = 1 = \mathcal{I}_{\text{MI}}(M')$.

$\mathcal{I}_{\text{MI}}$ +**EC** For any pair of minimal inconsistent sets M, M' (independently of whether they have the same cardinality) we always have $\mathcal{I}_{\text{MI}}(M) = 1 = \mathcal{I}_{\text{MI}}(M')$.

$\mathcal{I}_{\text{MI}^c}$ +**AT** Let M, M' be minimally inconsistent with $|M'| < |M|$. It follows directly that $\mathcal{I}_{\text{MI}^c}(M) = 1/|M| < 1/|M'| = \mathcal{I}_{\text{MI}^c}(M')$.

$\mathcal{I}_{\text{MI}^c}$ +**EC** Let M, M' be minimally inconsistent with $|M'| = |M|$. It follows directly that $\mathcal{I}_{\text{MI}^c}(M) = 1/|M| = 1/|M'| = \mathcal{I}_{\text{MI}^c}(M')$.

$\mathcal{I}_\eta$ +**AT** In [23] it has been shown (Theorem 2.12, slightly rephrased here) that for any minimal inconsistent M, $\mathcal{I}_\eta(M) = 1/|M|$. Then the proof of **AT** is analogous to the corresponding proof for $\mathcal{I}_{\text{MI}^c}$ (see above).

$\mathcal{I}_\eta$ +**EC** In [23] it has been shown (Theorem 2.12, slightly rephrased here) that for any minimal inconsistent M, $\mathcal{I}_\eta(M) = 1/|M|$. Then the proof of **EC** is analogous to the corresponding proof for $\mathcal{I}_{\text{MI}^c}$ (see above).

$\mathcal{I}_c$ −**AT** Consider $M = \{a, \neg a\}$ and $M' = \{\neg a \wedge a\}$. We have $|M| > |M'|$ but $\mathcal{I}_c(M) = 1 = \mathcal{I}_c(M')$.

$\mathcal{I}_c$ **−EC** Consider $M = \{a, \neg a\}$ and $M' = \{a \wedge b, \neg a \wedge \neg b\}$. We have $|M| = |M'|$ but $\mathcal{I}_c(M) = 1 \neq 2 = \mathcal{I}_c(M')$.

$\mathcal{I}_{mc}$ **−AT** Consider $M = \{a, \neg a\}$ and $M' = \{a \wedge \neg a\}$. We have $|M| > |M'|$ but $\mathcal{I}_{mc}(M) = 1 = \mathcal{I}_{mc}(M')$.

$\mathcal{I}_{mc}$ **+EC** Let $M \in \mathsf{MI}(\mathcal{K})$, if $|M| = 1$ then $\mathcal{I}_{mc}(M) = 1$ and if $|M| > 1$ then $\mathcal{I}_{mc}(M) = |M| - 1$.

$\mathcal{I}_p$ **−AT** Consider $M = \{a, b, \neg a \vee \neg b\}$ and $M' = \{\neg a, a\}$. We have $|M| > |M'|$ but $\mathcal{I}_p(M) = 3 > 2 = \mathcal{I}_p(M')$.

$\mathcal{I}_p$ **+EC** For any minimally inconsistent M, $\mathcal{I}_p(M) = |M|$.

$\mathcal{I}_{hs}$ **−AT** Consider $M = \{a, b, \neg a \vee \neg b\}$ and $M' = \{\neg a, a\}$. We have $|M| > |M'|$ but $\mathcal{I}_{hs}(M) = 1 = \mathcal{I}_{hs}(M')$.

$\mathcal{I}_{hs}$ **+EC** For $M \in \mathsf{MI}(\mathcal{K})$ observe that for $|M| = 1$, $\mathcal{I}_{hs}(M) = \infty$ and for $|M| > 1$ we have $\mathcal{I}_{hs}(M) = 1$.

$\mathcal{I}_{\text{dalal}}^{\Sigma}$ **−AT** Consider $M = \{a, b, \neg a \vee \neg b\}$ and $M' = \{\neg a, a\}$. We have $|M| > |M'|$ but $\mathcal{I}_{\text{dalal}}^{\Sigma}(M) = 1 = \mathcal{I}_{\text{dalal}}^{\Sigma}(M')$.

$\mathcal{I}_{\text{dalal}}^{\Sigma}$ **−EC** Consider $M = \{a, \neg a\}$ and $M' = \{a \wedge b, \neg a \wedge \neg b\}$. We have $|M| = |M'|$ but $\mathcal{I}_{\text{dalal}}^{\Sigma}(M) = 1 \neq 2 = \mathcal{I}_{\text{dalal}}^{\Sigma}(M')$.

$\mathcal{I}_{\text{dalal}}^{\max}$ **−AT** Consider $M = \{a, b, \neg a \vee \neg b\}$ and $M' = \{\neg a, a\}$. We have $|M| > |M'|$ but $\mathcal{I}_{\text{dalal}}^{\max}(M) = 1 = \mathcal{I}_{\text{dalal}}^{\max}(M')$.

$\mathcal{I}_{\text{dalal}}^{\max}$ **−EC** Consider $M = \{a, \neg a\}$ and $M' = \{a \wedge b \wedge c, \neg a \wedge \neg b \wedge \neg c\}$. We have $|M| = |M'|$ but $\mathcal{I}_{\text{dalal}}^{\max}(M) = 1 \neq 2 = \mathcal{I}_{\text{dalal}}^{\max}(M')$.

$\mathcal{I}_{\text{dalal}}^{\text{hit}}$ **−AT** Consider $M = \{a, b, \neg a \vee \neg b\}$ and $M' = \{\neg a, a\}$. We have $|M| > |M'|$ but $\mathcal{I}_{\text{dalal}}^{\text{hit}}(M) = 1 = \mathcal{I}_{\text{dalal}}^{\text{hit}}(M')$.

$\mathcal{I}_{\text{dalal}}^{\text{hit}}$ **+EC** Note that $\mathcal{I}_{\text{dalal}}^{\text{hit}}(M) = 1$ for every minimal inconsistent M.

$\mathcal{I}_{D_f}$ **+AT** For $M \in \mathsf{MI}(\mathcal{K})$ observe that $\mathcal{I}_{D_f}(M) = 1/|M|$. Then the proof of AT is analogous to the corresponding proof for $\mathcal{I}_{\mathsf{MI}^c}$ (see above).

$\mathcal{I}_{D_f}$ **+EC** For $M \in \mathsf{MI}(\mathcal{K})$ observe that $\mathcal{I}_{D_f}(M) = 1/|M|$. Then the proof of EC is analogous to the corresponding proof for $\mathcal{I}_{\mathsf{MI}^c}$ (see above).

$\mathcal{I}_{P_m}$ **−AT** Consider $M = \{\neg a, a\}$ and $M' = \{\neg a \wedge a\}$. We have $|M| > |M'|$ but $\mathcal{I}_{P_m}(M) = 1 = \mathcal{I}_{P_m}(M')$.

$\mathcal{I}_{P_m}$ **−EC** Consider $M = \{a, \neg a\}$ and $M' = \{a \wedge b, \neg a \wedge \neg b\}$. We have $|M| = |M'|$ but $\mathcal{I}_{P_m}(M) = 1 \neq 2 = \mathcal{I}_{P_m}(M')$.

$\mathcal{I}_{mv}$ −**AT** Consider $M = \{\neg a, a\}$ and $M' = \{\neg a \wedge a\}$. We have $|M| > |M'|$ but $\mathcal{I}_{mv}(M) = 1 = \mathcal{I}_{mv}(M')$.

$\mathcal{I}_{mv}$ +**EC** Observe $\mathcal{I}_{mv}(M) = 1$ for every minimal inconsistent set M.

$\mathcal{I}_{nc}$ −**AT** Consider $M = \{\neg a, a\}$ and $M' = \{\neg a \wedge a\}$. We have $|M| > |M'|$ but $\mathcal{I}_{nc}(M) = 1 = \mathcal{I}_{nc}(M')$.

$\mathcal{I}_{nc}$ +**EC** Observe $\mathcal{I}_{nc}(M) = 1$ for every minimal inconsistent set M.

$\mathcal{I}^{\text{fuz}}_{t_{\text{prod}}}$ −**AT** Consider $M = \{\neg a, a\}$ and $M' = \{\neg a \wedge a\}$. We have $|M| > |M'|$ but $\mathcal{I}^{\text{fuz}}_{t_{\text{prod}}}(M) = 0.75 = \mathcal{I}^{\text{fuz}}_{t_{\text{prod}}}(M')$.

$\mathcal{I}^{\text{fuz}}_{t_{\text{prod}}}$ −**EC** Consider $M = \{a, \neg a\}$ and $M' = \{a \wedge a, \neg a \wedge \neg a\}$. We have $|M| = |M'|$ but $\mathcal{I}^{\text{fuz}}_{t_{\text{prod}}}(M) = 0.75 \neq 0.9375 = \mathcal{I}^{\text{fuz}}_{t_{\text{prod}}}(M')$.

$\mathcal{I}^{\text{fuz},\Sigma}_{t_{\min}}$ −**AT** Consider $M = \{\neg a, a\}$ and $M' = \{\neg a \wedge a\}$. We have $|M| > |M'|$ but $\mathcal{I}^{\text{fuz},\Sigma}_{t_{\min}}(M) = 1 = \mathcal{I}^{\text{fuz},\Sigma}_{t_{\min}}(M')$.

$\mathcal{I}^{\text{fuz},\Sigma}_{t_{\min}}$ +**EC** Observe that for a minimal inconsistent M with $|M| = 1$, $\mathcal{I}^{\text{fuz},\Sigma}_{t_{\min}}(M) = 1/2$ due to Proposition 3 in [40]. Furthermore, for a minimal inconsistent M with $|M| > 1$ one can see that $\mathcal{I}^{\text{fuz},\Sigma}_{t_{\min}}(M) = 1$ (one can always define a fuzzy minimum interpretation ω in such a way that all but one formula of M are satisfied, i.e., $\omega(\alpha) = 1$, and exactly one formula β has $\omega(\beta) = 0$.

$\mathcal{I}^{\text{fuz},\Sigma}_{t_{\text{prod}}}$ −**AT** Consider $M = \{\neg a, a\}$ and $M' = \{\neg a \wedge a\}$. We have $|M| > |M'|$ but $\mathcal{I}^{\text{fuz},\Sigma}_{t_{\text{prod}}}(M) = 1 = \mathcal{I}^{\text{fuz},\Sigma}_{t_{\text{prod}}}(M')$.

$\mathcal{I}^{\text{fuz},\Sigma}_{t_{\text{prod}}}$ −**EC** Consider $M = \{a, \neg a\}$ and $M' = \{a \wedge b, \neg a \wedge \neg b\}$. We have $|M| = |M'|$ but $\mathcal{I}^{\text{fuz},\Sigma}_{t_{\text{prod}}}(M) = 1 \neq 1.5 = \mathcal{I}^{\text{fuz},\Sigma}_{t_{\text{prod}}}(M')$.

$\square$

We now provide proofs for the missing statements regarding expressivity of the measures $\mathcal{I}_{mcsc}$, $\mathcal{I}_{\text{forget}}$, $\mathcal{I}_{CC}$, and $\mathcal{I}_{\text{is}}$, see Table 4.

Theorem 7. $\mathcal{C}^v(\mathcal{I}_{mcsc}, n) = \mathcal{C}^p(\mathcal{I}_{mcsc}, n) = \infty$, $\mathcal{C}^f(\mathcal{I}_{mcsc}, n) = n + 1$. For $n > 1$, $\mathcal{C}^l(\mathcal{I}_{mcsc}, n) = \infty$.

Proof. Regarding $\mathcal{C}^v(\mathcal{I}_{mcsc}, n) = \mathcal{C}^p(\mathcal{I}_{mcsc}, n) = \infty$, consider the family

$$\mathcal{K}_i = \{\neg a, a, a \wedge a, \ldots, \underbrace{a \wedge \ldots \wedge a}_{i \text{ times}}\}$$

where each $\mathcal{K}_i$ and each formula in $\mathcal{K}_i$ mentions only a single atom, $i \in \mathbb{N}$. Note that each $\mathcal{K}_i$ contains exactly two maximal consistent subsets, namely

$\{\neg a\}$ and $\{a, a \wedge a, \ldots\}$. Those two also comprise the single maximal MC cover (which has an empty intersection). It follows that $\mathcal{I}_{mcsc}(\mathcal{K}_i) = |\mathcal{K}_i| - 0 = i+1$.

Regarding $\mathcal{C}^f(\mathcal{I}_{mcsc}, n) = n + 1$ note that $\mathcal{I}_{mcsc}$ is integer-valued and $\mathcal{I}_{mcsc}(\mathcal{K}) \leq |\mathcal{K}|$ by definition, showing that $\mathcal{C}^f(\mathcal{I}_{mcsc}, n) \leq n + 1$. To see that $\mathcal{C}^f(\mathcal{I}_{mcsc}, n) \geq n+1$ consider $\mathcal{K}_1, \ldots, \mathcal{K}_{n-1}$ from above showing that $\{2, \ldots, n\}$ are possible values for $\mathcal{I}_{mcsc}$ on knowledge bases of size n or smaller. Furthermore, we have $\mathcal{I}_{mcsc}(\emptyset) = 0$ and $\mathcal{I}_{mcsc}(\{a \wedge \neg a\}) = 1$, yielding $\mathcal{C}^f(\mathcal{I}_{mcsc}, n) \geq n + 1$.

Regarding $\mathcal{C}^l(\mathcal{I}_{mcsc}, n) = \infty$ for $n > 1$, consider the family

$$\mathcal{K}'_i = \{a_1, \neg a_1, \ldots, a_i, \neg a_i\}$$

and observe that $\mathcal{I}_{mcsc}(\mathcal{K}'_i) = 2i$. □

Theorem 8. $\mathcal{C}^v(\mathcal{I}_{\text{forget}}, n) = \mathcal{C}^f(\mathcal{I}_{\text{forget}}, n) = \mathcal{C}^p(\mathcal{I}_{\text{forget}}, n) = \infty$. For $n > 1$, $\mathcal{C}^l(\mathcal{I}_{\text{forget}}, n) = \infty$.

Proof. Regarding $\mathcal{C}^v(\mathcal{I}_{\text{forget}}, n) = \mathcal{C}^f(\mathcal{I}_{\text{forget}}, n) = \mathcal{C}^p(\mathcal{I}_{\text{forget}}, n) = \infty$, consider the family

$$\mathcal{K}_i = \{\underbrace{a \wedge \ldots \wedge a}_{i \text{ times}} \wedge \underbrace{\neg a \wedge \ldots \wedge \neg a}_{i \text{ times}}\}$$

where each $\mathcal{K}_i$ mentions only a single atom and consists of a single formula, $i \in \mathbb{N}$. Observe $\mathcal{I}_{\text{forget}}(\mathcal{K}_i) = i$.

Regarding $\mathcal{C}^l(\mathcal{I}_{\text{forget}}, n) = \infty$ for $n > 1$, consider the family

$$\mathcal{K}'_i = \{a_1, \neg a_1, \ldots, a_i, \neg a_i\}$$

and observe that $\mathcal{I}_{\text{forget}}(\mathcal{K}'_i) = i$. □

Theorem 9. $\mathcal{C}^v(\mathcal{I}_{CC}, n) = \mathcal{C}^p(\mathcal{I}_{CC}, n) = \infty$, $\mathcal{C}^f(\mathcal{I}_{CC}, n) = n + 1$. For $n > 1$, $\mathcal{C}^l(\mathcal{I}_{CC}, n) = \infty$.

Proof. Regarding $\mathcal{C}^v(\mathcal{I}_{CC}, n) = \infty$, consider the family

$$\mathcal{K}_i = \{a \wedge \neg a, a \wedge a \wedge \neg a, \ldots, \underbrace{a \wedge \ldots \wedge a}_{i \text{ times}} \wedge \neg a\}$$

with $\mathcal{I}_{CC}(\mathcal{K}_i) = i$.

Regarding $\mathcal{C}^f(\mathcal{I}_{CC}, n) = n + 1$, observe that $\mathcal{I}_{CC}$ is integer-valued. Furthermore, $\mathcal{I}_{CC}(\mathcal{K}) \leq |\mathcal{K}|$ as any CI partition $\{K_1, \ldots, K_n\}$ of $\mathcal{K}$ must satisfy $K_i \cap K_j \neq \emptyset$ for all i, j and therefore $n \leq |\mathcal{K}|$. It follows that $\mathcal{C}^f(\mathcal{I}_{CC}, n) \leq n+1$. For $\mathcal{C}^f(\mathcal{I}_{CC}, n) \geq n + 1$ consider for $i = 0, \ldots, n$ the family

$$\mathcal{K}'_i = \{a_1 \wedge \neg a_1, \ldots, a_i \wedge \neg a_i\}$$

with $|\mathcal{K}'_i| = \mathcal{I}_{CC}(\mathcal{K}'_i) = i$.

Regarding $\mathcal{C}^p(\mathcal{I}_{CC}, n) = \infty$ and $\mathcal{C}^l(\mathcal{I}_{CC}, n) = \infty$ for $n > 1$, consider the family

$$\mathcal{K}''_i = \{a_1, \neg a_1, \ldots, a_i, \neg a_i\}$$

and observe that $\mathcal{I}_{CC}(\mathcal{K}''_i) = i$. □

Theorem 10. $\mathcal{C}^v(\mathcal{I}_{is}, n) = \mathcal{C}^p(\mathcal{I}_{is}, n) = \infty$, $\mathcal{C}^f(\mathcal{I}_{is}, n) \leq 2^{\binom{n}{\lfloor n/2 \rfloor}} + 1$. For $n > 1$, $\mathcal{C}^l(\mathcal{I}_{is}, n) = \infty$

Proof. Regarding $\mathcal{C}^v(\mathcal{I}_{is}, n) = \mathcal{C}^p(\mathcal{I}_{is}, n) = \infty$, consider the family

$$\mathcal{K}_i = \{\neg a, a, a \wedge a, \ldots, \underbrace{a \wedge \ldots \wedge a}_{i \text{ times}}\}$$

where each $\mathcal{K}_i$ and each formula in $\mathcal{K}_i$ mentions only a single atom, $i \in \mathbb{N}$. Note

$$\mathsf{MI}(\mathcal{K}_i) = \{\{\neg a, a\}, \{\neg a, a \wedge a\}, \ldots, \{\neg a, \underbrace{a \wedge \ldots \wedge a}_{i \text{ times}}\}\}$$

It follows that every singleton subset of $\mathsf{MI}(\mathcal{K}_i)$ and the empty set are the only sets of pairwise disjoint subsets of $\mathsf{MI}(\mathcal{K}_i)$. Therefore $\mathcal{I}_{is}(\mathcal{K}_i) = \ln(i+1)$.

Regarding $\mathcal{C}^f(\mathcal{I}_{is}, n) \leq 2^{\binom{n}{\lfloor n/2 \rfloor}} + 1$, recall that $\mathcal{C}^f(\mathcal{I}_{\mathsf{MI}}, n) = \binom{n}{\lfloor n/2 \rfloor} + 1$ [38]. More specifically, the number of minimal inconsistent subsets of a knowledge base with at most n formulas is in $\{0, 1, \ldots, \binom{n}{\lfloor n/2 \rfloor}\}$. If a knowledge base has k minimal inconsistent subsets, i.e. $|\mathsf{MI}(\mathcal{K})| = k$, then there are at most 2^k sets of pairwise disjoint subsets of $\mathsf{MI}(\mathcal{K})$ (if all minimal inconsistent subsets are pairwise disjoint). Furthermore, the empty set is always a set of pairwise disjoint subsets of $\mathsf{MI}(\mathcal{K})$. Therefore, there are between 1 and 2^k sets of pairwise disjoint subsets of $\mathsf{MI}(\mathcal{K})$ (possibly not all values in-between are attained due to combinatorial reasons, but we are only interested in an upper bound here). Taking the case of a consistent knowledge base into account this shows that $\mathcal{C}^f(\mathcal{I}_{is}, n) \leq 2^{\binom{n}{\lfloor n/2 \rfloor}} + 1$.

Regarding $\mathcal{C}^l(\mathcal{I}_{is}, n) = \infty$ for $n > 1$, consider the family

$$\mathcal{K}'_i = \{a_1, \neg a_1, \ldots, a_i, \neg a_i\}$$

and observe that $\mathcal{I}_{is}(\mathcal{K}'_i) = \ln 2^i = i \ln 2$. □

We now provide proofs for the missing statements regarding the computational complexity of the measures $\mathcal{I}_{D_f}$ and $\mathcal{I}_{P_m}$, see Table 5.

Theorem 11. $\mathrm{EXACT}_{\mathcal{I}_{D_f}}$, $\mathrm{UPPER}_{\mathcal{I}_{D_f}}$, $\mathrm{LOWER}_{\mathcal{I}_{D_f}}$ are in **PSPACE** and $\mathrm{VALUE}_{\mathcal{I}_{D_f}}$ is in **FPSPACE**.

Proof. It suffices to show that $\mathcal{I}_{D_f}(\mathcal{K})$ can be computed in polynomial space for all $\mathcal{K}$. Note that the set of all values $|\mathsf{MI}^{(i)}(\mathcal{K})|$ and $|\mathsf{CN}^{(i)}(\mathcal{K})|$ for $i = 1,\ldots,|\mathcal{K}|$ can be stored in polynomial space and that $\mathcal{I}_{D_f}(\mathcal{K})$ can be computed from those values in polynomial space. As we can reuse space, we only need to show that computing each $|\mathsf{MI}^{(i)}(\mathcal{K})|$ and $|\mathsf{CN}^{(i)}(\mathcal{K})|$ for each $i = 1,\ldots,|\mathcal{K}|$ needs at most polynomial space. But this is clear, as we can enumerate each subset S of cardinality i (again reusing space), perform a check whether $S \in \mathsf{MI}^{(i)}(\mathcal{K})$ (or $S \in \mathsf{CN}^{(i)}(\mathcal{K})$) and update some counter. Note that $S \in \mathsf{MI}^{(i)}(\mathcal{K})$ and $S \in \mathsf{CN}^{(i)}(\mathcal{K})$ can be verified by enumerating all interpretations and checking for satisfiability (and additionally for $S \in \mathsf{MI}^{(i)}(\mathcal{K})$ checking each subset with one element less for satisfiability). This can all be done in polynomial space. □

Theorem 12. $\text{EXACT}_{\mathcal{I}_{P_m}}$, $\text{UPPER}_{\mathcal{I}_{P_m}}$, $\text{LOWER}_{\mathcal{I}_{P_m}}$ are in **PSPACE** and $\text{VALUE}_{\mathcal{I}_{P_m}}$ is in **FPSPACE**.

Proof. It suffices to show that $\mathcal{I}_{P_m}(\mathcal{K})$ can be computed in polynomial space for all $\mathcal{K}$. We now sketch an algorithm for computing $\mathcal{I}_{P_m}(\mathcal{K})$ running in polynomial space. For each proposition a we keep two counters c_a and $c_{\neg a}$ that keeps track of the number of minimal proofs we encountered for a and $\neg a$, respectively. Note that we only need polynomial space to store these counters. Then by reusing space we enumerate each subset S of $\mathcal{K}$ and check for each proposition a whether S is a minimal proof for a and/or $\neg a$, and update the corresponding counter. Note that checking whether a set S is a minimal proof for some α can be done by enumerating all interpretations (one after the other) and checking for entailment. □

	$\mathcal{C}^v(\mathcal{I},n)$	$\mathcal{C}^f(\mathcal{I},n)$	$\mathcal{C}^l(\mathcal{I},n)$	$\mathcal{C}^p(\mathcal{I},n)$
$\mathcal{I}_d$	2	2	2*	2
$\mathcal{I}_{\mathsf{MI}}$	∞	$\binom{n}{\lfloor n/2 \rfloor}+1$	∞^*	∞
$\mathcal{I}_{\mathsf{MI}^\mathsf{C}}$	∞	$\leq \Psi(n)^\ddagger$	∞^*	∞
$\mathcal{I}_\eta$	$\Phi(2^n)^\dagger$	$\leq \Phi(\binom{n}{\lfloor n/2 \rfloor}))^\dagger$	∞^{**}	∞^*
$\mathcal{I}_c$	$n+1$	∞	∞^*	∞
$\mathcal{I}_{mc}$	∞	$\binom{n}{\lfloor n/2 \rfloor}^{**}$	∞^*	∞
$\mathcal{I}_p$	∞	$n+1$	∞^*	∞
$\mathcal{I}_{hs}$	2^n+1	$n+1$	∞^{**}	∞^*
$\mathcal{I}_{\text{dalal}}^{\Sigma}$	∞	∞^*	∞^*	∞
$\mathcal{I}_{\text{dalal}}^{\max}$	$n+2$	∞^*	$\lfloor (n+7)/3 \rfloor^{**}$	$n+2$
$\mathcal{I}_{\text{dalal}}^{\text{hit}}$	∞	$n+1$	∞^*	∞
$\mathcal{I}_{D_f}$	∞	$\leq \Psi(n)^\ddagger$	∞^*	∞
$\mathcal{I}_{P_m}$	∞	∞	∞^*	∞
$\mathcal{I}_{mv}$	$n+1$	∞^*	∞^*	∞
$\mathcal{I}_{nc}$	∞	$n+1$	∞^*	∞
$\mathcal{I}_{t_{\text{prod}}}^{\text{fuz}}$	∞	∞	∞^*	∞
$\mathcal{I}_{t_{\min}}^{\text{fuz},\Sigma}$	∞	$n+1$	∞^*	∞
$\mathcal{I}_{t_{\text{prod}}}^{\text{fuz},\Sigma}$	∞	∞	∞^*	∞
$\mathcal{I}_{mcsc}$	∞	$n+1$	∞^*	∞
$\mathcal{I}_{\text{forget}}$	∞	∞	∞^*	∞
$\mathcal{I}_{CC}$	∞	$n+1$	∞^*	∞
$\mathcal{I}_{\text{is}}$	∞	$\leq 2^{\binom{n}{\lfloor n/2 \rfloor}}+1$	∞^*	∞

Table 4: Characteristics of inconsistency measures ($n \geq 1$);
*only for $n > 1$; **only for $n > 3$
†$\Phi(x)$ is the number of fractions in the Farey series of order x and can be defined as $\Phi(x) = |\{k/l \mid l = 1,\ldots,x, k = 0,\ldots,l\}|$, see e.g. http://oeis.org/A005728
‡$\Psi(n)$ is the number of profiles of monotone Boolean functions of n variables, see e.g. http://oeis.org/A220880

	EXACT$_\mathcal{I}$	UPPER$_\mathcal{I}$	LOWER$_\mathcal{I}$	VALUE$_\mathcal{I}$
$\mathcal{I}_d$	$D_1^p \cap \mathrm{co}D_1^p$	NP-c	coNP-c	FNP-c
$\mathcal{I}_{MI}$	$C_=$NP-h	CNP-c	CNP-c	$\#\cdot$coNP-c
$\mathcal{I}_{MI^C}$	$C_=$NP-h	CNP-h	CNP-h	$P^{\#\cdot\mathrm{coNP}}$
$\mathcal{I}_\eta$	D_1^p-c	NP-c	coNP-c	$FP^{NP[n]}$
$\mathcal{I}_c$	D_1^p-c	NP-c	coNP-c	$FP^{NP[\log n]}$-c
$\mathcal{I}_{mc}$	$C_=$NP-h	CNP-c	CNP-c	$\#\cdot$coNP-c†
$\mathcal{I}_p$	D_2^p-c	Π_2^p-c	Σ_2^p-c	$FP^{\Sigma_2^p[\log n]}$
$\mathcal{I}_{hs}$	D_1^p-c	NP-c	coNP-c	$FP^{NP[\log n]}$
$\mathcal{I}_{\mathrm{dalal}}^\Sigma$	D_1^p-c	NP-c	coNP-c	$FP^{NP[\log n]}$-c
$\mathcal{I}_{\mathrm{dalal}}^{\max}$	D_1^p-c	NP-c	coNP-c	$FP^{NP[\log n]}$
$\mathcal{I}_{\mathrm{dalal}}^{\mathrm{hit}}$	D_1^p-c	NP-c	coNP-c	$FP^{NP[\log n]}$-c
$\mathcal{I}_{D_f}$	PSPACE	PSPACE	PSPACE	FPSPACE
$\mathcal{I}_{P_m}$	PSPACE	PSPACE	PSPACE	FPSPACE
$\mathcal{I}_{mv}$	D_2^p-c	Π_2^p-c	Σ_2^p-c	$FP^{\Sigma_2^p[\log n]}$
$\mathcal{I}_{nc}$	D_2^p	Π_2^p-c	Σ_2^p-c	$FP^{\Sigma_2^p[\log n]}$
$\mathcal{I}_{t_{\mathrm{prod}}}^{\mathrm{fuz}}$	D_1^p	NP-c	coNP-c	?
$\mathcal{I}_{t_{\mathrm{min}}}^{\mathrm{fuz},\Sigma}$	D_1^p	NP-c	coNP-c	?
$\mathcal{I}_{t_{\mathrm{prod}}}^{\mathrm{fuz},\Sigma}$	D_1^p	NP-c	coNP-c	?
$\mathcal{I}_{mcsc}$	D_1^p-c	NP-c	coNP-c	$FP^{NP[\log n]}$
$\mathcal{I}_{\mathrm{forget}}$	D_1^p-c	NP-c	coNP-c	$FP^{NP[\log n]}$-c
$\mathcal{I}_{CC}$	D_3^p	Π_3^p	Σ_3^p	$FP^{\Sigma_3^p[\log n]}$
$\mathcal{I}_{is}$	$C_=$NP-h	CNP-c	CNP-c	$\#\cdot$coNP-c‡

Table 5: Computational complexity of the considered inconsistency measures (all statements are membership statements, an additionally attached "-c" ("-h") also indicates completeness (hardness) for the class); we note that all hardness results for $\#\cdot$coNP are under subtractive reductions; †we show complexity of the (minor) variation that omits subtracting one from the result; ‡we consider here the problem variant that does not apply a logarithm on the result; "?" indicates unknown results

Measuring Inconsistency in Argument Graphs

Anthony Hunter

Department of Computer Science
University College London, London, UK
anthony.hunter@ucl.ac.uk

Abstract

There have been a number of developments in measuring inconsistency in logic-based representations of knowledge. In constrast, the development of inconsistency measures for computational models of argument has been limited. To address this shortcoming, this paper provides a general framework for measuring inconsistency in abstract argumentation, together with some proposals for specific measures, and a consideration of measuring inconsistency in logic-based instantiations of argument graphs, including a review of some existing proposals and a consideration of how existing logic-based measures of inconsistency can be applied.

1 Introduction

Argumentation is an important cognitive ability for handling conflicting and incomplete information such as beliefs, assumptions, opinions, and goals. When we are faced with a situation where we find that our information is incomplete or inconsistent, we often resort to the use of arguments for and against a given position in order to make sense of the situation. Furthermore, when we interact with other people we often exchange arguments to reach a final agreement and/or to defend and promote an individual position.

In recent years, there has been substantial interest in the development of computational models of argument for capturing aspects of this cognitive ability (for reviews see [6, 13, 36]). This has led to the development of a number of directions including: (1) abstract argument models where arguments are atomic, and the emphasis is on the relationships between arguments; (2) logic-based (or structured) argument models where the emphasis is on the logical structure of the premises and claim of the arguments, and the logical definition of relations between arguments; and (3) dialogical argument models where the emphasis is on the protocols (i.e. allowed and obligatory moves that can be taken at each step of the dialogue) and strategies (i.e. mechanisms used by each participant to make the best choice of move at each step of the dialogue).

At the core of computational models of argument is the ability to represent and reason with inconsistency. So it is perhaps surprising that relatively little consideration has been given to measuring inconsistency in these models, particularly given the number of developments in measuring inconsistency in logic-based knowledgebases (see for example [31, 28, 22, 29, 25, 32, 30, 9, 37]).

A couple of exceptions are the consideration of the degree of undercut between an argument and counterargument [12, 13], and measuring inconsistency through argumentation [35]. Note, the approach of weighted argumentation frameworks [20] is not a measure of inconsistency as the approach assumes extra information (weights) to label each arc, and an inconsistency budget that allows arcs that sum to no more than the budget to be ignored.

There are a number of reasons why it is useful to investigate the measurement of inconsistency in argumentation: (1) to better characterize the nature of inconsistency in argumentation; (2) to analyse the inconsistency arising in specific argumentation situations; and (3) to direct the resolution of inconsistency as arising in argumentation. We will consider contributions to these three areas during the course of this chapter.

Given the central role of argument graphs (where each node is an argument and each arc denotes one argument attacking another) in modelling argumentation, we will consider the inconsistency of an argument graph. This is useful if we want to assess the overall conflict that is manifested by an argument graph, and we want to focus on actions that may allow us to decrease the graph inconsistency.

Consider for example some security analysts who are analyzing some conflicting reports concerning a foreign country that may be descending into civil war. These analysts may enter into a process as follows: (1) they collect relevant information concerning the political and security situation in the country; (2) they construct arguments from this information that draw tentative hypotheses about the situation in the country; (3) they compose these arguments into an argument graph; (4) they measure the inconsistency of the argument graph; (5) they use the measure of inconsistency to identify information requirements (i.e. queries to ascertain whether a particular argument should be accepted or rejected) that would result in commitments being made for some of these arguments; (6) they seek the answers to these queries; (7) they use these commitments to reduce the overall inconsistency of the graph; and (8) they terminate this process when sufficient commitments have been made so as to reduce the inconsistency to a sufficiently low level.

This kind of process may be of relevance to security analysts to augment recent proposals for argument-based security analysis technology such as by Toniolo et al [39]. Furthermore, this kind of process may be replicated in roles such as business intelligence analysis, policy planning, political planning, and science research.

Before we proceed, we need to consider why we are using the term inconsistency, and what we mean by it. When we use argumentation, we are normally considering situations where there exist both arguments and counterarguments. Suppose we have just two arguments A and B where B attacks A. Here, we regard this as an inconsistent situation because accepting both argument would not be conflict-free. In other words, there is an incompatibility between the arguments. One could argue that if we use the dialectical semantics proposed by Dung (as we will review in the next section), we obviate this problem since we would accept B and not accept A, and so in a sense, the inconsistency is resolved. But that does not mean that there is no inconsistency. Indeed, if A has no role, then why even present A in the argument graph? Yet we do present A because part of the role of argumentation is to consider the conflicts that arise, and then make sense of those conflicts by choosing for instance which argument to accept. Choosing which arguments to accept may be by using dialectical semantics. But that is only one course of action. For instance, we may have reasons to doubt B, and so it may then be reasonable to seek further information about B to either confirm it or to reject it (as indicated by the intelligence analysis scenario discussed above). So in general, we see a key part of argumentation is to identify inconsistencies, and then to have mechanisms for dealing with those inconsistencies. Hence, the purpose of this paper is to consider how we might measure these inconsistencies.

We proceed as follows: (Section 2) We review the basic definitions of abstract argumentation, considering both extension-based and label-based approaches; (Section 3) We investigate a general framework for measuring inconsistency in abstract argumentation, together with some proposals for specific measures; (Section 4) We review deductive argumentation for instantiating abstract argument graphs, we review an existing proposal for measuring inconsistency in deductive argumentation called degree of undercut, and we investigate a new approach that harnesses existing logic-based measures; (Section 5) We consider how we can use measures of inconsistency to direct the resolution of inconsistency in argumentation; (Section 6) We conclude with a discussion of the proposals in the chapter and of future work.

2 Review of Abstract Argumentation

Our framework builds on more general developments in the area of computational models of argument. These models aim to reflect how human argumentation uses conflicting information to construct and analyze arguments.

There is a number of frameworks for computational models of argumentation. They incorporate a formal representation of individual arguments and techniques for comparing conflicting arguments (for reviews see [6, 13, 36, 1, 10]). By basing our framework on these models, we can harness theory and tools as the basis of our solution.

2.1 Extension-based Semantics

We start with a brief review of abstract argumentation as proposed by Dung [23]. In this approach, each argument is treated as an atom, and so no internal structure of the argument needs to be identified.

Definition 1. *An **argument graph** is a pair $G = (\mathcal{A}, \mathcal{R})$ where $\mathcal{A}$ is a set and $\mathcal{R}$ is a binary relation over $\mathcal{A}$ (in symbols, $\mathcal{R} \subseteq \mathcal{A} \times \mathcal{A}$). Let $\mathsf{Nodes}(G)$ be the set of nodes in G and let $\mathsf{Arcs}(G) \subseteq \mathsf{Nodes}(G) \times \mathsf{Nodes}(G)$ be the set of arcs in G.*

So an argument graph is a directed graph. Each element $A \in \mathcal{A}$ is called an **argument** and $(A_i, A_j) \in \mathcal{R}$ means that A_i **attacks** A_j (accordingly, A_i is said to be an **attacker** of A_j). So A_i is a **counterargument** for A_j when $(A_i, A_j) \in \mathcal{R}$ holds.

Example 1. *Consider arguments $A_1 =$ "Patient has hypertension so prescribe diuretics", $A_2 =$ "Patient has hypertension so prescribe beta-blockers", and $A_3 =$ "Patient has emphysema which is a contraindication for beta-blockers". Here, we assume that A_1 and A_2 attack each other because we should only give one treatment and so giving one precludes the other, and we assume that A_3 attacks A_2 because it provides a counterargument to A_2. Hence, we get the following abstract argument graph.*

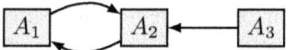

Arguments can work together as a coalition by attacking other arguments and by defending their members from attack as follows.

Definition 2. *Let $S \subseteq \mathcal{A}$ be a set of arguments.*

- *S **attacks** $A_j \in \mathcal{A}$ iff there is an argument $A_i \in S$ such that A_i attacks A_j.*

- *S **defends** $A_i \in S$ iff for each argument $A_j \in \mathcal{A}$, if A_j attacks A_i then S attacks A_j.*

The following gives a requirement that should hold for a coalition of arguments to make sense. If it holds, it means that the arguments in the set offer a consistent view on the topic of the argument graph.

Definition 3. *A set $S \subseteq \mathcal{A}$ of arguments is **conflict-free** iff there are no arguments A_i and A_j in S such that A_i attacks A_j.*

Now, we consider how we can find an acceptable set of arguments from an abstract argument graph. The simplest case of arguments that can be accepted is as follows.

Definition 4. *A set $S \subseteq \mathcal{A}$ of arguments is* **admissible** *iff S is conflict-free and defends all its arguments.*

The intuition here is that for a set of arguments to be accepted, we require that, if any one of them is challenged by a counterargument, then they offer grounds to challenge, in turn, the counterargument. There always exists at least one admissible set: The empty set is always admissible.

Clearly, the notion of admissible sets of arguments is the minimum requirement for a set of arguments to be accepted. We will focus on the following classes of acceptable arguments.

Definition 5. *Let Γ be a conflict-free set of arguments, and let* Defended $: \wp(\mathcal{A}) \mapsto \wp(\mathcal{A})$ *be a function such that* Defended$(\Gamma) = \{A \mid \Gamma \text{ defends } A\}$.

1. *Γ is a* **complete extension** *iff $\Gamma =$* Defended(Γ)

2. *Γ is a* **grounded extension** *iff it is the minimal (w.r.t. set inclusion) complete extension.*

3. *Γ is a* **preferred extension** *iff it is a maximal (w.r.t. set inclusion) complete extension.*

4. *Γ is a* **stable extension** *iff it is a preferred extension that attacks every argument that is not in the extension.*

The grounded extension is always unique, whereas there may be multiple preferred extensions. We illustrate these definitions with the following examples. As can be seen from the examples, the grounded extension provides a skeptical view on which arguments can be accepted, whereas each preferred extension takes a credulous view on which arguments can be accepted.

Example 2. *Continuing Example 1, there is only one complete set, and so this is both grounded and preferred. Note, $\{A_1, A_2\}$, $\{A_2, A_3\}$, and $\{A_1, A_2, A_3\}$ are not conflict-free subsets. Only the conflict-free subsets are given in the table.*

	Conflict-free	Admissible	Complete	Grounded	Preferred	Stable
$\{\}$	✓	✓	✗	✗	✗	✗
$\{A_1\}$	✓	✓	✗	✗	✗	✗
$\{A_2\}$	✓	✗	✗	✗	✗	✗
$\{A_3\}$	✓	✓	✗	✗	✗	✗
$\{A_1, A_3\}$	✓	✓	✓	✓	✓	✓

Example 3. *Consider the following argument graph.*

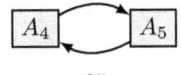

For this, there are two preferred sets, neither of which is grounded. Note $\{A_4, A_5\}$ is not conflict-free. Only the conflict-free subsets are given in the table.

	Conflict-free	Admissible	Complete	Grounded	Preferred	Stable
$\{\}$	✓	✓	✓	✓	×	×
$\{A_4\}$	✓	✓	✓	×	✓	✓
$\{A_5\}$	✓	✓	✓	×	✓	✓

The formalization we have reviewed in this section is abstract because both the nature of the arguments and the nature of the attack relation are ignored. In particular, the internal (logical) structure of each of the arguments is not made explicit. Nevertheless, Dung's proposal for abstract argumentation is ideal for clearly representing arguments and counterarguments, and for intuitively determining which arguments should be accepted (depending on whether we want to take a credulous or skeptical perspective).

Given an argument graph, let Extensions$_\sigma(G)$ denote the set of extensions according to σ where $\sigma = $ co denotes the complete extensions, $\sigma = $ pr denotes the preferred extensions, $\sigma = $ gr denotes the grounded extensions, and $\sigma = $ st denotes the stable extensions.

2.2 Labelling-based Semantics

We now review an alternative way of defining semantics for abstract argumentation proposed by Caminada and Gabbay [16]. A labelling L is a function $L : $ Nodes$(G) \to \{$in, out, undec$\}$ that assigns to each argument $A \in $ Nodes(G) either the value in, meaning that the argument is accepted, out, meaning that the argument is not accepted, or undec, meaning that the status of the argument is undecided. Let in$(L) = \{A \mid L(A) = $ in$\}$ and out(L) resp. undec(L) be defined analogously. The set in(L) for a labelling L is also called an *extension*. A labelling L is called *conflict-free* if for no $A, B \in $ in(L) we have that $(A, B) \in $ Arcs(G).

Definition 6. *A labelling L is called* **admissible** *if and only if for all arguments $A \in $ Nodes(G)*

1. *if $L(A) = $ out then there is $B \in $ Nodes(G) with $L(B) = $ in and $(B, A) \in $ Arcs(G), and*

2. *if $L(A) = $ in then $L(B) = $ out for all $B \in $ Nodes(G) with $(B, A) \in $ Arcs(G),*

and it is called **complete** *if, additionally, it satisfies*

3. *if* $L(A) = \mathsf{undec}$ *then there is no* $B \in \mathsf{Nodes}(G)$ *with* $(B, A) \in \mathsf{Arcs}(G)$ *and* $L(B) = \mathsf{in}$ *and there is a* $B' \in \mathsf{Nodes}(G)$ *with* $(B', A) \in \mathsf{Arcs}(G)$ *and* $L(B') \neq \mathsf{out}$.

The intuition behind admissibility is that an argument can only be accepted if there are no attackers that are accepted and if an argument is not accepted then there have to be some reasonable grounds. The idea behind the completeness property is that the status of an argument is only undec if it cannot be classified as in or out. Different types of classical semantics can be obtained by imposing further constraints.

Definition 7. *Let G be an argument graph, let $L : \mathsf{Nodes}(G) \to \{\mathsf{in}, \mathsf{out}, \mathsf{undec}\}$ be a complete labelling, and for all statements, minimality/maximality is with respect to set inclusion.*

- *L is* **grounded** *if and only if* $\mathsf{in}(L)$ *is minimal.*
- *L is* **preferred** *if and only if* $\mathsf{in}(L)$ *is maximal.*
- *L is* **stable** *if and only if* $\mathsf{undec}(L) = \emptyset$.

Example 4. *Continuing Example 2, there is one complete labelling.*

	A_1	A_2	A_3	Type
L_1	in	out	in	*grounded, preferred, stable*

Example 5. *Continuing Example 3, there are three complete labellings.*

	A_4	A_5	Type
L_1	undec	undec	*grounded*
L_2	in	out	*preferred, stable*
L_3	out	in	*preferred, stable*

The extension-based semantics and labelling-based semantics are equivalent. So for instance, for the grounded extension Γ for argument graph G, and the grounded labelling G, we have $\Gamma = \mathsf{in}(L)$. Similarly, each preferred extension (respectively stable) is represented by a preferred (respectively stable) labelling.

2.3 Subsidiary Definitions

We now consider some further simple definitions that we will use as subsidiary functions for our measures of inconsistency for abstract argument graphs.

Definition 8. *For a graph G, and an argument A, the indegree and outdegree functions are defined as follows.*

- $\mathsf{Indegree}(G, A) = |\{(B, A) \mid (B, A) \in \mathsf{Arcs}(G)\}|$

- Outdegree$(G, A) = |\{(A, B) \mid (A, B) \in \text{Arcs}(G)\}|$

Example 6. *For the graph below, we get the following indegree and outdegree values.*

$$\begin{array}{ll} \text{Indegree}(G, A_1) = 0 & \text{Outdegree}(G, A_1) = 2 \\ \text{Indegree}(G, A_2) = 3 & \text{Outdegree}(G, A_2) = 2 \\ \text{Indegree}(G, A_3) = 2 & \text{Outdegree}(G, A_3) = 3 \\ \text{Indegree}(G, A_4) = 2 & \text{Outdegree}(G, A_4) = 0 \end{array}$$

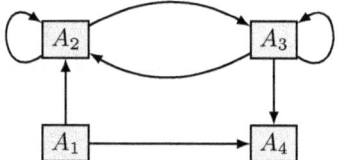

Definition 9. *For graphs G_1 and G_2, G_1 is a subgraph of G_2, denoted $G_1 \sqsubseteq G_2$, iff*

$\text{Nodes}(G_1) \subseteq \text{Nodes}(G_2)$ and $\text{Arcs}(G_1) \subseteq (\text{Arcs}(G_2) \cap (\text{Nodes}(G_1) \times \text{Nodes}(G_1)))$

Example 7. *For the following graphs G_1 (left) and G_2 (right), $G_2 \sqsubseteq G_1$ holds.*

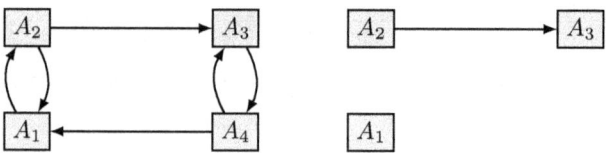

Definition 10. *For graph G, and a set of arguments $X \subseteq \text{Nodes}(G)$, the induced graph is a graph G' such that $G' \sqsubseteq G$ and $\text{Nodes}(G') = X$ and $\text{Arcs}(G') = \text{Arcs}(G) \cap (X \times X)$. Let $\text{Induce}(G, X)$ be the function that returns the induced graph for G and X.*

Example 8. *For the graph G (left), and $X = \{A_2, A_3, A_4\}$, the induced graph is $\text{Induce}(G, X)$ (right)*

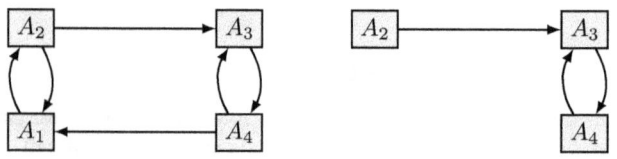

Definition 11. *For graphs G_1 and G_2, the graph union of G_1 and G_2, denoted $G_1 + G_2$, is*

$$(\text{Nodes}(G_1) \cup \text{Nodes}(G_2)), (\text{Arcs}(G_1) \cup \text{Arcs}(G_2))$$

Example 9. *For the graph G_1 (left), G_2 (middle), and $G_1 + G_2$ (right).*

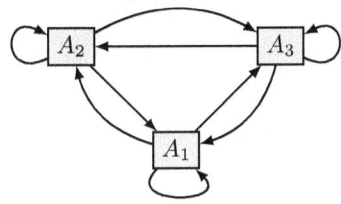

Definition 12. *A graph G is **complete** iff for all $A, B \in \mathsf{Nodes}(G)$, $(A, B) \in \mathsf{Arcs}(G)$.*

Example 10. *The following is a complete argument graph with three nodes.*

In the following definition, we define a cycle as a subset of nodes in the graph for which there is a circular path involving all these nodes.

Definition 13. *For an arc $E = (A, B)$, let $\mathsf{Source}(E) = A$ and $\mathsf{Target}(E) = B$. For a graph G, a **path** in G is a sequence of arcs $E_1, ...E_n$ such that for each E_i, where $1 \leq i < n$, $\mathsf{Target}(E_i) = \mathsf{Source}(E_{i+1})$, and for each E_i, E_j, where $1 \leq i, j \leq n$, if $i \neq j$, then $E_i \neq E_j$. A **cyclical path** in G is a path $E_1, ...E_n$ such that $\mathsf{Target}(E_n) = \mathsf{Source}(E_1)$. A **cycle** is the set of nodes appearing in the arcs in a cyclical path (i.e. if $E_1, \ldots, E_n$ is a cyclical path, then $\mathsf{Cycle}(\{E_1, \ldots, E_n\}) = \{\mathsf{Source}(E_i) \mid E_i \in \{E_1, \ldots, E_n\}\} \cup \{\mathsf{Target}(E_i) \mid E_i \in \{E_1, \ldots, E_n\}\}$). Let $\mathsf{Cycles}(G)$ denote the set of cycles in G. Finally, if $E_1, \ldots, E_n$ is a cyclical path, then the **length** of the cycle is n.*

Example 11. *In Example 10, there are 3 cycles of length 1 (i.e. each of the nodes has a self-attack), 3 cycles of length 2 (i.e. each pair of nodes has a cycle), and 1 cycle of length 3 (i.e. there is a cycle involving all three nodes). So $\mathsf{Cycles}(G) = 7$.*

Definition 14. *A graph G is **disjoint** with graph G' iff*

$$\mathsf{Nodes}(G) \cap \mathsf{Nodes}(G') = \emptyset$$

Definition 15. *A graph G has an **inverse graph** defined as*

$$\mathsf{Invert}(G) = (\mathsf{Nodes}(G), \mathsf{InverseArcs}(G))$$

where $\mathsf{InverseArcs}(G) = \{(B, A) \mid (A, B) \in \mathsf{Arcs}(G)\}$.

Definition 16. *A graph G is* **isomorphic** *with graph G' iff there is a bijection* $f : \mathsf{Nodes}(G) \to \mathsf{Nodes}(G')$ *such that* $(A, B) \in \mathsf{Arcs}(G)$ *iff* $(f(A), f(B)) \in \mathsf{Arcs}(G)$.

The following definition identifies a graph as being connected when for each pair of nodes in the graph, there is a sequence of arcs connecting them.

Definition 17. $G' \sqsubseteq G$ *is* **connected** *iff for all* $A_i, A_j \in \mathsf{Nodes}(G')$ *there is a path from A_i to A_j in the graph* $(\mathsf{Nodes}(G'), \mathsf{UndirectedArcs}(G'))$ *where* $\mathsf{UndirectedArcs}(G') = \{(A, B), (B, A) \mid (A, B) \in \mathsf{Arcs}(G')\}$.

The following definition for a multi-node component is a variant of the usual definition for a component. It is a component that has at least two elements.

Definition 18. G' *is a* **multi-node component** *of G iff the following three conditions hold.*

1. $G' \sqsubseteq G$ s.t. G' is connected

2. there is no G'' s.t. $G' \sqsubseteq G''$ and $G'' \sqsubseteq G$ and G'' is connected

3. $|\mathsf{Nodes}(G)| \geq 2$

Let $\mathsf{Components}(G) = \{G' \sqsubseteq G \mid G' \text{ is a multi-node component of } G\}$.

Example 12. *In the following graph, there are three multi-node components (left, middle, right).*

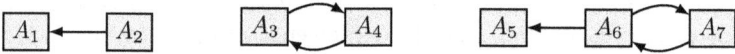

We introduce the definition for multi-node components because it forms the basis of a potentially useful measure of inconsistency that we will introduce in the next section. As we will see, we will not be interested in components that contain just 1 node.

3 Inconsistency Measures for Argument Graphs

Following developments in inconsistency measures for logical knowledgebases, we define a graph-based inconsistency measure as a function that assigns a real number to each graph such that the following constraints of consistency and freeness are satisfied. We explain these constraints as follows: (Consistency) If a graph has no arcs, then the graph contains no counterarguments, and hence it is consistent; and (Freeness) Adding an argument that does not attack any argument or is not attacked by an argument does not change the inconsistency of the graph.

Definition 19. *A* **graph-based inconsistency measure** *is a function* $I : \mathcal{G} \to \mathbb{R}$ *such that*

1. *(Consistency) If* $\mathsf{Arcs}(G) = \emptyset$, *then* $I(G) = 0$.

2. *(Freeness) If* $\mathsf{Nodes}(G) = \mathsf{Nodes}(G') \setminus \{A\}$ *and* $\mathsf{Arcs}(G) = \mathsf{Arcs}(G')$, *then* $I(G) = I(G')$.

The following are further optional properties of a graph-based inconsistency measure: (Monotonicity) Adding arguments and counterarguments cannot decrease inconsistency; (Inversion) If G and G' have the same arguments, but each attack in G is reversed in G', then they have the same inconsistency measure; (Isomorphism) If G and G' have the same structure, then they have the same inconsistency measure; (Disjoint additivity) If G_1 and G_2 are disjoint, then the inconsistency of $G_1 + G_2$ is the sum of the inconsistency of G_1 and G_2; and (Super-additivity) The inconsistency measure of the union of G_1 and G_2 is not less than the sum of the inconsistency measure of G_1 and the inconsistency measure of G_2.

Definition 20. *The following are further properties for a graph-based inconsistency measure.*

- *(Monotonicity) If* $G \sqsubseteq G'$, *then* $I(G) \leq I(G')$.

- *(Inversion) If* $G' = \mathsf{Invert}(G)$, *then* $I(G) = I(G')$.

- *(Isomorphic invariance) If* G *and* G' *are isomorphic, then* $I(G) = I(G')$.

- *(Disjoint additivity) If* G_1 *and* G_2 *are disjoint, then* $I(G_1 + G_2) = I(G_1) + I(G_2)$.

- *(Super-additivity)* $I(G_1 + G_2) \geq I(G_1) + I(G_2)$.

In the following subsections, we consider two classes of inconsistency measure for abstract argumentation, namely graph structure measures, and graph extension measures. We will compare and contrast them using the above properties.

3.1 Graph Structure Measures

We now provide some proposals for measures. We give examples of them, and then show that they are graph-based inconsistency measures.

Definition 21. *The following are measures* $I : \mathcal{G} \to \mathbb{R}$.

- *(Drastic)*

$$\text{If } \mathsf{Arcs}(G) \neq \emptyset, \text{ then } I_{dr}(G) = 1, \text{ otherwise } I_{dr}(G) = 0$$

- *(InSum)*
$$I_{in}(G) = \sum_{A \in \mathsf{Nodes}(G)} \mathsf{Indegree}(G, A)$$

- *(WeightedInSum)*
$$I_{win}(G) = \sum_{A \in \mathsf{Nodes}(G) \ s.t. \ \mathsf{Indegree}(G,A) \geq 1} \frac{1}{\mathsf{Indegree}(G, A)}$$

- *(WeightedOutSum)*
$$I_{wou}(G) = \sum_{A \in \mathsf{Nodes}(G) \ s.t. \ \mathsf{Outdegree}(G,A) \geq 1} \frac{1}{\mathsf{Outdegree}(G, A)}$$

- *(CycleCount)*
$$I_{cc}(G) = |\mathsf{Cycles}(G)|$$

- *(WeightedCycleCount)*
$$I_{wcc}(G) = \sum_{C \in \mathsf{Cycles}(G)} \frac{1}{|C|}$$

- *(WeightedComponentCount)*

If $\mathsf{Components}(G) = \emptyset$, then $I_{ic}(G) = 0$, otherwise
$$I_{ic}(G) = \sum_{X \in \mathsf{Components}(G)} (|X| - 1)^2,$$

We explain these measures as follows: (Drastic) If a graph has attacks (i.e. counterarguments), then the inconsistency measure is 1, otherwise the inconsistency measure is 0; (InSum) This is the sum of the indegrees for the nodes in the graph; (WeightedInSum) This is the sum of the inverse of indegrees for the nodes in the graph and so a node with a lower indegree has higher contribution to the inconsistency; (WeightedOutSum) This is the sum of the inverse of outdegrees for the nodes in the graph and so a node with a lower outdegree has a higher contribution to the inconsistency; (CycleCount) The is the number of the cycles in the graph; (WeightedCycleSum) This is the sum of the inverse of the number cycles in the graph and so a shorter cycle has a higher contribution to the inconsistency; and (WeightedComponentCount) This is the sum of the cardinality minus 1 squared of each component in the graph and so a larger component has a higher contribution to the inconsistency.

Example 13. *Consider the following graphs G_1 (left) and G_2 (right).*

Hence, we get the following inconsistency evaluations.

	I_{in}	I_{win}	I_{wou}	I_{cc}	I_{wcc}	I_{ic}
G_1	3	1/3	3	0	0	9
G_2	2	1/2	2	0	0	4

Example 14. *Consider the following graphs G_1 (left) and G_2 (right).*

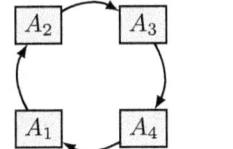 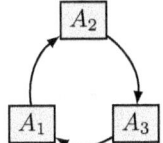

Hence, we get the following inconsistency evaluations.

	I_{in}	I_{win}	I_{wou}	I_{cc}	I_{wcc}	I_{ic}
G_1	4	4	4	1	1/4	9
G_2	3	3	3	1	1/3	4

The graph structure measures are graph-based inconsistency measures as shown in the following result.

Proposition 1. *The I_{dr}, I_{in}, I_{win}, I_{wou}, I_{cc}, I_{wcc} and I_{ic} measures are graph-based inconsistency measures according to Definition 19.*

Proof. Follows directly from definitions. □

The following result gives some idea of the difference of scale for each measure.

Proposition 2. *If G is a complete graph and $|\mathsf{Nodes}(G)| = n$ where $n > 0$, then*

$$I_{dr} = 1 \qquad I_{in} = n^2$$
$$I_{win} = 1 \qquad I_{wou} = 1$$
$$I_{cc} = 2^n - 1 \qquad I_{wcc} = \sum_{k=1}^{n} [\tfrac{1}{k} \times \tfrac{n!}{k!(n-k)!}]$$
$$I_{ic} = (n-1)^2$$

Proof. Assume G is a complete graph and $|\mathsf{Nodes}(G)| = n$. (I_{dr}) $\mathsf{Arcs}(G) \neq \emptyset$, and so $I_{dr} = 1$. (I_{in}) For every node A, $\mathsf{Indegree}(G, A) = n$, and so $I_{in} = n^2$. (I_{win}) For every node A, $\mathsf{Indegree}(G, A) = n$, and so $I_{win}(G) = \sum_{A \in \mathsf{Nodes}(G)} \frac{1}{\mathsf{Indegree}(G,A)} = \sum_{A \in \mathsf{Nodes}(G)} \frac{1}{n} = n \times \frac{1}{n} = 1$. ($I_{wou}$) Ditto. ($I_{cc}$) There are $2^n - 1$ subsets of nodes where each subset constitutes a cycle, and so $I_{cc} = 2^n - 1$. (I_{wcc}) For each non-empty subset of nodes constitute a cycle. Furthermore, for each $k \in \{1, \ldots, n\}$, there are $\frac{n!}{k!(n-k)!}$ subsets of cardinality k, and each of these contribute $\frac{1}{k}$ to the sum. Hence, the sum is $\sum_{k=1}^{n} [\frac{1}{k} \times \frac{n!}{k!(n-k)!}]$. ($I_{ic}$) There is a single component, and so $I_{ic} = (n-1)^2$. □

The following result shows which optional properties are satisfied by which measures.

Proposition 3. *For the graph structure measures in Definition 21, the adherence to the properties in Definition 20 is summarized in the following table.*

	I_{dr}	I_{in}	I_{win}	I_{wou}	I_{cc}	I_{wcc}	I_{ic}
Monotonicity	✓	✓	✓	✓	✓	✓	✓
Inversion	✓	✓	✗	✗	✓	✓	✓
Isomorphic invariance	✓	✓	✓	✓	✓	✓	✓
Disjoint additivity	✗	✓	✓	✓	✓	✓	✓
Super-additivity	✗	✓	✗	✗	✓	✓	✗

Proof. We consider each property as follows.

- Monotonicity. (I_{dr}, I_{in}, I_{win}, I_{wou}, I_{cc}, I_{wcc}, I_{ic}) Follows directly from the definition.

- Inversion. (I_{dr}, I_{in}) Follows directly from the definition. (I_{win}, I_{wou}) Consider G (left) and $\mathsf{Invert}(G)$ (right). Therefore, we obtain $I_{win}(G) = 1/2$, $I_{win}(\mathsf{Invert}(G)) = 2$, $I_{wou}(G) = 2$, and $I_{wou}(\mathsf{Invert}(G)) = 1/2$.

(I_{cc}, I_{wcc}, I_{ic}) Follows directly from the definition.

- Isomorphic invariance. (I_{dr}, I_{in}, I_{win}, I_{wou}, I_{cc}, I_{wcc}, I_{ic}) Follows directly from definition.

- Disjoint additivity. (I_{dr}) Consider the following graphs G_1 (left) and G_2 (right) where $I_{dr}(G_1) = 1$ and $I_{dr}(G_2) = 1$ but $I_{dr}(G_1 + G_2) = 1$.

(I_{in}, I_{win}, I_{wou}, I_{cc}, I_{wcc}, I_{ic}) Follows directly from the definition.

- Super-additivity. (I_{dr}) See counterexample for disjoint additivity. (I_{in}) Follows directly from the definition. (I_{win}) Consider the following graphs G_1 (left) and G_2 (right) where $I_{win}(G_1) = 1$ and $I_{win}(G_2) = 1$ but $I_{win}(G_1 + G_2) = 1/2$.

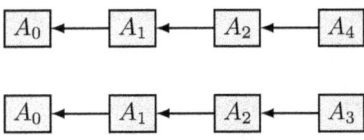

(I_{wou}) Use a similar counterexample to I_{win}. (I_{cc}, I_{wcc}) Follows directly from the definition. (I_{ic}) Consider the following graphs G_1 (top) and G_2 (bottom) where $I_{ic}(G_1) = 9$ and $I_{ic}(G_2) = 9$ but $I_{ic}(G_1 + G_2) = 16$.

$A_0 \leftarrow A_1 \leftarrow A_2 \leftarrow A_4$

$A_0 \leftarrow A_1 \leftarrow A_2 \leftarrow A_3$

□

The following property of order-compatibility holds when two measures give the same ranking to all the graphs. If this holds, then there is some overlap in what the two measures offer.

Definition 22. Measures I_x and I_y are **order-compatible** if for all G_1 and G_2,
$$I_x(G_1) < I_x(G_2) \text{ iff } I_y(G_1) < I_y(G_2)$$
otherwise I_x and I_y are **order-incompatible**.

Proposition 4. The I_{dr}, I_{in}, I_{win}, I_{wou}, I_{cc}, I_{wcc} and I_{ic} measures are pairwise order-incompatible,

Proof. From the differences in satisfaction of properties in Proposition 3, I_{dr} is pairwise incompatible with the other measures, I_{ic} is pairwise incompatible with the other measures, and I_{win} and I_{wou} are pairwise incompatible with the other measures, though from the properties in Proposition 3, we cannot discriminate I_{win} from I_{wou}. However, we can discriminate I_{win} from I_{wou} with graph G_1 (left) and G_2 (right), where $I_{win}(G_1) = 1/2$, $I_{win}(G_2) = 2$, $I_{wou}(G_1) = 2$, and $I_{wou}(G_2) = 1/2$.

Finally, from the properties in Proposition 3, we cannot discriminate between I_{in}, I_{wcc}, I_{cc}. So, from Proposition 2, assuming G is complete, if $|\mathsf{Nodes}(G)| \in \{2,3,4\}$, then $I_{in}(G) > I_{cc}(G)$, whereas if $|\mathsf{Nodes}(G)| \geq 5$, then $I_{in}(G) < I_{cc}(G)$. Similarly, if $|\mathsf{Nodes}(G)| = 2$, then $I_{in}(G) > I_{wcc}(G)$, whereas if $|\mathsf{Nodes}(G)| = 10$, then $I_{in}(G) < I_{wcc}(G)$. For I_{cc} and I_{wcc}, consider G_1 (left) and G_2 (right) which give $I_{cc}(G_1) = I_{cc}(G_2)$ and $I_{wcc}(G_1) > I_{wcc}(G_2)$.

□

In this section, we have considered a range of measures that take the structure of the argument graph into account. Each has its rationale, and any combination of them may provide useful insights into the nature of the conflict in an argumentation scenario.

3.2 Graph Extension Measures

We now consider measures of inconsistency that take the extensions of the graph into account. In the following, we consider three measures: (1) I_{pr} which gives the number of preferred extensions minus 1; (2) I_{ngr} which gives the number of arguments not in the grounded extension and not attacked by a member of the grounded extension; and (3) I_{nst} which gives the minimum number of arguments to be removed to get a stable extension.

Definition 23. *The following are measures* $I : \mathcal{G} \to \mathbb{R}$.

- *(PreferredCount)*

$$I_{pr}(G) = |\mathsf{Extensions}_{\mathsf{pr}}(G)| - 1$$

- *(NonGroundedCount)*

$$I_{ngr}(G) = |\mathsf{Nodes}(G) \setminus (\mathsf{Extensions}_{\mathsf{gr}}(G) \cup \mathsf{Attackees}(G))|$$

where $\mathsf{Attackees}(G) = \{B \mid (A, B) \in \mathsf{Arcs}(G) \text{ and } A \in \mathsf{Extensions}_{\mathsf{gr}}(G)\}$.

- *(UnstableCount)*

$$I_{ust}(G) = \min\{|X| \mid \mathsf{Extensions}_{\mathsf{st}}(\mathsf{Induce}(G, X)) \neq \emptyset \text{ s.t. } X \subseteq \mathsf{Nodes}(G)\}$$

Example 15. *For the following argument graph, we have the following extensions,*

- $\mathsf{Extensions}_{\mathsf{pr}}(G) = \{\{A_4, A_6, A_8\}, \{A_5, A_6, A_8\}\}$

- Extensions$_{gr}(G) = \{\{A_6, A_8\}\}$

and removing one argument from $\{A_1, A_2, A_3\}$ is the smallest number of arguments to be removed to get a stable extension. Hence, $I_{pr}(G) = 1$, $I_{ngr}(G) = 5$, and $I_{ust}(G) = 1$.

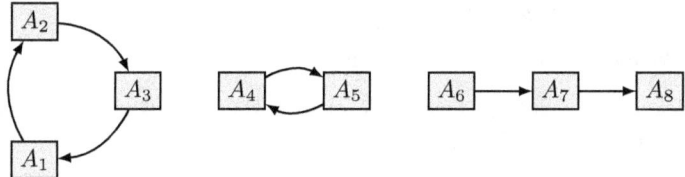

By way of further explanation for the I_{ust} measure, a stable extension is a preferred extension where every argument not in the extension is attacked by an argument that is in the extension. Not every graph has a stable extension. But if arguments and the attacks involving those arguments are removed, then a stable extension can be obtained from this subgraph. So this measure is based on the minimum number of arguments that need to be removed to get a stable extension.

The following result shows that the extension-based measures are indeed inconsistency measures but the subsequent result shows that most of the optional properties do not hold for these measures.

Proposition 5. *The I_{pr}, I_{ngr}, and I_{ust} measures are graph-based inconsistency measures according to Definition 19.*

Proof. We show the satisfaction of consistency by assuming that $\mathsf{Arcs}(G) = \emptyset$: ($I_{pr}$) $\mathsf{Extensions}_{pr}(G) = \{\mathsf{Nodes}(G)\}$, and so $I_{pr}(G) = 0$; (I_{ngr}) $\mathsf{Extensions}_{gr}(G) = \{\mathsf{Nodes}(G)\}$, and so $I_{ngr}(G) = 0$; (I_{ust}) $\mathsf{Extensions}_{st}(G) \neq \emptyset$, and so the smallest X such that $\mathsf{Extensions}_{st}(\mathsf{Induce}(G, X)) \neq \emptyset$ is $X = \emptyset$, and so $I_{ust}(G) = 0$. We show the satisfaction of freeness by assuming that $\mathsf{Nodes}(G) = \mathsf{Nodes}(G') \setminus \{A\}$ and $\mathsf{Arcs}(G) = \mathsf{Arcs}(G')$: ($I_{pr}$) $|\mathsf{Extensions}_{gr}(G)| = |\mathsf{Extensions}_{gr}(G')|$, and so $I_{pr}(G) = I_{pr}(G')$; (I_{ngr}) Let X be $\mathsf{Nodes}(G') \setminus \mathsf{Nodes}(G)$, $\mathsf{Extensions}_{pr}(G) = \{E\}$, and $\mathsf{Extensions}_{pr}(G') = \{E'\}$, and so from the assumptions, $E = E' \setminus X$, and hence $\mathsf{Nodes}(G) \setminus (E \cup \mathsf{Attackees}(G)) = \mathsf{Nodes}(G') \setminus (E' \cup \mathsf{Attackees}(G'))$, and so $I_{ngr}(G) = I_{ngr}(G')$; (I_{ust}) Let X_1 be the smallest subset of nodes such that $\mathsf{Extensions}_{st}(\mathsf{Induce}(G, X_1)) \neq \emptyset$. From the assumptions, X_1 is also the smallest subset of nodes such that $\mathsf{Extensions}_{st}(\mathsf{Induce}(G', X_1)) \neq \emptyset$. Hence, $I_{ust}(G) = I_{ust}(G')$. □

Proposition 6. *For the graph structure measures in Definition 21, the adherence to the properties in 20 is summarized in the following table.*

	I_{pr}	I_{ngr}	I_{ust}
Monotonicity	×	×	×
Inversion	×	×	×
Isomorphic invariance	✓	✓	✓
Disjoint additivity	×	✓	✓
Super-additivity	×	×	×

Proof. We consider each property as follows.

- Monotonicity. (I_{pr}) Consider G_1 (left) and G_2 (right) where $I_{pr}(G_1) = 1$, $I_{pr}(G_2) = 0$, and $G_1 \sqsubseteq G_2$.

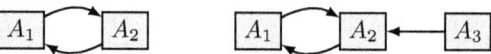

(I_{ngr}) Consider G_1 (above left) and G_2 (above right) where $I_{ngr}(G_1) = 2$, $I_{ngr}(G_2) = 0$, and $G_1 \sqsubseteq G_2$. (I_{ust}) Consider G_1 (left) and G_2 (right) where $I_{ust}(G_1) = 1$, $I_{ust}(G_2) = 0$, and $G_1 \sqsubseteq G_2$.

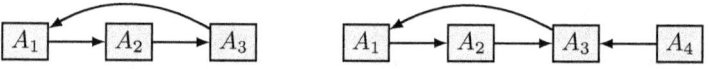

- Inversion. (I_{pr}) Consider G (left) and $\mathsf{Inverted}(G)$ with $I_{pr}(G) = 0$, and $I_{pr}(\mathsf{Inverted}(G)) = 1$.

(I_{ngr}) Consider G (above left) and $\mathsf{Inverted}(G)$ (above right) with $I_{ngr}(G) = 0$, and $I_{ngr}(\mathsf{Inverted}(G)) = 3$. ($I_{ust}$) Consider G (left) and $\mathsf{Inverted}(G)$ (right) with $I_{ust}(G) = 0$, and $I_{ust}(\mathsf{Inverted}(G)) = 1$.

- Isomorphic invariance. (I_{pr}, I_{ngr}, I_{ust}) Follows directly from definition.

- Disjoint additivity. (I_{pr}) Consider G_1 (left) and G_2 (right) where $I_{pr}(G_1) = 1$, $I_{pr}(G_2) = 1$, and $I_{pr}(G_1 + G_2) = 3$.

(I_{ngr}) If G_1 & G_2 are disjoint, then $\mathsf{Extensions}_{gr}(G_1) \cap \mathsf{Extensions}_{gr}(G_2) = \emptyset$, $\mathsf{Attackees}_{gr}(G_1) \cap \mathsf{Attackees}_{gr}(G_2) = \emptyset$, and $\mathsf{Nodes}_{gr}(G_1) \cap \mathsf{Nodes}_{gr}(G_2) = \emptyset$. So $I_{ngr}(G_1 + G_2) = I_{ngr}(G_1) + I_{ngr}(G_2)$. ($I_{ust}$) If G_1 and G_2 are disjoint, $\min\{|X| \mid \mathsf{Extensions}_{st}(\mathsf{Induce}(G_1 + G_2, X)) \neq \emptyset \text{ s.t. } X \subseteq \mathsf{Nodes}(G_1 + G_2)\} = \min\{|X| \mid \mathsf{Extensions}_{st}(\mathsf{Induce}(G_1, X)) \neq \emptyset \text{ s.t. } X \subseteq \mathsf{Nodes}(G_1)\} + \min\{|X| \mid \mathsf{Extensions}_{st}(\mathsf{Induce}(G_2, X)) \neq \emptyset \text{ such that } X \subseteq \mathsf{Nodes}(G_2)\}$. Therefore, $I_{ust}(G_1 + G_2) = I_{ust}(G_1) + I_{ust}(G_2)$.

- Super-additivity. (I_{pr}) Consider G_1 (left) and G_2 (right) where $I_{pr}(G_1) = 1$, $I_{pr}(G_2) = 0$, and $I_{pr}(G_1 + G_2) = 0$.

(I_{ngr}) Consider G_1 (above left) and G_2 (above right) where $I_{ngr}(G_1) = 2$, $I_{ngr}(G_2) = 0$, and $I_{ngr}(G_1 + G_2) = 0$. (I_{ust}) Consider G_1 (left) and G_2 (right) where $I_{ust}(G_1) = 1$, $I_{ust}(G_2) = 0$, and $I_{ust}(G_1 + G_2) = 0$.

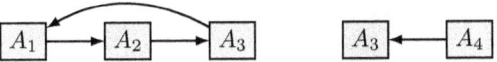

□

Proposition 7. *The I_{pr}, I_{ngr}, I_{ust} measures are pairwise order-incompatible. Each is also pairwise order-incompatible with each of the I_{dr}, I_{in}, I_{win}, I_{wou}, I_{cc}, I_{wcc} and I_{ic} measures.*

Proof. From the differences in satisfaction of properties in Proposition 3, and Proposition 6, the extension-based measures are pairwise incompatible with the structure-based measures. We now consider discriminating between the extension-based measures. We can discriminate I_{pr} from I_{ngr} with graph G_1 (left) and G_2 (right), where $I_{pr}(G_1) = 1$, $I_{pr}(G_2) = 0$, $I_{ngr}(G_1) = 2$, and $I_{ngr}(G_2) = 3$.

We can discriminate I_{pr} from I_{ust} with graph G_1 (left) and G_2 (right), where $I_{pr}(G_1) = 1$, $I_{pr}(G_2) = 3$, $I_{ust}(G_1) = 2$, and $I_{ust}(G_2) = 0$.

We can discriminate I_{pr} from I_{ust} with G_1 (left) and G_2 (right), where $I_{ngr}(G_1)$ = 1, $I_{ngr}(G_2) = 2$, $I_{ust}(G_1) = 1$, and $I_{ust}(G_2) = 1$.

□

In this subsection, we have considered an alternative class of measures based on the extensions of the graph. As can be seen from the properties that hold for them, they are very different in behaviour from the structured-based measures. On the one hand, measuring inconsistency in terms of the extensions seems a natural option, but on the other hand, they fail to satisfy most of the optional properties that we have considered, though this may be more a reflection of the nature of these properties, and that we need to consider a wider range of properties to justifiy and characterise extension-based measures.

4 Logic-based Instantiations

We now consider structured argumentation. This is where the arguments in the argument graph are instantiated with logical structure. We proceed by reviewing a specific approach to structured argumentation called deductive argumentation, and then we consider two approaches to measuring inconsistency in deductive argumentation. The first is the degree of undercut approach proposed in [12, 13] and the second is the application of existing logic-based measures of inconsistency to deductive arguments in an argument graph.

4.1 Review of Deductive Argumentation

In deductive argumentation, each argument is a pair where the first item is a set of premises that logically entails the second item according to some logical consequence relation (for a review see [14]). So we have a logical language to express the set of premises and the claim, and we have a logical consequence relation to relate the premises to the claim. So in order to construct argument graphs with deductive arguments, we need to specify the choice of logic (which we call the base logic) that we use to define arguments and counterarguments.

Classical logic is appealing as the choice of base logic as it reflects the richer deductive reasoning often seen in arguments arising in discussions and debates. We assume the usual propositional and predicate (first-order) languages for classical logic, and the usual **classical consequence relation**, denoted ⊢. A **knowledgebase** is a set of classical propositional or predicate formulae.

Definition 24. *For a classical knowledgebase Φ, and a classical formula α, $\langle\Phi,\alpha\rangle$ is a* **classical argument** *iff $\Phi \vdash \alpha$ and $\Phi \not\vdash \bot$ and there is no proper subset Φ' of Φ such that $\Phi' \vdash \alpha$. For an argument $A = \langle\Phi,\alpha\rangle$, the function* Support$(A)$ *returns Φ and the function* Claim(A) *returns α.*

Example 16. *The following classical argument uses a universally quantified formula in contrapositive reasoning to obtain the claim about number 77.*

$\langle\{\forall\texttt{X.multipleOfTen(X)} \rightarrow \texttt{even(X)}, \neg\texttt{even(77)}\}, \neg\texttt{multipleOfTen(77)}\rangle$

A counterargument is an argument that attacks another argument. In deductive argumentation, we define the notion of counterargument in terms of logical contradiction between the claim of the counterargument and the premises or claim of the attacked argument. Given the expressivity of classical logic (in terms of language and inferences), there are a number of natural ways to define counterarguments (as given in Definition 25 below). For instance, the definition of rebuttal captures the situation where two arguments have complementary claims, whereas the definition of defeater captures the situation where the claim of one argument negates the support of the other argument.

Definition 25. *Let A and B be two classical arguments. We define the following types of* **attack**. *In each case, A is the* **attacker** *and B is the* **attackee**.

- *A is a* **defeater** *of B if* Claim$(A) \vdash \neg \bigwedge$ Support(B).
- *A is a* **direct defeater** *of B if $\exists\phi \in$* Support(B) *s.t.* Claim$(A) \vdash \neg\phi$.
- *A is an* **undercut** *of B if $\exists\Psi \subseteq$* Support(B) *s.t.* Claim$(A) \equiv \neg \bigwedge \Psi$.
- *A is a* **direct undercut** *of B if $\exists\phi \in$* Support(B) *s.t.* Claim$(A) \equiv \neg\phi$.
- *A is a* **canonical undercut** *of B if* Claim$(A) \equiv \neg \bigwedge$ Support(B).
- *A is a* **rebuttal** *of B if* Claim$(A) \equiv \neg$Claim(B).
- *A is a* **defeating rebuttal** *of B if* Claim$(A) \vdash \neg$Claim(B).

To illustrate these different notions of counterargument, we consider the following examples, and we relate these definitions in Figure 1 where we show that defeaters are the most general of these definitions.

Example 17. *Let $\{a \vee b, a \leftrightarrow b, c \rightarrow a, \neg a \wedge \neg b, a, b, c, a \rightarrow b, \neg a, \neg b, \neg c\}$ be the knowledgebase from which the following are examples of attacks.*

$\langle\{a \vee b, c\}, (a \vee b) \wedge c\rangle$ *is a defeater of* $\langle\{\neg a, \neg b\}, \neg a \wedge \neg b\rangle$
$\langle\{a \vee b, c\}, (a \vee b) \wedge c\rangle$ *is a direct defeater of* $\langle\{\neg a \wedge \neg b\}, \neg a \wedge \neg b\rangle$
$\langle\{\neg a \wedge \neg b\}, \neg(a \wedge b)\rangle$ *is an undercut of* $\langle\{a, b, c\}, a \wedge b \wedge c\rangle$
$\langle\{\neg a \wedge \neg b\}, \neg a\rangle$ *is a direct undercut of* $\langle\{a, b, c\}, a \wedge b \wedge c\rangle$
$\langle\{\neg a \wedge \neg b\}, \neg(a \wedge b \wedge c)\rangle$ *is a canonical undercut of* $\langle\{a, b, c\}, a \wedge b \wedge c\rangle$
$\langle\{a, a \rightarrow b\}, b \vee c\rangle$ *is a rebuttal of* $\langle\{\neg a \wedge \neg b, \neg c\}, \neg(b \vee c)\rangle$
$\langle\{a, a \rightarrow b\}, b\rangle$ *is a defeating rebuttal of* $\langle\{\neg a \wedge \neg b, \neg c\}, \neg(b \vee c)\rangle$

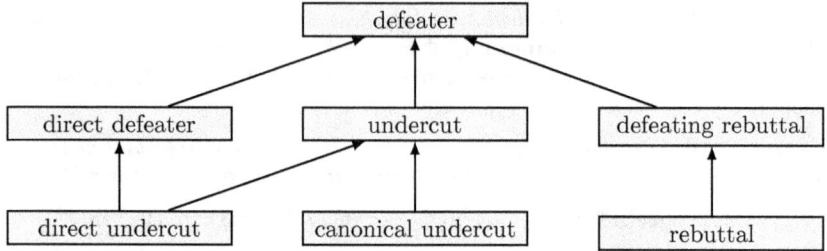

Figure 1: We can represent the containment between the attack relations as above where an arrow from R_1 to R_2 indicates that $R_1 \subseteq R_2$.

Example 18. *Consider the following arguments where A_2 is an undercut of A_1 though it is neither a direct undercut nor a canonical undercut. Essentially, the attack says that the flight cannot be both a low cost flight and a luxury flight.*

$A_1 = \langle \{\text{lowCostFly}, \text{luxFly}, \text{lowCostFly} \land \text{luxFly} \to \text{goodFly}\}, \text{goodFly} \rangle$
$A_2 = \langle \{\neg\text{lowCostFly} \lor \neg\text{luxFly}\}, \neg\text{lowCostFly} \lor \neg\text{luxFly} \rangle$

Trivially, undercuts are defeaters but it is also quite simple to establish that rebuttals are defeaters. Furthermore, if an argument has defeaters then it has undercuts. It may happen that an argument has defeaters but no rebuttals as illustrated next.

Example 19. *Consider $\{\neg\text{containsGarlic} \land \text{goodDish}, \neg\text{goodDish}\}$ as the knowledgebase. Then the following argument has at least one defeater but no rebuttal.*

$\langle \{\neg\text{containsGarlic} \land \text{goodDish}\}, \neg\text{containsGarlic} \rangle$

There are some important differences between rebuttals and undercuts that can be seen in the following examples.

Example 20. *Consider the following arguments. The first argument A_1 is a direct undercut to the second argument A_2, but neither rebuts each other. Furthermore, A_1 "agrees" with the claim of A_2 since the premises of A_1 could be used for an alternative argument with the same claim as A_2.*

$A_1 = \langle \{\neg\text{containsGarlic} \land \neg\text{goodDish}\}, \neg\text{containsGarlic} \rangle$
$A_2 = \langle \{\text{containsGarlic}, \text{containsGarlic} \to \neg\text{goodDish}\}, \neg\text{goodDish} \rangle$

Example 21. *Consider the following arguments. The first argument is a rebuttal of the second argument, but it is not an undercut because the claim of the*

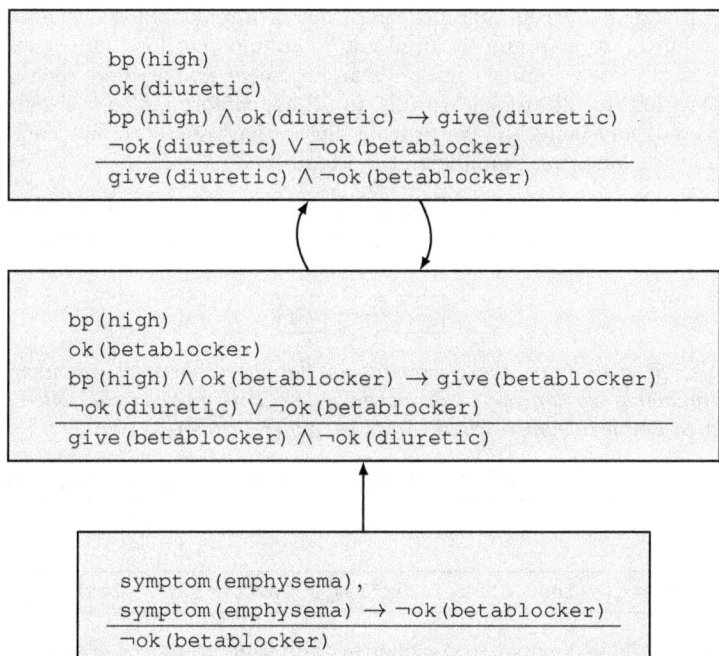

Figure 2: An instantiated argument graph for the abstract argument graph in Example 1. The atom **bp(high)** denotes that the patient has high blood pressure. The top two arguments rebut each other (i.e. the attack is defeating rebut). For this, each argument has an integrity constraint in the premises that says that it is not ok to give both betablocker and diuretic. So the top argument is attacked on the premise **ok(diuretic)** and the middle argument is attacked on the premise **ok(betablocker)**. So we are using the **ok** predicate as a normality condition for the rule to be applied.

first argument is not equivalent to the negation of some subset of the premises of the second argument.

$A_1 = \langle \{\texttt{goodDish}\}, \texttt{goodDish} \rangle$
$A_2 = \langle \{\texttt{containsGarlic}, \texttt{containsGarlic} \to \neg\texttt{goodDish}\}, \neg\texttt{goodDish} \rangle$

So an undercut for an argument need not be a rebuttal for that argument, and a rebuttal for an argument need not be an undercut for that argument.

An instantiated argument graph is an argument graph where each node is a classical argument, and each arc is an attack conforming to the definitions for attack (Definition 25). We provide illustrations of instantiated argument graphs in the following example and in Figure 2.

Example 22. *Consider the following argument graph where A_1 is "The flight is low cost and luxury, therefore it is a good flight", and A_2 is "A flight cannot be both low cost and luxury".*

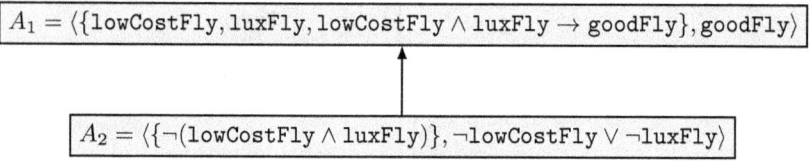

For this, we instantiate the arguments in the above abstract argument graph with arguments we considered in Example 18. This gives us the following instantiated argument graph where A_2 is an undercut to A_1.

$A_1 = \langle \{\texttt{lowCostFly}, \texttt{luxFly}, \texttt{lowCostFly} \land \texttt{luxFly} \to \texttt{goodFly}\}, \texttt{goodFly} \rangle$

$A_2 = \langle \{\neg(\texttt{lowCostFly} \land \texttt{luxFly})\}, \neg\texttt{lowCostFly} \lor \neg\texttt{luxFly} \rangle$

Perhaps the first paper to consider instantiating argument graphs with deductive arguments based on classical logic is by Cayrol [15] using direct undercut. For more details on deductive argumentation and how it can be used to instantiate argument graphs, see [14].

4.2 Degree of Undercut

An argument conflicts with each of its undercuts, by the very definition of an undercut. Now, some may conflict more than others, and some may conflict a little while others conflict a lot.

Example 23. *Consider the following argument graph G. Each undercut has a premise that negates some or all of the premises in the root. The left child has the weakest premise which can be read as saying that one of the premises in the root is false without saying which, the middle child says that one of the premises in the root is false and states which one, and the right child says that all of the premises are false.*

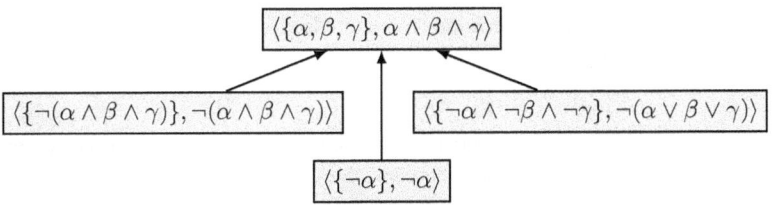

By these simple examples of undercuts, we see that there can be a difference in the amount of conflict between supports, and hence can be taken as an intuitive starting point for defining the degree of undercut that an argument has against its parent. To address this, the degree of undercut is a measure of the conflict between a pair of arguments based on the supports of these arguments [12, 13]. There are some alternatives for defining the degree of undercut, and we review one of these proposals in the next subsection.

4.2.1 Degree of Undercut based on Dalal Distance

In this section, we investigate a degree of undercut based on the distance between pairs of models. For this, we use the Dalal distance [18].

Definition 26. *For the language $\mathcal{L}$, let $\mathsf{Atoms}(\mathcal{L})$ be the set of atoms used in the language (and so the formulae in $\mathcal{L}$ are composed from $\mathsf{Atoms}(\mathcal{L})$ and the logical connectives using the usual inductive definition).*

Definition 27. *Let Π be a finite non-empty subset of $\mathsf{Atoms}(\mathcal{L})$ and let $w_i, w_j \in \wp(\Pi)$. The **Dalal distance** between w_i and w_j, denoted $\mathsf{Dalal}(w_i, w_j)$, is the difference in the number of atoms assigned true:*

$$\mathsf{Dalal}(w_i, w_j) = |w_i - w_j| + |w_j - w_i|$$

Example 24. *Let $w_1 = \{\alpha, \gamma, \delta\}$ and $w_2 = \{\beta, \gamma\}$ where $\{\alpha, \beta, \gamma, \delta\} \subseteq \Pi$. Then,*

$$\mathsf{Dalal}(w_1, w_2) = |\{\alpha, \delta\}| + |\{\beta\}| = 3$$

To evaluate the conflict between the support of an argument A and the support of an undercut A', we consider the models of $\mathsf{Support}(A)$ and $\mathsf{Support}(A')$, restricted to a set of atoms Π. For this, we require the following definition.

Definition 28. *Let Π is a finite non-empty subset of $\mathsf{Atoms}(\mathcal{L})$, let Φ be a set of formulae, and let $\models$ be the classical satisfaction relation.*

$$\mathsf{Models}(\Phi, \Pi) = \{w \in \wp(\Pi) \mid \forall \phi \in \Phi \text{ and } w \models \phi\}$$

Example 25. *Let $\Phi = \{\alpha \wedge \delta, \neg\phi, \gamma \vee \delta, \neg\psi, \beta \vee \gamma\} \subseteq \Delta$, and let Π be $\{\alpha, \beta, \gamma, \delta, \phi\} \subseteq \mathsf{Atoms}(\mathcal{L})$. Should it be the case that $\Pi \neq \mathsf{Atoms}(\mathcal{L})$, the set of* all the models of Φ is a proper superset of $\mathsf{Models}(\Phi, \Pi)$ as the latter consists exactly of the following models.

$$\{\alpha, \beta, \gamma, \delta\} \qquad \{\alpha, \beta, \delta\} \qquad \{\alpha, \gamma, \delta\}$$

To evaluate the conflict between two sets of formulae, we take a pair of models restricted to Π, one for each set, such that the Dalal distance is minimized. The degree of conflict is this distance divided by the maximum possible Dalal distance between a pair of models (i.e. $\log_2$ of the total number of models in $\wp(\Pi)$ which is $|\Pi|$).

Definition 29. *The* **degree of conflict** *wrt* Π, *denoted* $\mathsf{Conflict}(\Phi, \Psi, \Pi)$, *is:*

$$\frac{\min\{\mathsf{Dalal}(w_\Phi, w_\Psi) \mid w_\Phi \in \mathsf{Models}(\Phi, \Pi), w_\Psi \in \mathsf{Models}(\Psi, \Pi)\}}{|\Pi|}$$

Example 26. *Let* $\Pi = \{\alpha, \beta, \gamma, \delta\}$.

$\mathsf{Conflict}(\{\alpha \wedge \beta \wedge \gamma \wedge \delta\}, \{\neg\alpha \vee \neg\beta \vee \neg\gamma\}, \Pi) = 1/4$
$\mathsf{Conflict}(\{\alpha \wedge \beta \wedge \gamma \wedge \delta\}, \{\neg(\alpha \vee \beta)\}, \Pi) = 2/4$
$\mathsf{Conflict}(\{\alpha \wedge \beta \wedge \gamma \wedge \delta\}, \{\neg\alpha \wedge \neg\beta \wedge \neg\gamma\}, \Pi) = 3/4$

We obtain a degree of undercut by applying $\mathsf{Conflict}$ to supports as defined next.

Definition 30. *Let* $A_i = \langle \Phi_i, \alpha_i \rangle$ *and* $A_j = \langle \Phi_j, \alpha_j \rangle$ *be arguments.*

$$\mathsf{Degree}(A_i, A_j) = \mathsf{Conflict}(\Phi_i, \Phi_j, \mathsf{Atoms}(\Phi_i \cup \Phi_j))$$

Clearly, if A_i is an undercut for A_j, then $\mathsf{Degree}(A_i, A_j) > 0$.

Example 27. *Consider an argument with premises* $\{\alpha, \beta, \gamma\}$. *The claim is not important for the example. Values for degree of undercut by canonical undercuts are given below.*

$\mathsf{Degree}(\langle\{\alpha, \beta, \gamma\}, \ldots\rangle, \langle\{\neg\alpha \wedge \neg\beta \wedge \neg\gamma\}, \ldots\rangle) = 1$
$\mathsf{Degree}(\langle\{\alpha, \beta, \gamma\}, \ldots\rangle, \langle\{\neg\alpha \wedge \neg\beta\}, \ldots\rangle) = 2/3$
$\mathsf{Degree}(\langle\{\alpha, \beta, \gamma\}, \ldots\rangle, \langle\{\neg\alpha \vee \neg\beta \vee \neg\gamma\}, \ldots\rangle) = 1/3$
$\mathsf{Degree}(\langle\{\alpha, \beta, \gamma\}, \ldots\rangle, \langle\{\neg\alpha\}, \ldots\rangle) = 1/3$

Example 28. *Consider the following argument graph where* $\mathsf{Degree}(A_1, A_2) = 1/3$, $\mathsf{Degree}(A_1, A_3) = 2/3$, *and* $\mathsf{Degree}(A_1, A_4) = 3/4$.

$A_2 = \langle\{\neg\alpha \vee \neg\beta \vee \neg\gamma\}, \neg(\alpha \wedge \beta \wedge \gamma)\rangle \qquad A_3 = \langle\{\neg\alpha \wedge \neg\gamma\}, \neg(\alpha \wedge \beta \wedge \gamma)\rangle$

$A_1 = \langle\{\alpha \wedge \beta \wedge \gamma\}, \alpha\rangle$

$A_4 = \langle\{\neg\alpha \wedge \neg\beta \wedge \neg\gamma \wedge \neg\delta\}, \neg(\alpha \wedge \beta \wedge \gamma)\rangle$

Example 29. *As a more general example, let* $A_1 = \langle \{\neg(\alpha_1 \vee \ldots \vee \alpha_n)\}, \neg(\alpha_1 \wedge \ldots \wedge \alpha_n)\rangle$, $A_2 = \langle \{\neg\alpha_1 \vee \ldots \vee \neg\alpha_n\}, \neg(\alpha_1 \wedge \ldots \wedge \alpha_n)\rangle$, $A_3 = \langle \{\neg\alpha_1\}, \neg(\alpha_1 \wedge \ldots \wedge \alpha_n)\rangle$, $A_4 = \langle \{\alpha_1 \wedge \ldots \wedge \alpha_n\}, \alpha_1\rangle$.

$$\mathsf{Degree}(A_4, A_1) = n/n$$
$$\mathsf{Degree}(A_4, A_2) = 1/n$$
$$\mathsf{Degree}(A_4, A_3) = 1/n$$

The above examples indicate how the **Degree** measure differentiates between different kinds of attack, and the following result shows the **Degree** measure has the basic properties that we require.

Proposition 8. *Let* $A_i = \langle \Phi_i, \alpha_i\rangle$ *and* $A_j = \langle \Phi_j, \alpha_j\rangle$ *be arguments.*

(1) $0 \leq \mathsf{Degree}(A_i, A_j) \leq 1$
(2) $\mathsf{Degree}(A_i, A_j) = \mathsf{Degree}(A_j, A_i)$
(3) $\mathsf{Degree}(A_i, A_j) = 0$ *iff* $\Phi_i \cup \Phi_j \not\vdash \bot$

So the degree of undercut gives a value in the unit interval to represent how much two arguments differ in terms of their premises. In addition to this proposal, [13] presents some alternative proposals for defining degree of undercut.

4.2.2 Cumulative Degree of Undercut

We now consider how we can harness the notion of degree of undercut as an inconsistency measure for an argument graph.

Definition 31. *Let G be an argument graph. The* **cumulative degree of undercut** *in G, denoted $\mathsf{I_{cu}}(G)$, is given by*

$$I_{cu}(G) = \sum_{(A_i, A_j) \in \mathsf{Arcs}(G)} \mathsf{Degree}(A_i, A_j)$$

Example 30. *Consider the argument graph in Example 28. For this,* $I_{cu}(G) = 1/3 + 2/3 + 3/4 = 7/4$.

Proposition 9. *The I_{cu} measure is a graph-based inconsistency measure according to Definition 19.*

Proof. (Consistency) Assume $\mathsf{Arcs}(G) = \emptyset$. So $\sum_{(A_i, A_j) \in \mathsf{Arcs}(G)} \mathsf{Degree}(A_i, A_j) = 0$. So $I_{cu}(G) = 0$. (Freeness) Assume $\mathsf{Nodes}(G) = \mathsf{Nodes}(G') \setminus \{A\}$ and $\mathsf{Arcs}(G) = \mathsf{Arcs}(G')$. So

$$\sum_{(A_i, A_j) \in \mathsf{Arcs}(G)} \mathsf{Degree}(A_i, A_j) = \sum_{(A_i, A_j) \in \mathsf{Arcs}(G')} \mathsf{Degree}(A_i, A_j)$$

Therefore, $I_{cu}(G) = I_{cu}(G')$. □

Proposition 10. *The I_{cu} measure satisfies Monotonicity, Inversion, Isomorphic invariance, Disjoint additivity, and Super-additivity.*

Proof. (Monotonicity) Assume $G \sqsubseteq G'$. So $\mathsf{Arcs}(G) \subseteq \mathsf{Arcs}(G')$. So,

$$\sum\nolimits_{(A_i, A_j) \in \mathsf{Arcs}(G)} \mathsf{Degree}(A_i, A_j) \leq \sum\nolimits_{(A_i, A_j) \in \mathsf{Arcs}(G')} \mathsf{Degree}(A_i, A_j)$$

So, $I_{cu}(G) \leq I_{cu}(G')$. (Inversion) Assume $G' = \mathsf{Invert}(G)$. Since Degree is symmetric,

$$\sum\nolimits_{(A_i, A_j) \in \mathsf{Arcs}(G)} \mathsf{Degree}(A_i, A_j)$$
$$= \sum\nolimits_{(A_i, A_j) \in \mathsf{Arcs}(\mathsf{Invert}(G))} \mathsf{Degree}(A_j, A_i)$$

Therefore, $I_{cu}(G) = I_{cu}(\mathsf{Invert}(G))$. (Isomorphic invariance) Similar to proof for inversion. (Disjoint additivity) Assume G_1 and G_2 are disjoint. Therefore,

$$\sum\nolimits_{(A_i, A_j) \in \mathsf{Arcs}(G)} \mathsf{Degree}(A_i, A_j) =$$
$$\sum\nolimits_{(A_i, A_j) \in \mathsf{Arcs}(G_1)} \mathsf{Degree}(A_i, A_j)$$
$$+ \sum\nolimits_{(A_i, A_j) \in \mathsf{Arcs}((G_2)} \mathsf{Degree}(A_j, A_i)$$

Therefore, $I_{cu}(G) = I_{cu}(G_1) + I_{cu}(G_2)$. (Super-additivity) Similar to proof for disjoint additivity. $\square$

Proposition 11. *The I_{cu} measure is pairwise order-incomparable with each of the I_{dr}, I_{in}, I_{win}, I_{wou}, I_{cc}, I_{wcc}, I_{ic}, I_{pr}, I_{ngr}, and I_{ust} measures.*

Proof. From the differences in satisfaction of properties in Proposition 3, 6, and 10, I_{cu} is pairwise incompatible with the I_{win}, I_{wou}, I_{ic}, I_{pr}, I_{ngr}, and I_{ust} measures. However, from the properties in Proposition 3, we cannot discriminate I_{cu} from I_{in}, I_{wcc}, and I_{cc}. To discriminate I_{cu} from I_{in}, consider the following graphs G_1 (left) and G_2 (right) where $I_{in}(G_1) = 2$ and $I_{in}(G_2) = 2$, whereas $I_{cu}(G_1) = 2$ and $I_{cu}(G_2) = 2/3$.

$A_1 = \langle \{\alpha\}, \alpha \rangle$ ⇄ $A_2 = \langle \{\neg\alpha\}, \neg\alpha \rangle$

$A_5 = \langle \{\gamma, \neg\alpha \to \neg\gamma\}, \neg\alpha \rangle \to A_4 = \langle \{\alpha, \beta\}, \alpha \wedge \beta \rangle \leftarrow A_6 = \langle \{\delta, \delta \to \neg\beta\}, \neg\beta \rangle$

To discriminate I_{cu} from I_{cc}, consider the following graphs G_1 (left) and G_2 (right) where $I_{cc}(G_1) = 1$ and $I_{cc}(G_2) = 1$, whereas $I_{cu}(G_1) = 2$ and $I_{cu}(G_2) = 2/3$.

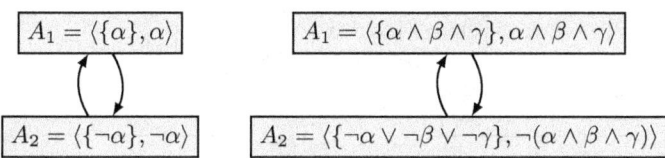

To discriminate I_{cu} from I_{wcc}, we can use a similar example to above. □

In the proposals for degree of undercut [12, 13], there are further options for degree of undercut, and these could be harnessed directly in the cumulative degree of undercut definition to provide potentially useful alternatives.

4.3 Application of Logic-based Measures of Inconsistency

In this section, we harness logic-based inconsistency measures to measure inconsistency in an argument graph instantiated with deductive arguments. We start by reviewing a couple of simple logic-based inconsistency measures. The first is the number of minimal inconsistent subsets of the knowledgebase, and the second is the sum of the inverse of the cardinality of each minimal inconsistent subset.

Definition 32. *Let K be a set of propositional formulae, and let $\mathsf{MinIncon}(K)$ be the set of minimal inconsistent subsets of K. The I_M measure and the $I_\#$ measure are defined as follows.*

$$I_M(K) = |\mathsf{MinIncon}(K)| \qquad I_\#(K) = \sum_{X \in \mathsf{MinIncon}(K)} \frac{1}{|X|}$$

Example 31. *Let $K = \{\alpha, \neg\alpha \vee \neg\beta, \beta, \neg\gamma, \neg\gamma \rightarrow \neg\alpha\}$. So $\mathsf{MinIncon}(K)$ is as below, $I_M(K) = 2$ and $I_\#(K) = 2/3$.*

$$\mathsf{MinIncon}(K) = \{\{\alpha, \neg\alpha \vee \neg\beta, \beta\}, \{\alpha, \neg\gamma, \neg\gamma \rightarrow \neg\alpha\}\}$$

The cumulative attack inconsistency measure takes the sum of the inconsistency measure of the premises of each attacker and attackee.

Definition 33. *The **cumulative attack inconsistency measure** w.r.t logic-based inconsistency measure I' is*

$$I_{I'}^C(G) = \sum_{(A_i, A_j) \in \mathsf{Arcs}(G)} I'(\mathsf{Support}(A_i) \cup \mathsf{Support}(A_j))$$

Example 32. *For the following graph G, $I_{I_M}^C(G) = 2$ and $I_{I_\#}^C(G) = 1$.*

$$A_2 = \langle \{\alpha, \beta, \alpha \wedge \beta \to \gamma\}, \gamma \rangle$$

$$A_1 = \langle \{\neg\alpha\}, \neg\alpha \rangle \qquad A_3 = \langle \{\neg\beta\}, \neg\beta \rangle$$

The support inconsistency measure, defined next, takes the inconsistency measure of the premises of all the arguments in the graph taken together.

Definition 34. *The* **support inconsistency measure** *w.r.t logical inconsistency measure I' is*

$$I^S_{I'}(G) = I'(\bigcup_{A \in \mathsf{Nodes}(G)} \mathsf{Support}(A))$$

Example 33. *Continuing Example 32, $I^S_{I_M}(G) = 2$ and $I^S_{I_\#}(G) = 1$.*

Example 34. *For the following graph G, $I^C_{I_M}(G) = 2$, $I^C_{I_\#}(G) = 1$, $I^S_{I_M}(G) = 3$, and $I^S_{I_\#}(G) = 3/2$.*

$$A_2 = \langle \{\alpha, \beta, \alpha \wedge \beta \to \gamma\}, \gamma \rangle$$

$$A_1 = \langle \{\neg\alpha \wedge \delta\}, \neg\alpha \rangle \qquad A_3 = \langle \{\neg\beta \wedge \neg\delta\}, \neg\beta \rangle$$

Proposition 12. *The $I^C_{I'}$ measure is a graph-based inconsistency measure according to Definition 19.*

Proof. (Consistency) Assume $\mathsf{Arcs}(G) = \emptyset$. So $\sum_{(A_i, A_j) \in \mathsf{Arcs}(G)} I'(\mathsf{Support}(A_i) \cup \mathsf{Support}(A_j)) = 0$. So $I^C_{I'}(G) = 0$. (Freeness) Assume $\mathsf{Nodes}(G) = \mathsf{Nodes}(G') \setminus \{A\}$ and $\mathsf{Arcs}(G) = \mathsf{Arcs}(G')$. So,

$$\sum_{(A_i, A_j) \in \mathsf{Arcs}(G)} I'(\mathsf{Support}(A_i) \cup \mathsf{Support}(A_j))$$
$$= \sum_{(A_i, A_j) \in \mathsf{Arcs}(G')} I'(\mathsf{Support}(A_i) \cup \mathsf{Support}(A_j))$$

So, $I^C_{I'}(G) = I^C_{I'}(G')$. □

Definition 35. *Let G be an argument graph instantiated with deductive arguments. G is* **reflective** *iff if $\bigcup_{A \in \mathsf{Nodes}(G)} \mathsf{Support}(A) \vdash \bot$, then $\mathsf{Arcs}(G) \neq \emptyset$.*

Assumption 1. *For the rest of the paper, we assume that all the argument graphs are reflective.*

Despite having an intuitive rationale, $I^S_{I'}$ is not a graph-based inconsistency measure according to Definition 19.

Proposition 13. *The $I_{I'}^S$ measure satisfies consistency but not freeness (as given in Definition 19).*

Proof. (Consistency) Assume $\mathsf{Arcs}(G) = \emptyset$. Therefore, there are no arguments A_i and A_j such that A_i attacks A_j. Therefore, $\bigcup_{A \in \mathsf{Nodes}(G)} \mathsf{Support}(A) \not\vdash \bot$. Therefore, $I_{I_M}^S(G) = 0$. (Freeness) Consider $A_1 = \langle \{\alpha \wedge \beta\}, \alpha \leftrightarrow \beta \rangle$, $A_2 = \langle \{\neg\alpha \wedge \gamma\}, \alpha \leftrightarrow \beta \rangle$, and $A_3 = \langle \{\neg\beta \vee \neg\gamma, \neg\beta \vee \neg\gamma \to \delta\}, \delta \rangle$. Let $\mathsf{Nodes}(G) = \{A_1, A_2\}$, $\mathsf{Arcs}(G) = \{(A_2, A_1)\}$, $\mathsf{Nodes}(G') = \{A_1, A_2, A_3\}$, and $\mathsf{Arcs}(G') = \{(A_2, A_1)\}$. So $I_{I_M}^S(G) = 1$ and $I_{I_M}^S(G') = 2$. □

Assumption 2. *We assume for the rest of this paper that when an argument A appears in $\mathsf{Nodes}(G)$ and in $\mathsf{Nodes}(G')$, then the logical argument associated with the node is the same (i.e. $\mathsf{Support}(A)$ is the same in both graphs, and $\mathsf{Claim}(A)$ is the same in both graphs). In addition, for argument $A \in \mathsf{Nodes}(G)$ and $A' \in \mathsf{Nodes}(G')$, if $\mathsf{Support}(A) = \mathsf{Support}(A')$ and $\mathsf{Claim}(A) = \mathsf{Claim}(A')$ then A and A' have the same name (i.e. $A = A'$).*

Proposition 14. *The $I_{I'}^C$ measure satisfies Monotonicity, Inversion, Isomorphic invariance, Disjoint additivity, and Super-additivity.*

Proof. (Monotonicity) Assume $G \sqsubseteq G'$. So $\mathsf{Arcs}(G) \subseteq \mathsf{Arcs}(G')$. So,

$$\sum_{(A_i, A_j) \in \mathsf{Arcs}(G)} I'(\mathsf{Support}(A_i) \cup \mathsf{Support}(A_j))$$
$$= \sum_{(A_i, A_j) \in \mathsf{Arcs}(G')} I'(\mathsf{Support}(A_i) \cup \mathsf{Support}(A_j))$$

So, $I_{I'}(G) \leq I_{I'}(G')$. (Inversion) Assume $G' = \mathsf{Invert}(G)$. So,

$$\sum_{(A_i, A_j) \in \mathsf{Arcs}(G)} I'(\mathsf{Support}(A_i) \cup \mathsf{Support}(A_j))$$
$$= \sum_{(A_i, A_j) \in \mathsf{Arcs}(\mathsf{Invert}(G))} I'(\mathsf{Support}(A_j) \cup \mathsf{Support}(A_i))$$

Therefore, $I_{I'}(G) = I_{I'}(\mathsf{Invert}(G))$. (Isomorphic invariance) Similar to proof for inversion. (Disjoint additivity) Assume G_1 and G_2 are disjoint. Therefore,

$$\sum_{(A_i, A_j) \in \mathsf{Arcs}(G)} I'(\mathsf{Support}(A_i) \cup \mathsf{Support}(A_j)) =$$
$$\sum_{(A_i, A_j) \in \mathsf{Arcs}(G_1)} I'(\mathsf{Support}(A_i) \cup \mathsf{Support}(A_j))$$
$$+ \sum_{(A_i, A_j) \in \mathsf{Arcs}((G_2)} I'(\mathsf{Support}(A_j) \cup \mathsf{Support}(A_i))$$

Therefore, $I_{I'}(G) = I_{I'}(G_1) + I_{I'}(G_2)$. (Super-additivity) Similar to proof for disjoint additivity. □

Proposition 15. *The $I_{I'}^S$ measure satisfies Monotonicity, Inversion, and Isomorphic invariance. However, $I_{I'}^S$ does not satisfy Disjoint additivity, or Super-additivity.*

Proof. (Monotonicity) Assume $G \sqsubseteq G'$. So $\mathsf{Arcs}(G) \subseteq \mathsf{Arcs}(G')$. So

$$\bigcup_{A \in \mathsf{Nodes}(G)} \mathsf{Support}(A) \subseteq \bigcup_{A \in \mathsf{Nodes}(G')} \mathsf{Support}(A)$$

So, $I_{I'}^S(G) \leq I_{I'}^S(G')$. (Inversion) Assume $G' = \mathsf{Invert}(G)$. So,

$$\bigcup_{A \in \mathsf{Nodes}(G)} \mathsf{Support}(A) = \bigcup_{A \in \mathsf{Nodes}(G')} \mathsf{Support}(A)$$

Therefore, $I_{I'}^S(G) \leq I_{I'}^S(\mathsf{Invert}(G))$. (Isomorphic invariance) Similar to proof for inversion. (Disjoint additivity) Consider G_1 (left) and G_2 (right) where $I_{I_M}(G_1 + G_2) = 2$ $I_{I_M}(G_1) = 2$ and $I_{I_M}(G_2) = 2$.

$A_1 = \langle \{\beta, \neg\beta \to \neg\alpha\}, \neg\alpha\rangle$		$A_4 = \langle \{\delta, \neg\delta \to \neg\epsilon\}, \neg\epsilon\rangle$
$A_2 = \langle \{\gamma, \neg\gamma \to \neg\beta\}, \neg\beta\rangle$		$A_5 = \langle \{\gamma, \neg\delta \to \neg\gamma\}, \neg\beta\rangle$
$A_3 = \langle \{\delta, \neg\delta \to \neg\gamma\}, \neg\gamma\rangle$		$A_6 = \langle \{\beta, \neg\gamma \to \neg\beta\}, \neg\gamma\rangle$

(Super-additivity) Similar to proof for disjoint additivity. □

For pairwise order-incompatibility, we consider $I' = I_M$ below. We can obtain similar results for other instantiations of $I_{I'}^C$ and $I_{I'}^S$.

Proposition 16. *The $I_{I_M}^C$ and $I_{I_M}^S$ measures are pairwise order-incompatible with each of the I_{dr}, I_{in}, I_{win}, I_{wou}, I_{cc}, I_{wcc}, I_{ic}, I_{pr}, I_{ngr}, I_{ust}, and I_{cu} measures.*

Proof. From the differences in satisfaction of properties in Propositions 3, 6, and 10, $I_{I_M}^C$ is pairwise incompatible with the I_{win}, I_{wou}, I_{ic}, I_{pr}, I_{ngr}, and I_{ust} measures. However, from the properties in Proposition 3, we cannot discriminate $I_{I_M}^C$ from I_{in}, I_{cc}, I_{wcc}, and I_{cu}. To discriminate $I_{I_M}^C$ from I_{in}, consider the following graphs G_1 (left) and G_2 (right) where $I_{in}(G_1) = 2$ and $I_{in}(G_2) = 1$, whereas $I_{I_M}^C(G_1) = 2$ and $I_{I_M}^C(G_2) = 2$.

$A_1 = \langle \{\alpha\}, \alpha\rangle$		$A_4 = \langle \{\alpha, \alpha \to \beta\}, \beta\rangle$
$A_2 = \langle \{\neg\alpha\}, \neg\alpha\rangle$		$A_5 = \langle \{\neg\beta, \neg\beta \to \neg\alpha\}, \neg\alpha\rangle$

To discriminate $I_{I_M}^C$ from I_{cc}, consider the above graphs G_1 (above left) and G_2 (above right) where $I_{cc}(G_1) = 1$ and $I_{cc}(G_2) = 0$, whereas $I_{I_M}^C(G_1) = 2$ and

$I^C_{I_M}(G_2) = 2$. To discriminate $I^C_{I_M}$ from I_{wcc} and from I_{cu}, we can use a similar example to above. To show the $I^S_{I_M}$ measure is pairwise incompatible with each of the $I_{dr}, I_{in}, I_{win}, I_{wou}, I_{cc}, I_{wcc}, I_{ic}, I_{pr}, I_{ngr}, I_{ust}$, and I_{cu} measures, we can use the failure of freeness to create examples where order-compatibility fails for each pairwise comparison. □

In this subsection, we have harnessed two existing logic-based inconsistency measures, I_M and $I_\#$, for measuring inconsistency in argument graphs instantiated with deductive arguments. There is a wide range of further measures of inconsistency that we could deploy in this role (for reviews see [25, 37]). In addition, we have only considered two ways of applying logic-based measures, namely $I^C_{I'}$ and $I^S_{I'}$. Further, ways of applying logic-based measures include analyzing the support in extensions of an argument graph to identify inconsistency. For instance, it is not necessarily the case that the union of the support of the arguments in an extension is consistent (for more discussion of this point, see [24]). Another option is to check the inconsistency measure of the premises of the defenders of an argument since it is not necessarily the case that these would be consistent. We leave investigation of these options to further work.

4.4 Related Work

In the converse of what we have considered in this section, deductive argumentation has been used for measuring inconsistency. For this, an argument tree (as defined by [11, 13]) is used. Each node in the tree is an argument. Each child is a canonical undercut of its parents. For each node in the tree, each canonical undercut of the node is a child of the node (except when the premises of the child have all occurred in the support of the argument in ancestor arguments). This exception prohibits infinite branches where each argument has the same premises that have already occurred on the branch.

Example 35. *Consider the knowledgebase* $K = \{\alpha, \neg\alpha, \neg\alpha \lor \beta, \beta, \neg\beta\}$. *The following is an argument tree with A_1 being the root.*

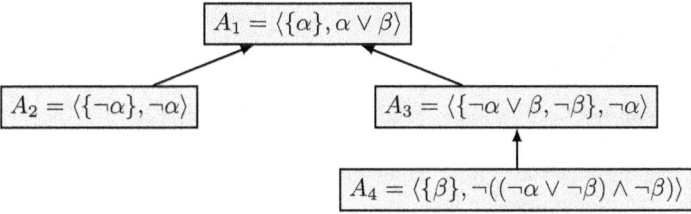

In [35], an argument tree is constructed where the argument at the root has a single premise. Then, three proposals are made for evaluating the inconsistency of this formula.

Definition 36. Let T be the argument tree with the argument A at the root having $\mathsf{Support}(A) = \phi$. An I_{ARG}^x measure is defined as follows

$$I_{ARG}^x(\phi, T) = |\mathsf{Undercuts}(\phi, T)| \times f^x(\phi, T)$$

where $\mathsf{Undercuts}(\phi, T)$ is the set of undercuts of the root argument in the tree T, $\mathsf{Height}(T)$ is the height of the tree (i.e. the number of edges in a path from root to leaf on the longest branch), and $\mathsf{Depth}(n, T)$ is the depth of node n in the tree (i.e. the number of edges in a path from root to node n).

$$f^1(\phi, T) = \frac{1}{\mathsf{Height}(T)}$$

$$f^2(\phi, T) = \frac{1}{\sum_{n \in \mathsf{Nodes}(T) \setminus \{\phi\}} \mathsf{Depth}(n, T)}$$

$$f^3(\phi, T) = \sum_{n \in \mathsf{Nodes}(T) \setminus \{\phi\}} \frac{1}{\mathsf{Depth}(n, T)}$$

So $I_{ARG}^1(\phi, T)$ takes the height of the tree into account, $I_{ARG}^2(\phi, T)$ takes the inverse of the sum of the depth of each node into account, and $I_{ARG}^3(\phi, T)$ takes the sum of the inverse of the depth of each node into account.

Example 36. *Continuing Example 35, the measures for three of the formulae are tabulated.*

	α	$\neg\alpha \vee \beta$	$\neg\alpha$
I_{ARG}^1	1	1/2	1/3
I_{ARG}^2	1/2	1/5	1/6
I_{ARG}^3	5	2	11/6

Whilst the proposal by Raddaoui [35] is for measuring inconsistency in a formula, it is possible that the ideas could be adapted for measuring the inconsistency of argument graphs.

5 Resolution through Commitment

An agent can commit to some arguments (i.e. declare whether they think an argument is acceptable or not), and s/he can be queried about those commitments. We assume that commitment by an agent is represented by the belief an agent has in the arguments.

In this section, we consider how we can model the agent. For this, we assume a labelling function as defined in Section 2.2. Initially, if we know nothing about the agent, we start with a uniform labelling that assigns undec to each argument. Then suppose the agent declares a commitment to an argument

— either by saying that the label for the argument is in or that it is out — we can consider what the ramifications are of that commitment on the other beliefs, and moreover, we can use it for resolving inconsistency (i.e. reducing the measure of inconsistency of the argument graph).

We proceed by introducing some subsidiary definitions for labellings, and for generating a subgraph of a graph based on a labelling. The first subsidiary definition specifies a labelling for which there is no undecided label.

Definition 37. *A labelling L is* **committed** *for graph G iff for all $A \in$ Nodes(G), $A \in$ in(G) or $A \in$ out(G).*

Example 37. *Consider the following graph. For this, the following labelling is committed: $L(A1) =$ out, $L(A2) =$ in, and $L(A3) =$ out.*

$$\boxed{A_1} \rightleftarrows \boxed{A_2} \leftarrow \boxed{A_3}$$

The next subsidiary definition constrains a labelling to take into account the attack relationship. As we illustrate in the subsequent example, a strict labelling is not necessarily an admissible labelling, though every admissible labelling is a strict labelling.

Definition 38. *A labelling L is* **strict** *for graph G iff for all $(A, B) \in$ Arcs(G), if $A \in$ in(G), then $B \in$ out(G).*

Example 38. *Consider the graph in Example 37. For this, the labellings that are committed and strict are tabulated below.*

	A_1	A_2	A_3
L_1	out	out	out
L_2	in	out	out
L_3	out	out	in
L_4	in	out	in
L_5	out	in	out

The following definition forms a subgraph from a graph and a labelling by deleting every node that is labelled out and deleting every arc that has either the source or the target labelled out. The reason we want this new graph is that if an agent commits to an argument being out, then that argument is no longer acceptable, and we can ignore it from further consideration.

Definition 39. *Given an argument graph G and labelling L, the* **new graph** *function is* NewGraph$(G, L) = G'$ *where*

- Nodes$(G') = \{A \in$ Nodes$(G) \mid L(A) \neq$ out$\}$
- Arcs$(G') = \{(A, B) \in$ Arcs$(G) \mid L(A) \neq$ out *and* $L(B) \neq$ out$\}$

Example 39. *Consider the following graph G with the labelling $L(A_1) =$ undec, $L(A_2) =$ undec, $L(A_3) =$ out, $L(A_4) =$ in, $L(A_5) =$ out, $L(A_6) =$ in, and $L(A_7) =$ undec.*

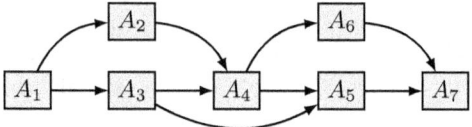

So the new graph G' for this graph and labelling is below.

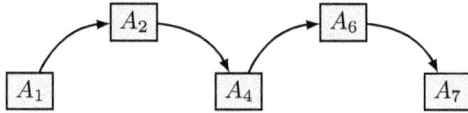

We can apply a measure of inconsistency to the graph G and G', and determine the reduction in inconsistency. For instance, $I_{in}(G) = 9$ and $I_{in}(G') = 4$. Similarly, $I_{win}(G) = 9/2$ and $I_{win}(G') = 4$.

When we query an agent about an argument A, we get a reply of either in or out. Given this information, we need to update the labelling to L' as follows.

Definition 40. *Let L be a labelling for graph G, and let $A \in \mathsf{Nodes}(G)$ be a query.*

- *If the answer for A is in, then*
  - $L'(A) =$ in
  - *for each $(A, B) \in \mathsf{Nodes}(G)$, $L'(B) =$ out,*
  - *for each $(B, A) \in \mathsf{Nodes}(G)$, $L'(B) =$ out,*
  - *for all other arguments C, $L'(C) = L(C)$.*

- *If the answer for A is out, then*
  - $L'(A) =$ out
  - *for all other arguments C, $L'(C) = L(C)$.*

Example 40. *Consider the following graph. Let L_1 be the original labelling, let L_2 be the new labelling obtained after the first query, and let L_3 be the new labelling obtained after the second query.*

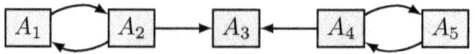

Suppose the first query concerns A_1, with the reply out, and the second query concerns A_4, with the reply in. The labellings are tabulated below.

	A_1	A_2	A_3	A_4	A_5
L_1	undec	undec	undec	undec	undec
L_2	out	undec	undec	undec	undec
L_3	out	undec	out	in	out

Now we can show how measures of inconsistency can help in deciding which arguments to query. We illustrate this in the following example.

Example 41. *Consider the following graph G where $I_{in}(G) = 6$ and $I_{cc}(G) = 2$. Suppose $L_0(A_i) = $ undec for all $A_i \in \{A_1, \ldots, A_5\}$. We could query any of these arguments.*

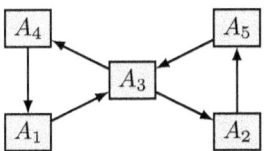

In the following, each bullet point concerns a specific query and specific answer to that query. In each case, given a graph G, and the revised labelling L, we obtain the new graph G' where $\mathsf{NewGraph}(G, L) = G'$.

- *If we query A_3, and we get the answer A_3 is in, then the resulting labelling is*

$$L(A_1) = \text{out} \quad L(A_2) = \text{out} \quad L(A_3) = \text{in}$$
$$L(A_4) = \text{out} \quad L(A_5) = \text{out}$$

  – *Hence, $\mathsf{NewGraph}(G, L) = (\{A_3\}, \{\})$.*
  – *So $I_{in}(G') = 0$ and $I_{cc}(G') = 0$.*

- *If we query A_3, and we get the answer A_3 is out, then the resulting labelling is*

$$L(A_1) = \text{undec} \quad L(A_2) = \text{undec} \quad L(A_3) = \text{out}$$
$$L(A_4) = \text{undec} \quad L(A_5) = \text{undec}$$

  – *Hence, $\mathsf{NewGraph}(G, L) = (\{A_1, A_2, A_4, A_5\}, \{(A_4, A_1), (A_2, A_5)\})$.*
  – *So $I_{in}(G') = 2$ and $I_{cc}(G') = 0$.*

- *If we query A_1, and we get the answer A_1 is in, then the resulting labelling is*

$$L(A_1) = \text{in} \quad L(A_2) = \text{undec} \quad L(A_3) = \text{out}$$
$$L(A_4) = \text{out} \quad L(A_5) = \text{undec}$$

- Hence, NewGraph$(G, L) = (\{A_1, A_2, A_5\}, \{(A_2, A_5)\})$.
- So $I_{in}(G') = 1$ and $I_{cc}(G') = 0$.

- If we query A_1, and we get the answer A_1 is out, then the resulting labelling is

$$L(A_1) = \text{out} \quad L(A_2) = \text{undec} \quad L(A_3) = \text{undec}$$
$$L(A_4) = \text{undec} \quad L(A_5) = \text{undec}$$

- Hence, NewGraph$(G, L) =$

$$(\{A_2, A_3, A_4, A_5\}, \{(A_3, A_4), (A_2, A_5), (A_3, A_2), (A_5, A_3)\})$$

- So $I_{in}(G') = 4$ and $I_{cc}(G') = 1$.

- If we query A_2, and we get the answer A_2 is in, then the resulting labelling is

$$L(A_1) = \text{undec} \quad L(A_2) = \text{in} \quad L(A_3) = \text{out}$$
$$L(A_4) = \text{undec} \quad L(A_5) = \text{out}$$

- Hence, NewGraph$(G, L) = (\{A_1, A_2, A_4\}, \{(A_4, A_1)\})$.
- So $I_{in}(G') = 1$ and $I_{cc}(G') = 0$.

- If we query A_2, and we get the answer A_2 is out, then the resulting labelling is

$$L(A_1) = \text{undec} \quad L(A_2) = \text{out} \quad L(A_3) = \text{undec}$$
$$L(A_4) = \text{undec} \quad L(A_5) = \text{undec}$$

- Hence, NewGraph$(G, L) =$

$$(\{A_1, A_3, A_4 A_5\}, \{(A_1, A_3), (A_3, A_4), (A_4, A_1), (A_5, A_3)\})$$

- So $I_{in}(G') = 4$ and $I_{cc}(G') = 1$.

Because of the graph structure, querying A4 is the same as querying A2 and querying A5 is the same as querying A1. Therefore, if we query A_3, we will have the maximum reduction in inconsistency if we use the I_{in} or I_{cc} measures where the reduction is the average of the reduction for the in and out answers.

In this section, we have seen how we can use the inconsistency measures for graphs as a way of guiding the selecting of arguments to query an agent. The answer from the agent is a commitment to accepting or not accepting the argument, and this can be used to resolve inconsistencies in the graph.

6 Discussion

This chapter makes the following contributions: (1) A proposal for a general framework of postulates for characterizing measures of inconsistency for argument graphs; (2) Proposals for graph structure measures and graph extension measures as instances of measures of inconsistency for argument graphs; (3) A review of the degree of undercut approach to measuring inconsistency for argument graphs instantiated with deductive arguments; (4) An investigation of the use of existing (logic-based) measures of inconsistency to measuring inconsistency for argument graphs instantiated with deductive arguments; and (5) An outline of how measures of inconsistency for argument graphs can be used as part of a process for inconsistency resolution in argumentation.

In future work, it would be good to give further properties for an inconsistency measure and relationships between them, further definitions for an inconsistency measure, further results for specific classes of graphs, and methods for resolution based on commitments.

The structure-based measures are in a sense measuring aspects of graph complexity. There are numerous options for features of argument graphs that could be considered (see for instance, features used for selecting argument solvers [41]). Further options for measuring graph complexity include measuring the sparseness of the graph (e.g. average indegree, average outdegree, indegree distribution, or outdegree distribution) which may give more recognition to unattacked arguments, radius of the graph (which is the maximum eccentricity of any node in the graph where the eccentricity of a node is the length of the shortest path to the node furthest away from that node), and dimensions of the graph (i.e. the minimum number of dimensions of Euclidean space required to represent all arcs with unit length). Another possible field for options for graph-based measures of inconsistency is graph entropy (see for example [21]).

A comparison with proposals for argument strength would be interesting. These consider a weight to individual arguments which can be affected by arguments that impinge upon it. A number of proposals have been made (e.g. [11, 17, 34, 2, 3, 4, 5, 26, 38, 7]), and comparisons undertake with a range of postulates (for a review see [8]). Possibly, analogous postulates can be proposed for inconsistency measures in graphs. Also it would be interesting to consider relationships with approaches to weighted labeling (e.g. [19]).

It would also be interesting to consider inconsistency measures for other forms of structured argumentation such as assumption-based argumentation (for a tutorial, see [40]), defeasible logic programming (for a tutorial, see [27]), and ASPIC+ (for a tutorial, see [33]). This would involve developing logic-based inconsistency measures for these non-standard logical formalisms.

Finally, it would be valuable to apply these techniques as part of the process for inconsistency resolution in argumentation for an application such as intelligence analysis to investigate the usability of the measures, and whether indeed there are tangible benefits to using these inconsistency measures.

Acknowledgements I am very grateful to John Grant and the anonymous reviewers for numerous comments for improving a draft of this paper. This research was partly funded by EPSRC grant EP/N008294/1 for the Framework for Computational Persuasion project.

References

[1] K. Atkinson, P. Baroni, M. Giacomin, A. Hunter, H. Prakken, C. Reed, G. Simari, M. Thimm, and S. Villata. Towards Artificial Argumentation. *AI Magazine*, 38(3):25–36, 2017.

[2] L. Amgoud and J. Ben-Naim. Ranking-based Semantics for Argumentation Frameworks. In *Proceedings of the 7th International Conference on Scalable Uncertainty Management (SUM'13)*, volume 8078 of *Lecture Notes In Artificial Intelligence*, pages 134–147, Washington, DC, USA, September 2013.

[3] L. Amgoud and J. Ben-Naim. Argumentation-based Ranking Logics. In *Proceedings of the 14th International Conference on Autonomous Agents and Multiagent Systems (AAMAS'15)*, pages 1511–1519, 2015.

[4] L. Amgoud and J. Ben-Naim. Axiomatic Foundations of Acceptability Semantics. In *Proceedings of the 15th International Conference on Principles of Knowledge Representation and Reasoning (KR'16)*, 2016.

[5] L. Amgoud, J. Ben-Naim, D. Doder, and S. Vesic. Ranking Arguments with Compensation-based Semantics. In *Proceedings of the 15th International Conference on Principles of Knowledge Representation and Reasoning (KR'16)*, pages 12–21, 2016.

[6] T. Bench-Capon and P. Dunne. Argumentation in Artificial Intelligence. *Artificial Intelligence*, 171(10-15):619–641, 2007.

[7] E. Bonzon, J. Delobelle, S. Konieczny, and N. Maudet. Argumentation Ranking Semantics based on Propagation. In *Proceedings of the Sixth International Conference on Computational Models of Argumentation (COMMA'16)*, pages 139–150, 2016.

[8] E. Bonzon, J. Delobelle, S. Konieczny, and N. Maudet. A comparative Study of Ranking-based Semantics for Abstract Argumentation. In *Proceedings of the 30th AAAI Conference on Artificial Intelligence (AAAI'16)*, pages 914–920, 2016.

[9] Ph. Besnard. Revisiting Postulates for Inconsistency Measures. In *Logics in Artificial Intelligence (JELIA'14)*, LNCS, pages 383–396. Springer, 2014.

[10] P. Baroni, D. Gabbay, M. Giacomin, and L. van der Torre, editors. *Handbook of Formal Argumentation*. College Publications, 2018. in preparation.

[11] Ph. Besnard and A. Hunter. A Logic-based Theory of Deductive Arguments. *Artificial Intelligence*, 128:203–235, 2001.

[12] Ph. Besnard and A. Hunter. Practical First-order Argumentation. In *Proceedings of the 20th American National Conference on Artificial Intelligence (AAAI'05)*, pages 590–595. MIT Press, 2005.

[13] Ph. Besnard and A. Hunter. *Elements of Argumentation*. MIT Press, 2008.

[14] Ph. Besnard and A. Hunter. Constructing Argument Graphs with Deductive Arguments: A Tutorial. *Argument and Computation*, 5(1):5–30, 2014.

[15] C. Cayrol. On the Relation between Argumentation and Non-monotonic Coherence-based Entailment. In *Proceedings of the 14th International Joint Conference on Artificial Intelligence (IJCAI'95)*, pages 1443–1448, 1995.

[16] M. Caminada and D. Gabbay. A Logical Account of Formal Argumentation. *Studia Logica*, 93(2-3):109–145, 2009.

[17] C. Cayrol and M.-C. Lagasquie-Schiex. Graduality in Argumentation. *Journal of Artificial Intelligence Research*, 23:245–297, 2005.

[18] M. Dalal. Investigations into a Theory of Knowledge base Revision. In *Proceedings of National Conference on Artificial Intelligence (AAAI'88)*, pages 475–479, 1988.

[19] C. da Costa Pereira, M. Dragoni, A. Tettamanzi, and S. Villata. Fuzzy Labeling for Abstract Argumentation: An Empirical Evaluation. In *Proceedings of the 10th International Conference on Scalable Uncertainty Management (SUM'13)*, volume 9858 of *Lecture Notes In Artificial Intelligence*, pages 126–139, 2016.

[20] P.E. Dunne, A. Hunter, P. McBurney, S. Parsons, and M. Wooldridge. Weighted Argument Systems: Basic Definitions, Algorithms, and Complexity Results. *Artificial Intelligence*, 175:457–486, 2011.

[21] M. Dehmer and A. Mowshowitz. A History of Graph Entropy Measures. *Information Sciences*, 181:5778, 2011.

[22] D. Doder, M. Rašković, Z. Marković, and Z. Ognjanović. Measures of Inconsistency and Defaults. *International Journal of Approximate Reasoning*, 51(7):832–845, 2010.

[23] P. Dung. On the Acceptability of Arguments and its Fundamental Role in Nonmonotonic Reasoning, Logic Programming, and n-person Games. *Artificial Intelligence*, 77:321–357, 1995.

[24] N. Gorogiannis and A. Hunter. Instantiating Abstract Argumentation with Classical Logic Arguments: Postulates and Properties. *Artificial Intelligence*, 175(9-10):1479–1497, 2011.

[25] J. Grant and A. Hunter. Measuring the Good and the Bad in Inconsistent Information. In *International Joint Conference on Artificial Intelligence (IJCAI'11)*, pages 2632–2637, 2011.

[26] D. Grossi and S. Modgil. On the Graded Acceptability of Arguments. In *Proceedings of the 24th International Joint Conference on Artificial Intelligence (IJCAI'15)*, pages 868–874, 2015.

[27] A. Garcia and G. Simari. Defeasible Logic Programming: Delp-servers, Contextual Queries, and Explanations for Answers. *Argument and Computation*, 5(1):63–88, 2014.

[28] A. Hunter and S. Konieczny. Approaches to Measuring Inconsistent Information. In *Inconsistency Tolerance*, volume 3300 of *LNCS*, pages 189–234. Springer, 2004.

[29] A. Hunter and S. Konieczny. On the Measure of Conflicts: Shapley Inconsistency Values. *Artificial Intelligence*, 174:1007–1026, 2010.

[30] S. Jabbour, Y. Ma, and B. Raddaoui. Inconsistency Measurement thanks to MUS Decomposition. In *Proceedings of the 2014 international conference onAutonomous agents and multi-agent systems*, pages 877–884. International Foundation for Autonomous Agents andMultiagent Systems, 2014.

[31] K. Knight. Measuring Inconsistency. *Journal of Philosophical Logic*, 31:77–98, 2001.

[32] K. Mu, W. Liu, and Z. Jin. Measuring the Blame of each Formula for Inconsistent Prioritized Knowledge bases. *Journal of Logic and Computation*, 22(3):481–516, 2012.

[33] S. Modgil and H. Prakken. The ASPIC+ Framework for Structured Argumentation. *Argumentation and Computation*, 5(1):31–62, 2014.

[34] P.A. Matt and F. Toni. A Game-theoretic Perspective on the Notion of Argument Strength in Abstract Argumentation. In *Proceedings of the 11th European Conference on Logics in Artificial Intelligence (JELIA'08)*, pages 285–297. Springer Berlin Heidelberg, 2008.

[35] B. Raddaoui. Computing Inconsistency using Logical Argumentation. In *Proceedings of the International Conference on Agents and Artificial Intelligence ICAART*, pages 164–172, 2015.

[36] I. Rahwan and G. Simari, editors. *Argumentation in Artificial Intelligence*. Springer, 2009.

[37] M. Thimm. On the Expressivity of Inconsistency Measures. *Artificial Intelligence*, 234:120–151, 2016.

[38] M. Thimm and G. Kern-Isberner. On Controversiality of Arguments and Stratified Labelings. In *Proceedings of the Fifth International Conference on Computational Models of Argumentation (COMMA'14)*, pages 413–420, September 2014.

[39] A. Toniolo, T. Norman, A. Etuk, F. Cerutti, R. Ouyang, M. Srivastava, N. Oren, T. Dropps, J. Allen, and P. Sullivan. Agent Support to Reasoning with Different Types of Evidence in Intelligence Analysis. In *Proceedings of the 14th International Conference on Autonomous Agents and Multiagent Systems (AAMAS 2015)*, pages 781–789, 2015.

[40] F. Toni. A Tutorial on Assumption-based Argumentation. *Argumentation and Computation*, 5(1):89–117, 2014.

[41] M. Vallati, F. Cerutti, and M. Giacomin. Argumentation Frameworks Features: An Initial Study. In *Proceedings of the European Conference on Artificial Intelligence (ECAI'14)*, pages 1117 – 1118, 2014.

Inconsistency Measures for Disjunctive Logic Programs Under Answer Set Semantics

Markus Ulbricht[1], Matthias Thimm[2],
Gerhard Brewka[1]

[1]Department of Computer Science
Leipzig University, Germany
{brewka,mulbricht}@informatik.uni-leipzig.de,
[2]Institute for Web Science and Technologies
University of Koblenz-Landau, Germany
thimm@uni-koblenz.de

Abstract

We address the issue of quantitatively assessing the severity of inconsistencies in disjunctive logic programs under the answer set semantics. Taking the non-monotonicity of answer set semantics into account brings new challenges that have to be addressed by reasonable accounts of inconsistency measures. We investigate the behaviour of inconsistency in logic programs by revisiting existing rationality postulates for inconsistency measurement and developing novel ones taking non-monotonicity into account. Further, we develop new measures for this setting and investigate their properties, in particular with respect to their compliance to these rationality postulates and their computational complexity.

1 Introduction

Inconsistency is an omnipresent phenomenon in logical accounts of knowledge representation and reasoning (KR) [9, 27, 23, 16, 12]. Classical logics usually suffer from the *principle of explosion* which renders reasoning meaningless, as everything can be derived from inconsistent theories. Therefore, reasoning under inconsistency [4, 32, 34] is an important research area in KR. In general, one can distinguish two paradigms in handling inconsistent information. The first paradigm advocates living with inconsistency but providing non-classical semantics that allow the derivation of non-trivial information, such as using paraconsistent reasoning [7], reasoning with possibilistic logic [16, 17], or formal argumentation [3]. The second paradigm is about explicitly restoring consistency, thus changing the theory itself, as it is done in e.g. belief revision [27] or belief merging [33]. A quantitative approach for *analyzing* inconsistencies is given by the field *inconsistency measurement* which investigates functions $\mathcal{I}$ that assign real numbers to knowledge bases, with the intuitive meaning that larger values indicate more severe inconsistency, see [46, 45, 48] for surveys and [47, 30, 26, 6, 1, 40] for some recent approaches.

Answer set programming (ASP, see [10] for an overview) is an emerging problem solving paradigm. It is based on logic programs under the answer set semantics [22, 21], a popular non-monotonic formalism for knowledge representation and reasoning which consists of rules possibly containing default-negated literals. Inconsistencies occur in ASP for two reasons, cf. [43]. First, the rules allow the derivation of two complementary literals l and $\neg l$—also called *incoherence* in e.g. [37]—thus producing inconsistencies similar to e.g. propositional logic. Second, due to the use of default negation it may happen that some literal assumed to be false is again derived (called *instability*). Hence, analyzing and handling inconsistency in ASP poses additional challenges (in comparison to monotonic logics) that need to be addressed, cf. [19, 13]. Some few works handle these challenges by adapting the classical techniques mentioned above to ASP, such as paraconsistent reasoning [8] or belief revision [15].

In this paper, we investigate the problem of measuring inconsistency in ASP. The issue of measuring inconsistency in logic programs is more challenging compared to the setting of propositional knowledge bases due to the non-monotonicity of answer set semantics. This becomes apparent when considering the *Monotonicity* postulate which is usually satisfied by inconsistency measures for propositional knowledge bases. It demands $\mathcal{I}(P') \geq \mathcal{I}(P)$ whenever $P \subseteq P'$ for any logical theories P and P', i.e., the severity of inconsistency cannot be decreased by adding new information. Consider now the two logic programs P and P' given as follows:

$P: b \leftarrow \text{not } a.$ $P': b \leftarrow \text{not } a.$
$\quad \neg b \leftarrow \text{not } a.$ $\quad \neg b \leftarrow \text{not } a.$
$\quad\quad\quad\quad\quad\quad\quad\quad\quad\quad\quad a.$

We have $P \subseteq P'$ but P is inconsistent while P' is not, so we would expect $\mathcal{I}(P') < \mathcal{I}(P')$ for any reasonable measure $\mathcal{I}$. Therefore, simply taking classical inconsistency measures and applying them to the setting of logic programs does not yield the desired behavior. Many rationality postulates such as *Monotonicity* are already disputed in the case of propositional knowledge bases, cf. [5]. Taking non-monotonicity of the knowledge representation formalism into account, a rational account of the severity of inconsistency calls for a specific investigation, which we will undertake in the remainder of this paper.

The main contributions of this paper are as follows:

1. We revisit the notion of inconsistency measures for ASP and develop

six new inconsistency measures for this setting (Section 3). Several of the measures are based on the effort it takes to repair an inconsistent program, others measure various kinds of distances between the actual and the intended outcome.

2. We critically examine existing rationality postulates, and develop novel ones taking non-monotonicity into account (Section 4). Several of our postulates aim to replace the unintended monotonicity postulate by weaker variants.

3. We analyze our new measures by checking their compliance with the rationality postulates (Section 5). In a nutshell, the main outcome of this analysis is that our new measures are well-behaved in the light of the postulates.

4. We finally perform an in-depth complexity analysis of computational problems related to our measures (Section 6).

Furthermore, we will give necessary preliminaries in Section 2 and conclude in Section 7. Readers familiar with logic programming and thus tempted to skip the preliminaries section should be aware that—for reasons explained in Section 2—we use a slightly nonstandard definition of answer sets which allows programs to have multiple inconsistent answer sets.

A brief description of a proper subset of the postulates and measures investigated in this paper was presented in the extended abstract [50]. The abstract did not cover disjunctive programs, and no complexity analysis was given.

2 Preliminaries

In this paper, we consider logic programs with disjunction in the head of rules and two kinds of negation, namely strong negation "¬" and default negation "**not**", under the answer set semantics [22, 21]. In [22] such programs were called extended disjunctive databases, whereas Gelfond and Leone [21] simply speak of logic programs or A-Prolog programs. We will call these programs *extended disjunctive logic programs*.

For the remainder of this paper, we assume we are (implicitly) given an infinite set $\mathcal{L}$ of literals. Now, an extended disjunctive logic program P is a finite set of rules r of the form

$$l_0 \vee \ldots \vee l_k \leftarrow l_{k+1}, \ldots, l_m, \text{not } l_{m+1}, \ldots, \text{not } l_n. \tag{1}$$

where $l_0, \ldots, l_n \in \mathcal{L}$ and $0 \leq k \leq m \leq n$. In particular, no function symbols or variables occur in r.

For a rule r of the form (1) we write $head(r) = \{l_0, ..., l_k\}$, $body(r) = \{l_{k+1}, \ldots, l_m, \text{not } l_{m+1}, \ldots, \text{not } l_n\}$, $pos(r) = \{l_{k+1}, \ldots, l_m\}$ and $neg(r) =$

$\{l_{m+1}, \ldots, l_n\}$. For a set M of literals, let $\mathcal{A}(M)$ be the set of all atoms occurring in M. We let $\mathcal{A}(r)$ and $\mathcal{L}(r)$ be the set of all atoms and literals occurring in r, respectively. Similarly, let $\mathcal{A}(P)$ and $\mathcal{L}(P)$ be the set of all atoms and literals that occur in a program P, respectively. Further, let $body(P) = \cup_{r \in P} body(r)$, and analogously for $pos(P)$ and $neg(P)$.

We write "$l_0 \vee \ldots \vee l_k.$" instead of "$l_0 \vee \ldots \vee l_k \leftarrow .$" for rules with a trivial body. If in addition $k = 0$ holds, i.e., the rule is of the form "$l_0.$", we call it a *fact*.

The set of all disjunctive logic programs is denoted by $\mathcal{P}$. Throughout the paper, we distinguish the following subclasses of $\mathcal{P}$:

- *Extended disjunctive logic programs*, i.e., programs that consist of rules of the form (1). Since we do not particularly focus on programs without the occurrence of strong negation "¬", we will call such programs simply *disjunctive logic programs* in most cases.

- *Extended logic programs*, i.e., programs that consist of rules of the form (1) with $k = 0$, i.e., with no occurrence of the disjunction "∨" in the head of rules. In order to emphasize the lack of disjunction, we will also call such programs *extended disjunction-free logic programs* or *disjunction-free logic programs* for short. Analogously, we call rules of the form (1) with $k = 0$ *disjunction-free rules*.

- *Extended disjunctive classical logic programs*, i.e., programs that consist of rules of the form (1) with $n = m$, i.e., with no occurrence of the default negation "not" in the body of rules. We will call such programs simply *classical logic programs*. Analogously, we call rules of the form (1) with $m = n$ *classical rules*.

- *Extended classical disjunction-free logic programs*, i.e., programs that consist of rules of the form (1) with $k = 0$ and $n = m$, i.e., neither "∨" nor "not" occurs in a rule. We will call such programs simply *classical disjunction-free logic programs*. Analogously, we call rules of the form (1) with $m = n$ *classical disjunction-free rules*.

We now turn to the semantics, i.e., the definition of answer sets. There are actually variants of the definition in the literature which differ in whether inconsistent answer sets are admitted or not. The original definition in [22] allows for a single inconsistent answer set, namely $\mathcal{L}$, in cases where a subset of rules without default negation generates an inconsistency. In this paper, we are interested in more fine-grained distinctions than the single inconsistent answer set $\mathcal{L}$ would allow. For this reason answer sets in this paper can be arbitrary subsets of $\mathcal{L}$. For a set M of literals and a literal l we say M satisfies l ($M \models l$) iff $l \in M$. If L is a set of literals, then $M \models L$ iff $M \models l$ for all $l \in L$. Now consider a classical rule, i.e., a rule r of the form (1) with $m = n$. We say M satisfies r, denoted by $M \models r$ iff $\exists i \in \{0,\ldots,k\} : M \models l_i$ whenever $M \models body(r)$.

Now we are ready to define answer sets of a given program.

Definition 1. Let P be a classical disjunctive logic program. A set M of literals is called an *answer set* of P if $M \models r$ for all rules $r \in P$ (denoted by $M \models P$) and there is no $M' \subsetneq M$ with $M' \models P$. For an arbitrary program P, M is an answer set of P iff M is an answer set of P^M, where P^M is the *reduct* of P with respect to M, i.e.,

$$P^M = \{head(r) \leftarrow pos(r) \mid r \in P, \; neg(r) \cap M = \emptyset\}.$$

Note that, so far, we defined what an *answer set* is no matter whether it is consistent or not. In principle, *any* set of literals can be an answer set of a given program. We now distinguish consistent and inconsistent answer sets.

Definition 2. A set M of literals is called *consistent* if it does not contain both a and $\neg a$ for an atom a. A program P is called *consistent* if it has at least one consistent answer set, otherwise it is called *inconsistent*. Let $\text{Ans}(P)$ denote the set of all answer sets of P and $\text{Ans}_{Inc}(P)$ and $\text{Ans}_{Con}(P)$ the inconsistent and consistent ones, respectively. So we have $\text{Ans}(P) = \text{Ans}_{Inc}(P) \cup \text{Ans}_{Con}(P)$.

Hence, a program P can be inconsistent, because

- it has no answer set, i.e., $\text{Ans}(P) = \emptyset$ or
- it only has inconsistent answer sets, i.e., $\text{Ans}(P) = \text{Ans}_{Inc}(P)$.

Note that in particular, P is inconsistent iff $\text{Ans}_{Con}(P) = \emptyset$.

Example 1. The program

$$P_1: \qquad a \vee \neg a. \qquad\qquad \neg a \leftarrow a.$$

has two answer sets, $\{a, \neg a\}$ and $\{\neg a\}$. The latter is consistent and so is the program. The same program with "$a \leftarrow \neg a.$" as additional rule, i.e., the program

$$P_2: \qquad a \vee \neg a. \qquad \neg a \leftarrow a. \qquad a \leftarrow \neg a.$$

has $\{a, \neg a\}$ as unique answer set and is therefore inconsistent. Now consider the program

$$P_3: \qquad b \leftarrow \text{not } c. \qquad c \leftarrow \text{not } d. \qquad d \leftarrow \text{not } b.$$

One can check that P_3 does not have an answer set. In fact, it is a quite simple example of an inconsistency that stems from an odd loop in the dependency graph (cf. Definition 15 below). In contrast,

$$P_3': \qquad b \leftarrow \text{not } c. \qquad c \leftarrow \text{not } d. \qquad d \leftarrow \text{not } b.$$
$$\qquad\qquad d \leftarrow \text{not } e.$$

has $\{b, d\}$ as answer set, because d can inferred due to the added rule "$d \leftarrow$ not e.". However, the program

$$P_3'': \quad b \leftarrow \text{not } c. \quad\quad c \leftarrow \text{not } d. \quad\quad d \leftarrow \text{not } b.$$
$$ d \leftarrow \text{not } e. \quad\quad e.$$

again has no answer set.

3 Inconsistency Measures

In the literature on inconsistency measurement—see e.g. [28, 24, 46]—inconsistency measures are functions that aim at assessing the severity of the inconsistency in knowledge bases formalized in propositional logic. Here, we are interested in measuring inconsistency for logic programs and only consider measures defined on those. Let $\mathbb{R}^\infty_{\geq 0}$ be the set of non-negative real values including ∞.

Definition 3. An *inconsistency measure* $\mathcal{I}$ is a function $\mathcal{I} : \mathcal{P} \to \mathbb{R}^\infty_{\geq 0}$.

The basic intuition behind an inconsistency measure $\mathcal{I}$ is that the larger the inconsistency in P the larger the value $\mathcal{I}(P)$. We now propose concrete inconsistency measures for logic programs. Inconsistency of logic programs can occur due to two different reasons, namely because the program has no answer set at all or because all answer sets are inconsistent, cf. [43]. Different measures should assess those reasons differently.

We start with a very simple measure which just indicates whether a given program is consistent or not [29].

Definition 4. Define $\mathcal{I}_{01} : \mathcal{P} \to \{0, 1\}$ via

$$\mathcal{I}_{01}(P) = \begin{cases} 0 & \text{if } P \text{ is consistent,} \\ 1 & \text{otherwise} \end{cases}$$

for all $P \in \mathcal{P}$.

We call $\mathcal{I}_{01}$ the *drastic inconsistency measure*. Of course, this measure fails to provide the distinction we are aiming for, namely the distinction between less and more inconsistent programs. We will therefore introduce various more fine-grained measures in the remainder of this section.

In Section 3.1, we introduce a generalization of the measure $\mathcal{I}_{\text{MI}}$ [29] which, in its original definition, counts the number of minimal inconsistent subsets of a knowledge base $\mathcal{K}$ (a set of propositional formulas). For that, we utilize a notion of inconsistency for nonmonotonic logics developed in [11]. We continue with measures that are based on the distance of inconsistent answer sets to consistent ones (Section 3.2). Then, we consider syntactic approaches that are based on the effort needed to turn an inconsistent program into a consistent one (Section 3.3).

3.1 Measures based on Strong Inconsistency

One of the most basic inconsistency measures for a propositional knowledge base $\mathcal{K}$ is $\mathcal{I}_{\mathsf{MI}}$ [29] defined via $\mathcal{I}_{\mathsf{MI}}(\mathcal{K}) = |\mathsf{MI}(\mathcal{K})|$ where $\mathsf{MI}(\mathcal{K})$ is the set of minimal inconsistent subsets of $\mathcal{K}$. As consistent programs may contain inconsistent subsets—cf. e.g., P_3' from above—this measure is quite meaningless in ASP. In [11], a refined notion of inconsistency for nonmonotonic logics has been proposed which does not have this drawback. We give the definition for the special case of ASP.

Definition 5. Let P be a logic program. A subset $H \subseteq P$, is called *strongly P-inconsistent* if $H \subseteq H' \subseteq P$ implies H' is inconsistent. The set H is *minimal strongly P-inconsistent* if H is strongly P-inconsistent and $H' \subsetneq H$ implies that H' is not strongly P-inconsistent. Let $SI_{min}(P)$ be the set of all minimal strongly P-inconsistent subsets of P.

The main motivation for this notion of inconsistency is that a generalization of Reiter's hitting set duality [42] can be proved (cf. [11] for more details). Now, we can define our generalized measure as follows.

Definition 6. Define $\mathcal{I}_{\mathsf{MSI}} : \mathcal{P} \to \mathbb{R}_{\geq 0}^{\infty}$ via

$$\mathcal{I}_{\mathsf{MSI}}(P) = |SI_{min}(P)|$$

i.e., the measure outputs the number of minimal strongly P-inconsistent subsets of P.

Given a propositional knowledge base $\mathcal{K}$, the minimal inconsistent subsets $\mathsf{MI}(\mathcal{K})$ are interpreted as the "raw" conflicts within $\mathcal{K}$. Similarly, $\mathcal{I}_{\mathsf{MSI}}$ counts the number of "raw" conflicts within a program P.

Example 2. Consider

$$P_2: \quad a \vee \neg a. \quad\quad \neg a \leftarrow a. \quad\quad a \leftarrow \neg a.$$

again. Any proper subset H of the program is consistent. Hence, P_2 itself is the only strongly P_2-inconsistent subset. We obtain

$$\mathcal{I}_{\mathsf{MSI}}(P_2) = 1.$$

Now consider

$$P_3'': \quad b \leftarrow \text{not } c. \quad\quad c \leftarrow \text{not } d. \quad\quad d \leftarrow \text{not } b.$$
$$ d \leftarrow \text{not } e. \quad\quad\quad\quad\quad e.$$

The subset

$$H = \{b \leftarrow \text{not } c., \quad c \leftarrow \text{not } d., \quad d \leftarrow \text{not } b.\}$$

is inconsistent, but it is *not* strongly P_3''-inconsistent as it is contained in the consistent set

$$P_3': \quad b \leftarrow \text{not } c. \quad c \leftarrow \text{not } d. \quad d \leftarrow \text{not } b.$$
$$d \leftarrow \text{not } e.$$

One can verifiy that

$$SI_{min}(P_3'') = \{\{b \leftarrow \text{not } c., c \leftarrow \text{not } d., d \leftarrow \text{not } b., e.\}\}$$

and hence,

$$\mathcal{I}_{\mathsf{MSI}}(P_3'') = 1.$$

3.2 Distance-based Measures

We now consider measures that focus on the (possibly inconsistent) answer sets of a program instead of the program itself. To this end, we now introduce inconsistency measures that make use of distance measures to assess the inconsistency of answer sets. Note that distance-based measures have also been used in the setting of propositional logic, see for instance [26].

Observe that, while a program P might not have an answer set, for any set M of literals the reduct P^M is a classical logic program and thus has a nonempty set of answer sets. So, one could consider the distance between a set M and consistent answer sets in $\text{Ans}_{Con}(P^M)$.

Definition 7. A mapping $d : 2^{\mathcal{L}} \times 2^{\mathcal{L}} \to [0, \infty)$ is called a *distance* if it satisfies

- $d(X, Y) = 0$ if and only if $X = Y$,
- $d(X, Y) = d(Y, X)$,
- $d(X, Y) \leq d(X, Z) + d(Z, Y)$

for any $X, Y, Z \subseteq \mathcal{L}$.

In the following, we only consider the number of literals in the symmetric difference of two sets as an example of a distance measure between sets. Investigating other distances is left for future work.

Definition 8. Let M and M' be two sets of literals. The *sd-distance* (sd="symmetric difference") $d_{sd}(M, M')$ between M and M' is defined via $d_{sd}(M, M') = |(M \cup M') \setminus (M \cap M')|$. If $\mathcal{M}$ is a set of sets of literals, we let

$$d_{sd}(M, \mathcal{M}) = \min_{M' \in \mathcal{M}} d(M, M').$$

It should be obvious that d_{sd} is indeed a distance function on sets according to Definition 7.

Now we can consider measures of the following kind.

Definition 9. Let d be a distance measure. Define $\mathcal{I}_d : \mathcal{P} \to \mathbb{R}_{\geq 0}^{\infty}$ via

$$\mathcal{I}_d(P) = \min_{\substack{M \subseteq \mathcal{L} \\ M \text{ consistent}}} \{d(M, M') \mid M' \in \text{Ans}_{Con}(P^M)\}$$

for all $P \in \mathcal{P}$ with $\min \emptyset = \infty$.

Note that we only allow consistent sets of literals because we want to measure the distance between a potential consistent answer set M of P and the consistent models of P^M. In the following, we abbreviate the inconsistency measure $\mathcal{I}_{d_{sd}}$ simply by $\mathcal{I}_{sd}$ and focus on this instance.

Example 3. Consider the simple case

$$P_3: \qquad b \leftarrow \text{not } c. \qquad c \leftarrow \text{not } d. \qquad d \leftarrow \text{not } b.$$

We formally show that $\mathcal{I}_{sd}(P_3) = 1$. As the program is inconsistent, it is quite easy to see that $\mathcal{I}_{sd}(P_3) \geq 1$ holds. Now let $M = \{b\}$. Then the reduct is given via

$$P_3^M: \qquad\qquad b. \qquad\qquad c.$$

with minimal model $\{b, c\}$. We thus found a set M with

$$\exists M' \in \text{Ans}_{Con}(P_3^M) : d_{sd}(M, M') = 1.$$

Thus, $\mathcal{I}_{sd}(P_3) \leq 1$. For the program

$$P_2: \qquad a \vee \neg a. \qquad \neg a \leftarrow a. \qquad a \leftarrow \neg a.$$

we see that $\mathcal{I}_{sd}(P_2) = \infty$ because there is no set M of literals such that P_2^M has a consistent answer set.

Remark 1. Note that $\mathcal{I}_{sd}(P) \in \{0, \infty\}$ if P is a classical program.

Another approach that is based on the semantics of a program P rather than the syntax is considering all answer sets of P and measuring the distance to a consistent one. Again, one could do this with arbitrary distances d. However, we will again only look at the symmetric difference d_{sd}. Given an inconsistent set M of literals, the minimal distance $d_{sd}(M, M')$ between M and a consistent set M' is simply the number of complementary literals in M. Let $\mathbb{N}_0$ denote the set of natural numbers including zero.

Definition 10. A set M of literals is called k-inconsistent, $k \in \mathbb{N}_0$, if there are exactly k atoms a such that $a \in M$ and $\neg a \in M$.

Further, we have to take into account that a program might be inconsistent due to having no answer set. We assign ∞ to such programs as they are a special case for this measure.

Definition 11. Define $\mathcal{I}_\# : \mathcal{P} \to \mathbb{R}_{\geq 0}^\infty$ via

$$\mathcal{I}_\#(P) = \min_{M \in \text{Ans}(P)} \{k \mid M \text{ is } k\text{-inconsistent}\}$$

with $\min \emptyset = \infty$.

Example 4. Since

$$P_3: \quad b \leftarrow \text{not } c. \quad\quad c \leftarrow \text{not } d. \quad\quad d \leftarrow \text{not } b.$$

has no answer set, $\mathcal{I}_\#(P_3) = \infty$. For the program

$$P_2: \quad a \vee \neg a. \quad\quad \neg a \leftarrow a. \quad\quad a \leftarrow \neg a.$$

we obtain $\mathcal{I}_\#(P_2) = 1$ due to the inconsistent answer set $M = \{a, \neg a.\}$.

This concludes our discussion on distance-based measures.

3.3 Modification-based Measures

Our next measure $\mathcal{I}_\pm$ aims at measuring the effort needed to turn an inconsistent program into a consistent one. More specifically, it quantifies the number of modifications in terms of deleting and adding rules, necessary in order to restore consistency. Deleting certain rules can surely be sufficient to prevent P from entailing contradictions, but as already pointed out before, adding rules can also resolve inconsistency.

Definition 12. Define $\mathcal{I}_\pm : \mathcal{P} \to \mathbb{R}_{\geq 0}^\infty$ via

$$\mathcal{I}_\pm(P) = \min\{|A| + |D| \mid A, D \in \mathcal{P} \text{ such that } (P \cup A) \setminus D \text{ is consistent}\}$$

for all $P \in \mathcal{P}$.

Example 5. Let us consider our examples from above again. Since deleting any rule of

$$P_2: \quad a \vee \neg a. \quad\quad \neg a \leftarrow a. \quad\quad a \leftarrow \neg a.$$

renders the program consistent, $\mathcal{I}_\pm(P_2) = 1$. The same is true for

$$P_3'': \quad b \leftarrow \text{not } c. \quad\quad c \leftarrow \text{not } d. \quad\quad d \leftarrow \text{not } b.$$
$$ d \leftarrow \text{not } e. \quad\quad e.$$

In the latter case, however, one could also *add* rules. For example, $P_3'' \cup \{d.\}$ is consistent.

The definition of $\mathcal{I}_\pm$ allows the addition of any rule in order to restore consistency. But in fact, it is sufficient to only consider addition of facts instead of general rules. First, we show that adding rules with disjunction in the head is not necessary, which is intuitive since ASP requires minimality anyway.

Proposition 1. *Let P be an inconsistent program. If r is a rule such that $P \cup \{r\}$ is consistent, then there is a literal $a \in head(r)$ such that $P \cup \{a \leftarrow body(r).\}$ is consistent.*

Proof. Let M be a consistent answer set of $P \cup \{r\}$. P being inconsistent implies that M is not an answer set of P. Thus, $M \models body(\{r\}^M)$ and $\{r\}^M \neq \emptyset$ because otherwise one could delete the rule while maintaining M as answer set. It follows that $head(r) \cap M \neq \emptyset$ since M is a model of $(P \cup \{r\})^M$. Let $a \in head(r) \cap M$. We show that M is an answer set of $P \cup \{a \leftarrow body(r).\}$ as well. By definition, M is a minimal model of $(P \cup \{r\})^M$. Since $a \in M$, M is a model of $(P \cup \{a \leftarrow body(r).\})^M$ as well. Now assume M is not a minimal model and let $M' \subsetneq M$ be a model of $(P \cup \{a \leftarrow body(r).\})^M$. Due to $a \in head(r)$ this implies that M' is a model of $(P \cup \{r\})^M$, too. Since M was assumed to be an answer set of $(P \cup \{r\})^M$, this yields a contradiction. $\square$

Now we are ready to show that facts are sufficient.

Proposition 2. *Let P be an inconsistent program. If r is a rule such that $P \cup \{r\}$ is consistent, then there is a literal $a \in head(r)$ such that $P \cup \{a.\}$ is also consistent.*

Proof. Using Proposition 1, we assume that $head(r)$ contains only one literal a. As in the proof of Proposition 1, we see that $M \models body(\{r\}^M)$ and $\{r\}^M \neq \emptyset$. Since M is a model of P^M, we obtain $a \in M$. However, this means M is an answer set of $P \cup \{a.\}$. $\square$

We now consider a simplified version of $\mathcal{I}_\pm$ that focuses entirely on additions of rules to restore consistency. This measure is adequate whenever the information in the given program is considered highly reliable and when it is thus more likely that information was forgotten rather than represented incorrectly. Adding additional assumptions seems to be a reasonable solution to resolve all the inconsistencies in such situations. This motivates the measure $\mathcal{I}_+$ which applies this solution.

Definition 13. Let $\mathcal{I}_+ : \mathcal{P} \to \mathbb{R}_{\geq 0}^\infty$ be the measure given via

$$\mathcal{I}_+(P) = \min\{|A| \mid A \in \mathcal{P} \text{ such that } P \cup A \text{ is consistent}\}$$

with $\min \emptyset = \infty$.

The case $\mathcal{I}_+(P) = \infty$ occurs whenever adding rules cannot resolve inconsistency, e.g., if a program contains two contradicting facts. Note that Proposition 2 applies to $\mathcal{I}_+$ as it does to $\mathcal{I}_\pm$. Hence, we can also w.l.o.g. assume the set A in the definition of $\mathcal{I}_+$ to be a set of facts.

Example 6. The program

$$P_2: \qquad a \vee \neg a. \qquad \neg a \leftarrow a. \qquad a \leftarrow \neg a.$$

cannot be repaired by adding rules. Hence, $\mathcal{I}_+(P_2) = \infty$. As argued before, the inconsistency of

$$P_3'': \qquad b \leftarrow \text{not } c. \qquad c \leftarrow \text{not } d. \qquad d \leftarrow \text{not } b.$$
$$ d \leftarrow \text{not } e. \qquad e.$$

can be resolved by adding e.g. "d.". Thus, $\mathcal{I}_+(P_3'') = 1$.

The following observation is obvious.

Proposition 3. *For any program $P \in \mathcal{P}$, $\mathcal{I}_\pm(P) \leq \mathcal{I}_+(P)$.*

Remark 2. If P is an inconsistent classical program, then adding rules will never resolve inconsistency. Hence $\mathcal{I}_+(P) \in \{0, \infty\}$ if P is a classical program.

Of course, it is also possible to focus entirely on deletions of rules. Since the empty program is obviously consistent, it is always possible to restore consistency via deletions alone. Deletions are the obvious choice when modeling errors may have occurred.

Definition 14. Let $\mathcal{I}_- : \mathcal{P} \to \mathbb{R}_{\geq 0}^\infty$ be the measure given via

$$\mathcal{I}_-(P) = \min\{|D| \mid D \in \mathcal{P} \text{ such that } P \setminus D \text{ is consistent}\}$$

for any $P \in \mathcal{P}$.

The following observations also follow directly by definition.

Proposition 4. *For any $P \in \mathcal{P}$, $\mathcal{I}_\pm(P) \leq \mathcal{I}_-(P)$. If P is a classical program, $\mathcal{I}_\pm(P) = \mathcal{I}_-(P)$.*

4 Rationality Postulates for Inconsistency Measures

Research in inconsistency measurement is driven by *rationality postulates*, i.e., desirable properties that should hold for concrete approaches. There is a growing number of rationality postulates for inconsistency measurement but not every postulate is generally accepted, see [5] for a recent discussion on this topic.

In the following, we revisit a selection of the most popular postulates—see [45] for a recent survey—and phrase them within our context of logic programs.

Let $\mathcal{I} : \mathcal{P} \to \mathbb{R}_{\geq 0}^{\infty}$ be some inconsistency measure and $P, P' \in \mathcal{P}$ some disjunctive logic programs. The most central property of any inconsistency measure is that it is able to distinguish consistency from inconsistency.

Consistency P is consistent iff $\mathcal{I}(P) = 0$.

The above postulate establishes that 0 is the minimal inconsistency value and that it is reserved for consistent programs.

We have already mentioned *Monotonicity* as a desirable property for inconsistency measures in monotonic logics in the introduction.

Monotonicity $\mathcal{I}(P) \leq \mathcal{I}(P')$ whenever $P \subseteq P'$.

Satisfaction of this postulate is generally *not* desirable for ASP, as we discussed before.

In the rest of this section we will propose new postulates for logic programs. We first discuss weaker variants of monotonicity in Section 4.1 and then discuss some further postulates in Section 4.2.

4.1 Weakening Monotonicity

Although monotonicity as a general property is undesired, as we have seen, we still wish to require some form of monotonicity in special cases. First, if a program P does not contain any default negation, no additional information can resolve any conflicts in P.

CLP-Monotonicity If P is a classical logic program and P' an arbitrary one, then $\mathcal{I}(P) \leq \mathcal{I}(P \cup P')$.

In the above postulate, CLP stands for "classical logic program". We can further elaborate on the idea of *CLP-Monotonicity*: Whenever P' has no influence on P, then $\mathcal{I}(P) \leq \mathcal{I}(P \cup P')$ should hold as well. To make this precise, we need the notion of the dependency graph of a program [2].

Definition 15. Let P be an extended logic program. The *dependency graph* D_P of the program P is a labeled directed graph having $\mathcal{L}(P)$ as vertices and there is an edge (l_i, l_j, s) iff P contains a rule r such that $l_j \in head(r)$ and $l_i \in pos(r) \cup neg(r)$. The label $s \in \{+, -\}$ indicates whether $l_i \in pos(r)$ or $l_i \in neg(r)$. For any $l \in \mathcal{L}(P)$, let $\mathsf{Path}(P, l)$ be the set of all literals l' such that there is a directed path (with any labels) from l to l' in D_P (including l itself). For a set M of literals, let

$$\mathsf{Path}(P, M) = \bigcup_{l \in M} \mathsf{Path}(P, l).$$

Example 7. Consider the union of the programs considered in Example 1 except the fact "e." i.e., the program P_4 given as follows:

$$P_4: \qquad a \vee \neg a. \qquad \neg a \leftarrow a. \qquad a \leftarrow \neg a..$$
$$b \leftarrow \text{not } c. \qquad c \leftarrow \text{not } d. \qquad d \leftarrow \text{not } b.$$
$$d \leftarrow \text{not } e.$$

The dependency graph of P_4 is depicted in Figure 1. For example, $\mathsf{Path}(P_4, b) = \{b, c, d\}$.

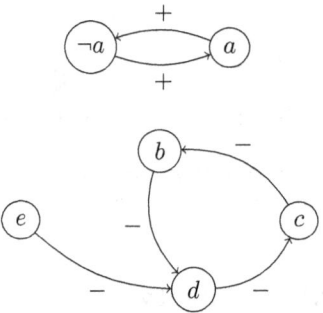

Figure 1: Dependency graph of P_4

Now we are ready to describe the notion of *splitting* logic programs [36].

Definition 16. Let P be an extended logic program. A set U of literals is called a *splitting set* for P, if $head(r) \cap U \neq \emptyset$ implies $\mathcal{L}(r) \subseteq U$ for any rule $r \in P$. For a splitting set U, let $bot_U(P)$ be the set of all rules $r \in P$ with $head(r) \subseteq U$. This set of rules is called the *bottom part* of P with respect to U.

In other words, if l is a literal that is not contained in U, i.e., $l \in \mathcal{L}(P) \setminus U$, then it cannot be in the head of any rule in $bot_U(P)$. For the dependency graph D_P, this means that while there could be a path from a literal $l' \in U$ to l, the converse is not true. Splitting is used, for example, to effectively compute answer sets because it allows handling $bot_U(P)$ without taking the rest of the program into account, see [36] for more details. However, since splitting is generally useful to examine the structure of a program, we are also interested in this notion here.

Example 8. Consider program P_4 again.

$$P_4: \quad a \vee \neg a. \qquad \neg a \leftarrow a. \qquad a \leftarrow \neg a.$$
$$b \leftarrow \text{not } c. \qquad c \leftarrow \text{not } d. \qquad d \leftarrow \text{not } b.$$
$$d \leftarrow \text{not } e.$$

For the splitting set $U = \{a, \neg a, e\}$, $bot_U(P_4)$ is the program

$$bot_U(P_4): \quad a \vee \neg a. \qquad \neg a \leftarrow a. \qquad a \leftarrow \neg a.$$

The definition of a splitting set ensures that the set of nodes that corresponds to $U = \{a, \neg a, e\}$ has no in-going edge from outside in the dependency graph, cf. Figure 1.

Theorem 1 ([36]). *Let U be a splitting set for a program P. Every answer set M of P is of the form $M = X \cup Y$ with an answer set X of $bot_U(P)$ and a set Y of literals.*

Corollary 1. *Let U be a splitting set of P. If P is consistent, then so is $bot_U(P)$.*

Proof. Let U be a splitting set of P and M a consistent answer set. Due to Theorem 1, there is a subset X of M such that X is an answer set of $bot_U(P)$. M being consistent implies X is consistent. Thus, $bot_U(P)$ is consistent. □

Example 9. We continue Example 8. The bottom part $bot_U(P_4)$ has one answer set, namely $\{a, \neg a\}$. Due to Theorem 1, all answer sets of P_4 contain both a and $\neg a$. Indeed, P_4 has the single answer set $\{a, \neg a, b, d, \}$. In particular, it is impossible to resolve the inconsistency of the program without changing the bottom part.

Intuitively, if $bot_U(P)$ is inconsistent, changing the rest of the program will not remove the reason why the bottom part is inconsistent; it is imposed on P. So, it is reasonable to assume that P is at least as inconsistent as $bot_U(P)$.

Split-Monotonicity If U is a splitting set for P, then $\mathcal{I}(bot_U(P)) \leq \mathcal{I}(P)$.

In the above notions of monotonicity, we always require properties ensuring that, once a program entails a contradiction, the added rules do not fix this. In the case of *CLP-Monotonicity*, this is done by requiring P to be a classical program and in the presence of a splitting set U, this property is inherent since the bottom part is just independent from the remainder of the program. However, we also achieve this goal if we "avoid" non-monotonicity: Adding a rule r to a program P will not resolve inconsistency as long as we make sure that the rule is not "involved in the non-monotonicity" of the program. This is the case if there are no paths from any literal of the head of r to default-negated literals. This ensures that the rule is meaningless for the derivation of such literals. To motivate this postulate, we make the following observations.

Lemma 1. *Let P be a disjunctive logic program and $r \notin P$ a rule with*

$$\mathsf{Path}(P \cup \{r\}, head(r)) \cap neg(P \cup \{r\}) = \emptyset. \tag{2}$$

Then, for all answer sets M^* of $P \cup \{r\}$, there is an answer set M of P with $M \subseteq M^*$.

Proof. For notational convenience, we let

$$P^* = P \cup \{r\}.$$

Let M^* be an answer set of P^*, i.e., $M^* \in \text{Ans}_{Con}((P^*)^{M^*})$. The outline of the proof is as follows: From M^* we find a minimal model M of P^{M^*} with $M \subseteq M^*$. Using (2), we will see that $P^{M^*} = P^M$ and hence, M being a minimal model of P^{M^*} implies it is a minimal model of P^M as well. However, this means that M is an answer set of P.

So let M^* be a minimal model of $(P^*)^{M^*}$. Of course, M^* is a model of P^{M^*} as well, but it might not be minimal. Since P^{M^*} is a classical program, it has a minimal model. So, we can remove literals from M^* until we obtain a minimal model $M \subseteq M^*$ from P^{M^*}. We now examine the set $X = M^* \setminus M$. We assume $X \neq \emptyset$, i.e., M is a proper subset of M^*. Otherwise, the claim follows trivially. Hence, M is not a model of $(P^*)^{M^*}$, because M^* was assumed to be minimal. However, the only rule in $(P^*)^{M^*}$ that is not contained in P^{M^*} is $\{r\}^{M^*}$. Since M is a model of P^{M^*}, this means that M does not satisfy $\{r\}^{M^*}$.

As a minimal model of $(P^*)^{M^*}$, M^* satisfies $\{r\}^{M^*}$. Thus, we can find a literal $l_1 \in X = M^* \setminus M$ with $l_1 \in head(r)$. Now, $M_1 = M \cup \{l_1\}$ satisfies $\{r\}^{M^*}$. Clearly, $M_1 \subseteq M^*$. If $M_1 = M^*$, then we found that l_1 is the only literal that we need to add, i.e., that it is the only literal in $X = M^* \setminus M$. Otherwise, M_1 cannot be a model of $(P^*)^{M^*}$, because M^* was assumed to be minimal. In this case, we find a rule that is not satisfied and continue as above to obtain a set $M_2 = M_1 \cup \{l_2\} \subseteq M^*$. Since M^* is a model of $(P^*)^{M^*}$, we will never encounter a situation, where at some point M_i cannot be augmented with an additional literal to satisfy an unsatisfied rule. Since M^* is minimal, the procedure will not stop until we constructed M^*, i.e., found an n such that $M_n = M^*$. In particular, this means that we have to augment M with all literals in X.

Recall that the first literal we added, l_1, was contained in $head(r)$. Since M is a model of P^{M^*}, M satisfies all rules in P^{M^*}. Thus, the fact that we might have to add further literals stems from augmenting M with l_1. Using this observation, one can easily inductively verify that all literals in $X = M^* \setminus M$, i.e., all literals we added in the procedure described above, are contained in $\mathsf{Path}(P \cup \{r\}, head(r))$.

Now, we are ready to show that $P^{M^*} = P^M$. As pointed out, M^* is of the form

$$M^* = M \cup X \tag{3}$$

with
$$X \subseteq \mathsf{Path}(P^*, head(r)). \qquad (4)$$
Now, (4) and (2) imply
$$X \cap neg(P^*) \subseteq \mathsf{Path}(P^*, head(r)) \cap neg(P^*) = \emptyset.$$
Since the construction of the reduct depends on literals in $neg(P)$ only, the literals in X can be ignored and thus we obtain
$$P^{M^*} = P^{M \cup X} = P^M. \qquad (5)$$
As already mentioned, this implies that M is a minimal model of P^M as well and thus an answer set of P. □

In particular, this means that moving from $P \cup \{r\}$ to P cannot introduce inconsistency.

Corollary 2. *Let P be a disjunctive logic program and $r \notin P$ a rule with*
$$\mathsf{Path}(P \cup \{r\}, head(r)) \cap neg(P \cup \{r\}) = \emptyset.$$
If $P \cup \{r\}$ is consistent, then so is P.

Proof. Let M^* be a consistent answer set of $P \cup \{r\}$. Due to Lemma 1, there is a subset $M \subseteq M^*$ such that M is an answer set of P. M^* being consistent implies M is consistent. So, P has a consistent answer set. □

Example 10. Consider again the inconsistent program P_3''.

$$P_3'': \qquad b \leftarrow \text{not } c. \qquad c \leftarrow \text{not } d. \qquad d \leftarrow \text{not } b.$$
$$ d \leftarrow \text{not } e. \qquad \phantom{c \leftarrow \text{not } d.} e.$$

If we add the fact "$r = d.$", the program becomes consistent, having $\{b, d\}$ as the only answer set. However, we have $d \in neg(P)$ and in particular,
$$\mathsf{Path}(P_3'' \cup \{r\}, head(r)) \cap neg(P_3'' \cup \{r\}) \neq \emptyset$$
and thus, this example does not satisfy the premise of Corollary 2.

Due to the above considerations, we deem the following postulate as desirable (I="Independence").

I-Monotonicity If r is a rule with $\mathsf{Path}(P \cup \{r\}, head(r)) \cap neg(P \cup \{r\}) = \emptyset$, then $\mathcal{I}(P) \leq \mathcal{I}(P \cup \{r\})$.

So far, we considered situations where augmenting a program P with rules cannot prevent the entailment of contradictions. However, there is another situation where a rule r should never decrease the severity of inconsistency—namely, if r is a constraint:

Definition 17. Let P be a disjunctive logic program. A rule of the form

$$a \leftarrow l_1, \ldots, l_k, \text{not } l_{k+1}, \ldots, \text{not } l_m, \text{not } a. \tag{6}$$

where a is an atom which does not occur elsewhere in the program is called a *constraint*.

Intuitively, a constraint $r \in P$ of the form (6) ensures that a set M of literals cannot be an answer set of P if M contains the literals $\{l_1, \ldots, l_k\}$, but none of $l_{k+1}, \ldots, l_m$. Hence, constraints reduce the amount of answer sets of a program. It is straightforward to see that the inconsistency of P implies the inconsistency of $P \cup \{r\}$ for a constraint r.

This motivates the following postulate (Con="Constraint").

Con-Monotonicity If r is a constraint, then $\mathcal{I}(P) \leq \mathcal{I}(P \cup \{r\})$ for any $P \in \mathcal{P}$.

This concludes our discussion on weaker versions of monotonicity.

4.2 Further Postulates

We will now discuss postulates which consider cases where the inconsistency value should *not* change. For this reason, we consider established notions of equivalence for logic programs, cf. [35, 49].

Definition 18. Two logic programs P, P' are *equivalent*, denoted by $P \equiv P'$, if $\text{Ans}(P) = \text{Ans}(P')$. They are *strongly equivalent*, denoted by $P \equiv_s P'$, if $\text{Ans}(P \cup H) = \text{Ans}(P' \cup H)$ for every $H \in \mathcal{P}$.

The notion of equivalence is only useful if we are interested in the answer sets of a given program. Two programs can be equivalent while encoding different information, e.g.,

$$P: \quad b \leftarrow \text{not } a. \qquad\qquad P': \quad b.$$
$$\neg b \leftarrow \text{not } a. \qquad\qquad \neg b.$$

Furthermore, the above programs behave differently if we consider them as parts of a larger program. Even though this shows that this notion of equivalence is weak, respecting equivalent programs still seems to be a desirable property for inconsistency measures that are determined by the answer sets of a program rather than other aspects.

E-Indifference If $P \equiv P'$, then $\mathcal{I}(P) = \mathcal{I}(P')$.

The analogous postulate for strongly equivalent programs is obviously even more desirable.

SE-Indifference If $P \equiv_s P'$, then $\mathcal{I}(P) = \mathcal{I}(P')$.

Due to the definition of $\equiv_s$, the postulate *Exchange* from [5], which is advocated by Besnard as desirable for propositional logics, seems reasonable for strongly equivalent programs P and P', too.

Exchange If $P \equiv_s P'$, then for all $H \in \mathcal{P}$, $\mathcal{I}(P \cup H) = \mathcal{I}(P' \cup H)$.

However, *Exchange* and *SE-Indifference* coincide.

Proposition 5. *Let $\mathcal{I}$ be an inconsistency measure. If $\mathcal{I}$ satisfies* E-Indifference, *then it satisfies* SE-Indifference *as well. Furthermore, it satisfies* SE-Indifference *if and only if it satisfies* Exchange.

Proof. The first statement is clear. The satisfaction of *Exchange* implies the satisfaction of *SE-Indifference*, because we can choose $H = \emptyset$. The converse holds since $P \cup H \equiv_s P' \cup H$ for any program H. □

Our final two postulates are concerned with the language used in a program. Intuitively, parts of a program which do not share any vocabulary elements with the rest of the program should be assessed separately with respect to inconsistency.

Language Separability If $\mathcal{A}(P) \cap \mathcal{A}(P') = \emptyset$, then $\mathcal{I}(P \cup P') = \mathcal{I}(P) + \mathcal{I}(P')$.

Example 11. Consider again the following programs from Example 1

$$P_2: \quad a \vee \neg a. \quad\quad \neg a \leftarrow a. \quad\quad a \leftarrow \neg a..$$
$$P_3'': \quad b \leftarrow \text{not } c. \quad\quad c \leftarrow \text{not } d. \quad\quad d \leftarrow \text{not } b.$$
$$ \quad d \leftarrow \text{not } e. \quad\quad e.$$

As they do not share any atoms, their conflicts should be considered independent and hence,

$$\mathcal{I}(P_2 \cup P_3'') = \mathcal{I}(P_2) + \mathcal{I}(P_3'')$$

should hold.

Another approach builds on the notion of *safe* formulas [44]. A consistent formula α is *safe* with respect to a set $\mathcal{K}$ of propositional formulas if α and $\mathcal{K}$ do not share any atoms. So, adding α to $\mathcal{K}$ will never introduce inconsistency. The corresponding postulate *safe-formula independence* requires $\mathcal{I}(\mathcal{K}) = \mathcal{I}(\mathcal{K} \cup \{\alpha\})$ whenever α is safe with respect to $\mathcal{K}$.

Due to the directedness of rules we can be somewhat more liberal in defining a corresponding notion of safeness for the ASP-setting and do not need to require completely disjoint languages for r and P. There are two different ways in which a single rule r can have an effect on the consistency of a program P. First of all, the literals in the head of r may interact with the the program and thus introduce or eliminate inconsistency. To make sure this kind of interaction cannot happen, we have to require that the atoms in the head of r are disjoint from the atoms appearing in P. But there is another, more indirect way in which a new rule can cause contradiction, namely if r is self-contradictory, in the sense that a literal is derivable if and only if it is not derivable. The simplest example is the rule "$a \leftarrow$ not a." which leads to the non-existence of answer sets if added to a program P whenever atom a does not appear in P. To eliminate the possibility of this phenomenon we require that the literals in the head of r not occur default-negated in the body of r. This leads to the following definition.

Definition 19. Let P be a disjunctive logic program. A rule r is called *safe* with respect to P if $\mathcal{A}(head(r)) \cap \mathcal{A}(P) = \emptyset$ and $head(r) \cap neg(r) = \emptyset$.

Our discussion motivates the following postulate.

Safe-Rule Independence If P is a logic program and r safe with respect to P, then $\mathcal{I}(P) = \mathcal{I}(P \cup \{r\})$.

Remark 3. Consider a disjunctive logic program P and a rule r that is safe with respect to P. We can view $U = \mathcal{L}(P)$ as a splitting set of $P \cup \{r\}$ and obtain P as a bottom part. Then, for any measure $\mathcal{I}$ that satisfies Split-Monotonicity, we have $\mathcal{I}(P) \leq \mathcal{I}(P \cup \{r\})$. The same holds for any measure $\mathcal{I}$ satisfying I-Monotonicity, since $\mathsf{Path}(P \cup \{r\}, head(r)) \cap neg(P \cup \{r\}) = \emptyset$ is clear for a safe rule r. Neither Split-Monotonicity nor I-Monotonicity implies Safe-Rule Independence, though, since we do not obtain $\mathcal{I}(P) \geq \mathcal{I}(P \cup \{r\})$.

5 Compliance with Rationality Postulates

Table 1 gives an overview of the compliance of our measures with respect to the rationality postulates from Section 4 and thus summarizes Propositions 6-12 and Examples 12-14 below. Note that, naturally, none of our measures satisfies the ordinary *Monotonicity* postulate which is also not desired for ASP, cf. Section 1.

Proposition 6. $\mathcal{I}_{01}$ *satisfies* Consistency, CLP-Monotonicity, Split-Monotonicity, I-Monotonicity, Con-Monotonicity, E-Indifference, SE-Indifference, Exchange *and* Safe-Rule Independence

Proof. Split-Monotonicity follows from Corollary 1 and *I-Monotonicity* from Corollary 2. The rest is clear. □

Naturally, $\mathcal{I}_{01}$ does not satisfy *Language Separability* since $\mathcal{I}_{01}(P) \in \{0, 1\}$ for any program $P \in \mathcal{P}$.

	$\mathcal{I}_{01}$	$\mathcal{I}_\pm$	$\mathcal{I}_+$	$\mathcal{I}_-$	$\mathcal{I}_{\mathsf{MSI}}$	$\mathcal{I}_\#$	$\mathcal{I}_{sd}$
Consistency	✓	✓	✓	✓	✓	✓	✓
Monotonicity	✗	✗	✗	✗	✗	✗	✗
CLP-Monotonicity	✓	✓	✓	✓	✓	✓	✓
Split-Monotonicity	✓	✓	✓	✓	✓	✓	✓
I-Monotonicity	✓	✓	✓	✓	✓	✓	✓
Con-Monotonicity	✓	✓	✓	✓	✓	✓	✓
E-Indifference	✓	✗	✗	✗	✗	✓	✗
SE-Indifference	✓	✗	✓	✗	✗	✓	✗
Exchange	✓	✗	✓	✗	✗	✓	✗
Language Separability	✗	✓	✓	✓	✓	✓	✓
Safe-Rule Independence	✓	✓	✓	✓	✓	✓	✓

Table 1: Compliance of inconsistency measures with respect to our rationality postulates

Proposition 7. $\mathcal{I}_\pm$ *satisfies* Consistency, CLP-Monotonicity, Split-Monotonicity, I-Monotonicity, Con-Monotonicity, Language Separability *and* Safe-Rule Independence.

Proof. **Consistency** Clear from the definition of $\mathcal{I}_\pm$.

CLP-Monotonicity Let P be a classical program and P' an arbitrary one. Assume $\mathcal{I}_\pm(P \cup P') = k$. Let $(P \cup P' \cup A) \setminus D$ be consistent with $|A| + |D| = k$. Since P is a classical logic program, $P \setminus D$ must be consistent as well since adding rules cannot restore consistency. Hence, $\mathcal{I}_\pm(P) \leq |D| \leq |A| + |D| = k$.

Split-Monotonicity Let P be a program and U a splitting set. Let $\mathcal{I}_\pm(P) = k$ and let $(P \cup A) \setminus D$ be consistent with $|A| + |D| = k$. We use Proposition 2 to assume that A is a set of facts. Thus, $(P \cup A) \setminus D$ contains no additional edges in the dependency graph compared to P which implies that U is a splitting set of $(P \cup A) \setminus D$ as well. In particular, if we let A_U be the subset of A such that $r \in A_U$ if and only if $head(r) \subseteq U$, then $(bot_U(P)) \cup A_U) \setminus D$ is the corresponding bottom program. Now let M be a consistent answer set of $(P \cup A) \setminus D$. Due to Theorem 1, there is a subset $X \subseteq M$ such that X is an answer set of $(bot_U(P)) \cup A_U) \setminus D$. As a subset of the consistent set M of literals, X is consistent. Therefore, $(bot_U(P)) \cup A_U) \setminus D$ is consistent. Hence,

$$\mathcal{I}_\pm(bot_U(P)) \leq |D| + |A_U| \leq |D| + |A| = k.$$

I-Monotonicity Let $\mathcal{I}_\pm(P \cup \{r\}) = k$. Let $(P \cup \{r\} \cup A) \setminus D$ be consistent with $|A| + |D| = k$. Let M^* be a consistent answer set of $(P \cup \{r\} \cup A) \setminus D$.

Using Proposition 2, we assume that A is a set of facts. So, since

$$\mathsf{Path}(P \cup \{r\}, head(r)) \cap neg(P \cup \{r\}) = \emptyset,$$

we also obtain

$$\mathsf{Path}(((P \cup A) \setminus D) \cup \{r\}, head(r)) \cap neg(P \cup \{r\}) = \emptyset,$$

because adding facts (and deleting rules) does not extend the dependency graph. Furthermore, we have

$$neg(((P \cup A) \setminus D) \cup \{r\}) = neg((P \setminus D) \cup \{r\}) \subseteq neg(P \cup \{r\})$$

and hence,

$$\mathsf{Path}(((P \cup A) \setminus D) \cup \{r\}, head(r)) \cap neg(P \cup \{r\}) = \emptyset$$

also implies

$$\mathsf{Path}(((P \cup A) \setminus D) \cup \{r\}, head(r)) \cap neg(((P \cup A) \setminus D) \cup \{r\}) = \emptyset.$$

So, we can apply Lemma 1 to the program $(P \cup A) \setminus D$ and the additional rule r. Since M^* was assumed to be an answer set of $(P \cup \{r\} \cup A) \setminus D$, we obtain that $(P \cup A) \setminus D$ has an answer set M with $M \subseteq M^*$. With M^* being consistent, M is consistent, too. Likewise, $(P \cup A) \setminus D$ is consistent. Thus, $\mathcal{I}_\pm(P) \le |A| + |D| = k = \mathcal{I}_\pm(P \cup \{r\})$.

Con-Monotonicity If r is a constraint and $(P \cup \{r\} \cup A) \setminus D$ is consistent, then $(P \cup A) \setminus D$ is consistent as well.

Language Separability This is clear since P and P' need to be considered independently.

Safe-Rule Independence That is clear, since no matter which facts A we add or rules D we delete, whether or not $(P \cup A) \setminus D$ is consistent will never depend on a safe rule r. The only exception would be adding a fact "$\neg a.$" for an atom $a \in head(r)$ which, however, will never be beneficial to resolve the inconsistency of P anyway. □

It is clear that $\mathcal{I}_\pm$ does not satisfy the remaining postulates *(S)E-Indifference* and *Exchange* since the notion of (strong) equivalence does not take into account *how many* rules a program contains to ensure the entailment of certain atoms. The measure $\mathcal{I}_\pm$, however, is tailored to assess this. Consider, for example

Example 12.

$$P: \quad a. \quad \neg a.$$
$$P': \quad a. \quad \neg a.$$
$$\quad b. \quad a \leftarrow b. \quad \neg a \leftarrow a.$$

with $\mathcal{I}_\pm(P) = 1$ and $\mathcal{I}_\pm(P') = 2$.

In contrast, the measure $\mathcal{I}_+$ satisfies *SE-Indifference* and *Exchange* as only adding rules is considered here. The rest is rather similar to $\mathcal{I}_\pm$.

Proposition 8. $\mathcal{I}_+$ *satisfies* Consistency, CLP-Monotonicity, Split-Monotonicity, I-Monotonicity, Con-Monotonicity, SE-Indifference, Exchange, Language Separability *and* Safe-Rule Independence.

Proof. **Consistency** This is clear.

CLP-Monotonicity If P is consistent, then $\mathcal{I}_+(P) = 0$. If P is inconsistent, then $\mathcal{I}_+(P) = \infty$ since adding rules cannot resolve inconsistency. For the same reason, P being inconsistent implies that $P \cup P'$ is inconsistent with $\mathcal{I}_+(P \cup P') = \infty$. Hence, in both cases, we obtain $\mathcal{I}_+(P) \leq \mathcal{I}_+(P \cup P')$.

Split-Monotonicity, I-Monotonicity and **Con-Monotonicity** Similar to $\mathcal{I}_\pm$.

SE-Indifference *Exchange* and Proposition 5.

Exchange Let $P \equiv_s P'$ and let $H \in \mathcal{P}$. Let $\mathcal{I}_+(P \cup H) = k$. This means that there is a set A of facts with $|A| = k$ such that $P \cup H \cup A$ is consistent. $P \equiv_s P'$ implies that $P' \cup H \cup A$ is consistent as well, yielding $\mathcal{I}_+(P' \cup H) \leq k = \mathcal{I}_+(P \cup H)$. Of course, we obtain $\mathcal{I}_+(P \cup H) \leq \mathcal{I}_+(P' \cup H)$ similarly.

Language Separability and **Safe-Rule Independence** Similar to $\mathcal{I}_\pm$. □

The postulate *E-Indifference* is not satisfied by $\mathcal{I}_+$.

Example 13. The programs

$$P: \ b \leftarrow \text{not } a. \qquad P': \ b.$$
$$\neg b \leftarrow \text{not } a. \qquad \neg b.$$

are equivalent. However, $\mathcal{I}_+(P) = 1$ and $\mathcal{I}_+(P') = \infty$.

Proposition 9. $\mathcal{I}_-$ *satisfies* Consistency, CLP-Monotonicity, Split-Monotonicity, I-Monotonicity, Con-Monotonicity, Language Separability *and* Safe-Rule Independence.

Proof. Similar to $\mathcal{I}_\pm$. □

For the same reason as $\mathcal{I}_\pm$, the measure $\mathcal{I}_-$ does not satisfy *(S)E-Indifference* and *Exchange*.

We now turn to the measure $\mathcal{I}_{\mathsf{MSI}}$.

Proposition 10. $\mathcal{I}_{\mathsf{MSI}}$ *satisfies* Consistency, CLP-Monotonicity, Split-Monotonicity, I-Monotonicity, Con-Monotonicity, Language Separability *and* Safe-Rule Independence.

Proof. **Consistency** As already pointed out in [11], $SI_{min}(P) = \emptyset$ iff P is consistent.

CLP-Monotonicity Let P be a classical program and P' an arbitrary one. If H is strongly P-inconsistent, then H is strongly $P \cup P'$-inconsistent as well, because conflicts within P cannot be resolved by adding rules. Similarly, one can see that the minimality of H is preserved. Hence, $H \in SI_{min}(P)$ implies $H \in SI_{min}(P \cup P')$. Hence, $SI_{min}(P) \subseteq SI_{min}(P \cup P')$ which implies $\mathcal{I}_{\mathsf{MSI}}(P) \leq \mathcal{I}_{\mathsf{MSI}}(P \cup P')$.

Split-Monotonicity Let P be a program and U a splitting set. By Theorem 1 (and similar considerations as above) we obtain $SI_{min}(bot_U(P)) \subseteq SI_{min}(P)$. The claim follows.

I-Monotonicity Similar, utilizing Lemma 1.

Con-Monotonicity Similar as constraints cannot restore consistency.

Language Separability and **Safe-Rule Independence** are clear. □

Regarding *(S)E-Indifference* and *Exchange*, the situation is similar as in the case of $\mathcal{I}_\pm$.

Proposition 11. $\mathcal{I}_{sd}$ *satisfies* Consistency, CLP-Monotonicity, Split-Monotonicity, I-Monotonicity, Con-Monotonicity, Language Separability *and* Safe-Rule Independence.

Proof. **Consistency** This is clear by definition.

CLP-Monotonicity If P is inconsistent, then it has only inconsistent answer sets. In this case, the reduct of $P \cup P'$ with respect to any set of literals has only inconsistent answer sets as well. Hence, $\mathcal{I}_{sd}(P \cup P') = \infty$. If P is consistent, then $\mathcal{I}_{sd}(P) = 0$. In both cases, $\mathcal{I}_{sd}(P) \leq \mathcal{I}_{sd}(P \cup P')$ holds.

Split-Monotonicity Let P be a disjunctive logic program, U a splitting set and $bot_U(P)$ the corresponding bottom program. The case $\mathcal{I}_{sd}(P) = \infty$ is clear. We consider $\mathcal{I}_{sd}(P) = k < \infty$. Let M be a consistent set of literals with $d_{sd}(M, M') = k$ for an $M' \in \text{Ans}_{Con}(P^M)$. Note that U is also a splitting set of P^M with $(bot_U(P))^M$ as the corresponding bottom part. So, due to Theorem 1, M' is of the form $X' \cup Y'$ with

$$X' \in \text{Ans}_{Con}((bot_U(P))^M).$$

We can w.l.o.g. assume $X' \cap Y' = \emptyset$.
Now consider $X = M \cap U$ and $Y = M \setminus X$. Then, similarly to X' and Y' we have $X \cup Y = M$ and $X \cap Y = \emptyset$.
We obtain $(bot_U(P))^M = (bot_U(P))^X$ due to the construction of the reduct and U being a splitting set. Thus,

$$X' \in \text{Ans}_{Con}((bot_U(P))^X).$$

The last step argues that the constructed sets are disjoint: We have $X, X' \subseteq U$ and $Y \cap U = Y' \cap U = \emptyset$. So, $X \cap Y' = X' \cap Y = \emptyset$. Furthermore, $X \cap Y = X' \cap Y' = \emptyset$ was already mentioned. Thus, we can calculate

$$d_{sd}(X, X') \leq d_{sd}(X \cup Y, X' \cup Y') = d_{sd}(M, M') = k.$$

To summarize, we found two sets X, X' of literals with $X' \in \text{Ans}_{Con}((bot_U(P))^X)$ and $d_{sd}(X, X') \leq k$. Hence, $\mathcal{I}_{sd}(bot_U(P)) \leq k$.

I-Monotonicity Again, we only have to consider the case $\mathcal{I}_{sd}(P \cup \{r\}) = k < \infty$. Let M^* be a consistent set of literals such that $(M^*)' \in \text{Ans}_{Con}((P \cup \{r\})^{M^*})$ is consistent with $d_{sd}(M^*, (M^*)') = k$.
Consider $P^{(M^*)}$. As seen in the proof of Lemma 1, there is a set $M' \in \text{Ans}_{Con}(P^{M^*})$ of literals such that $(M^*)'$ is of the form $(M^*)' = M' \cup X'$ for a set X' of literals with $P^{(M^*)'} = P^{M' \cup X'} = P^{M'}$ (cf. (3) and (5)). Furthermore, M' and X' are disjoint.
Now consider $M = M^* \setminus X'$. We have $M' \in \text{Ans}_{Con}(P^{M^*}) = \text{Ans}_{Con}(P^{M^* \setminus X'})$ and

$$d_{sd}(M, M') = d_{sd}(M^* \setminus X', (M^*)' \setminus X') \leq d_{sd}(M^*, (M^*)') = k.$$

This implies $\mathcal{I}_{sd}(P) \leq k$.

Con-Monotonicity If $M' \in \text{Ans}_{Con}((P \cup \{r\})^M)$ with $d_{sd}(M, M') = k$, then we have $M' \in \text{Ans}_{Con}(P^M)$ as well and hence $\mathcal{I}_{sd}(P) \leq k$.

Language Separability and **Safe-Rule Independence** are clear. □

The following example shows that $\mathcal{I}_{sd}$ does not satisfy *SE-Indifference*. It makes use of the following observation: Even for two strongly equivalent logic programs P and P', the corresponding reducts P^M and P'^M with respect to to a given set M of literals might not be equivalent. It follows that $\mathcal{I}_{sd}$ does not satisfy *E-Indifference* or *Exchange*, either.

Example 14. Consider the following program P_5.

$$P_5: \quad a. \qquad b_1 \leftarrow \text{not } a. \qquad b_2 \leftarrow \text{not } a.$$
$$c_1 \leftarrow \text{not } b_1, \text{not } c_1. \qquad c_2 \leftarrow \text{not } b_2, \text{not } c_2.$$

The program is inconsistent since b_1 and b_2 are not entailed. We obtain the best set of literals if we get rid of a (to be able to use the default "not a"). The corresponding reduct is

$$P_5^{\{b_1, b_2\}}: \qquad a. \qquad b_1. \qquad b_2.$$

with $d_{sd}(\{b_1, b_2\}, \text{Ans}_{Con}(P_5^{\{b_1, b_2\}})) = 1$. Since the program is inconsistent (i.e., $\mathcal{I}_{sd}(P_5) \geq 1$), $\mathcal{I}_{sd}(P_5) = 1$. Now consider the same program with two additional rules that prevent us from considering sets of literals without a.

$$P_6: \quad a. \qquad b_1 \leftarrow \text{not } a. \qquad b_2 \leftarrow \text{not } a.$$
$$c_1 \leftarrow \text{not } b_1, \text{not } c_1. \qquad c_2 \leftarrow \text{not } b_2, \text{not } c_2.$$
$$d \leftarrow \text{not } a. \qquad \neg d \leftarrow \text{not } a.$$

For any set M of literals with $a \notin M$, the reduct P_6^M contains "$d.$" and "$\neg d.$". Hence, we have to consider sets containing a and thus, the reduct will never contain "$b_1.$" or "$b_2.$". Now one can check that $\mathcal{I}_{sd}(P_6) = 2$: For example, if we use the set $\{a, b_1\}$ we obtain the reduct

$$P_6^{\{a, b_1\}}: \qquad a. \qquad c_2.$$

with minimal model $\{a, c_2\}$, i.e., the distance is 2.

However, the programs P_5 and P_6 are strongly equivalent since the fact "$a.$" renders the rules "$d \leftarrow \text{not } a.$" and "$\neg d \leftarrow \text{not } a.$" meaningless. Thus, *SE-Indifference* is not satisfied.

Proposition 12. $\mathcal{I}_\#$ *satisfies* Consistency, CLP-Monotonicity, Split-Monotonicity, I-Monotonicity, Con-Monotonicity, E-Indifference, SE-Indifference, Exchange, Language Separability *and* Safe-Rule Independence.

Proof. **Consistency** Clear due to the definition.

CLP-Monotonicity Clear due to the monotonicity of classical logic programs.

Split-Monotonicity Let P be a program and U a splitting set. Let $\tilde{X}$ be an answer set of $bot_U(P)$ with a minimal amount of complementary literals, say $2k$. Thus, $\mathcal{I}_\#(bot_U(P)) = k$. Now let M be an answer set of P. Due to Theorem 1, $X \subseteq M$ for an answer set X of $bot_U(P)$. Thus, M contains at least as many complementary literals as X, which itself contains at least as many as $\tilde{X}$. Hence, $\mathcal{I}_\#(P) \geq k$.

I-Monotonicity Let $\mathcal{I}_\#(P \cup \{r\}) = k^*$ and let M^* be a k^*-inconsistent answer set of $P \cup \{r\}$. By Lemma 1, there is an answer set M of P with $M \subseteq M^*$. So, if M is k-consistent, then $k \leq k^*$ and we obtain $\mathcal{I}_\#(P) \leq k \leq k^* = \mathcal{I}_\#(P \cup \{r\})$.

Con-Monotonicity If $\mathcal{I}_\#(P \cup \{r\}) = k$ and M is a k-inconsistent answer set of $P \cup \{r\}$, then M is a k-inconsistent answer set of P as well and hence, $\mathcal{I}_\#(P) \leq k$.

E-Indifference Clear due to the definition of $\mathcal{I}_\#$.

SE-Indifference *E-Indifference* and Proposition 5.

Exchange *E-Indifference* and Proposition 5.

Language Separability The number of complementary literals adds up.

Safe-Rule Independence This is clear. □

Our analysis shows that the measures investigated in this paper satisfy most of the postulates we discussed—with the exception of monotonicity, which, as explained, is not desirable in our context anyway. This provides our measures with some basic justification and shows that they behave in an expected way. The proposed postulates were adapted from existing ones for propositional logics and only modifications necessary to account for the non-monotonicity of ASP were made.

6 Computational Complexity

We now address the computational complexity of the measures we developed in this paper. After establishing some notation and making some general observations we investigate the complexity of our measures on disjunctive logic programs in Section 6.1. Afterwards, we have a look at some special cases. In particular, we discuss the complexity of disjunction-free programs in Section 6.2, stratified programs in Section 6.3, and classical disjunction-free programs in Section 6.4. The techniques we are going to use for our investigation are not applicable for the measure $\mathcal{I}_\mathsf{MSI}$, though, and we leave a rigorous investigation of its computational complexity for future work.

Following [48], we consider the three decision problems $\textsc{Exact}_\mathcal{I}$, $\textsc{Upper}_\mathcal{I}$, $\textsc{Lower}_\mathcal{I}$, and the natural function problem $\textsc{Value}_\mathcal{I}$. Let $\mathcal{I}$ be an arbitrary inconsistency measure on logic programs. As we will investigate complexity with respect to different subclasses of logic programs, let $X \subseteq \mathcal{P}$ be some set of logic programs.

$\textsc{Exact}_\mathcal{I}^X$ **Input:** $P \in X$, $x \in [0, \infty]$
            Output: TRUE iff $\mathcal{I}(P) = x$

$\textsc{Upper}_\mathcal{I}^X$ **Input:** $P \in X$, $x \in [0, \infty]$
            Output: TRUE iff $\mathcal{I}(P) \leq x$

$\textsc{Lower}_\mathcal{I}^X$ **Input:** $P \in X$, $x \in (0, \infty]$
            Output: TRUE iff $\mathcal{I}(P) \geq x$

$\textsc{Value}_\mathcal{I}^X$ **Input:** $P \in X$
            Output: The value of $\mathcal{I}(P)$

We assume the reader to be familiar with the complexity classes P, NP and coNP. We make use of the polynomial hierarchy, defined using oracle machines as $\Sigma_0^p := \Pi_0^p := \mathsf{P}$ and $\Sigma_{i+1}^p := \mathsf{NP}^{\Sigma_i^p}$ and $\Pi_{i+1}^p := \mathsf{coNP}^{\Sigma_i^p}$ for $i \geq 0$. Here, $\mathcal{C}^\mathcal{D}$ is the class of decision problems solvable in $\mathcal{C}$ having access to an oracle for some problem that is complete in $\mathcal{D}$. The class D_i^p consists of all languages that are the intersection of a language in Σ_i^p and a language in Π_i^p. We also consider the functional complexity classes $\mathsf{FP}^{\mathsf{NP}[\log n]}$ and $\mathsf{FP}^{\Sigma_2^p[\log n]}$, i.e., classes that contain problems whose solution can be computed in P with access to a logarithmically bounded number of calls to an NP resp. Σ_2^p oracle.

In order to discuss computational complexity we need a measure of the size of the input to our problems. For that we use the following straightforward notion of the length of a logic program.

Definition 20. Let r be a rule of the form

$$l_0 \vee \ldots \vee l_k \leftarrow l_{k+1}, \ldots, l_m, \text{not } l_{m+1}, \ldots, \text{not } l_n.$$

Then the length $\mathrm{len}(r)$ of r is defined as $\mathrm{len}(r) = n + 1$. For a program P, its length is defined via $\mathrm{len}(P) = \sum_{r \in P} \mathrm{len}(r)$.

The work [48] already established some relationships between the individual computational problems for the propositional case, in particular, results pertaining to the bounds for the complexity of other problems if the complexity of one problem is known. We will now extend these results to the case of logic programs. For that, we introduce a simple notion of *expressivity* for our measures, cf. [46].

Definition 21. Let $\mathcal{I}$ be an inconsistency measure. The *expressivity* $\mathcal{C}_{\mathcal{I}}$ of $\mathcal{I}$ is the function $\mathcal{C}_{\mathcal{I}} : \mathbb{N} \to 2^{\mathbb{R}}$ defined via $\mathcal{C}_{\mathcal{I}}(n) = \{\mathcal{I}(P) \mid \mathrm{len}(P) \leq n\}$ for all $n \in \mathbb{N}$.

In other words, $\mathcal{C}_{\mathcal{I}}(n)$ is the set of values $\mathcal{I}$ can attain on logic programs of length n or smaller. For our measures we can see that the size of this set is bounded linearly by the length of the programs.

Lemma 2. For $\mathcal{I} \in \{\mathcal{I}_{01}, \mathcal{I}_\pm, \mathcal{I}_+, \mathcal{I}_-, \mathcal{I}_{sd}, \mathcal{I}_\#\}$, $\mathcal{C}_{\mathcal{I}}(n) \subseteq \{0, \ldots, n, \infty\}$.

Proof. For $\mathcal{I} \in \{\mathcal{I}_\pm, \mathcal{I}_-\}$, the claim follows from $|P| \leq |\mathrm{len}(P)|$ and the observation that the empty program is consistent. Since we can consider adding facts only in the case of $\mathcal{I}_+$, at most $|\mathcal{L}(P)| \leq |\mathrm{len}(P)|$ rules can be added. If adding rules cannot restore consistency, the value is ∞. Hence, $\mathcal{I}_+(P) \leq n+1$. The case $\mathcal{I} \in \{\mathcal{I}_{sd}, \mathcal{I}_\#\}$ is similar using $|\mathcal{L}(P)| \leq |\mathrm{len}(P)|$. □

The above results allow us to adopt Lemma 2 from [48] which provides an upper bound for the complexity of $\mathrm{VALUE}_{\mathcal{I}}^X$ given the complexity of $\mathrm{UPPER}_{\mathcal{I}}^X$. As there are only linearly many different values of $\mathcal{I}$ one can perform a binary search on these values in order to realize an algorithm for $\mathrm{VALUE}_{\mathcal{I}}^X$ while using only logarithmically many calls to an algorithm for $\mathrm{UPPER}_{\mathcal{I}}^X$.

Lemma 3 (Lemma 2 in [48]). *Let $\mathcal{I}$ be an inconsistency measure, $i > 0$ an integer, and $X \subseteq \mathcal{P}$. If $\mathrm{UPPER}_{\mathcal{I}}^X$ is in Σ_i^p or Π_i^p and $|\mathcal{C}_{\mathcal{I}}(n)| \in \mathcal{O}(n^k)$ for some $k \in \mathbb{N}$, then $\mathrm{VALUE}_{\mathcal{I}}^X$ is in $\mathsf{FP}^{\Sigma_i^p[\log n]}$.*

Naturally, we expect $\mathrm{UPPER}_{\mathcal{I}}^X$ and $\mathrm{LOWER}_{\mathcal{I}}^X$ to be complementary problems and $\mathrm{EXACT}_{\mathcal{I}}^X$ a combination of both. However, in order to see this, we adopt the following notion from [48].

Definition 22. An inconsistency measure $\mathcal{I}$ is called *well-serializable* if the following problems are in P:

- Given $n \in \mathbb{N}$ and $x \in \mathcal{C}_{\mathcal{I}}(n)$, find $y \in \mathcal{C}_{\mathcal{I}}(n)$ such that $x < y$ and there is no $y' \in \mathcal{C}_{\mathcal{I}}(n)$ with $x < y' < y$.

- Given $n \in \mathbb{N}$ and $x \in \mathcal{C}_{\mathcal{I}}(n)$, find $y \in \mathcal{C}_{\mathcal{I}}(n)$ such that $y < x$ and there is no $y' \in \mathcal{C}_{\mathcal{I}}(n)$ with $y < y' < x$.

In other words, finding the immediate successor and predecessor of a value $x \in \mathcal{C}_{\mathcal{I}}(n)$ is tractable for a well-serializable measure $\mathcal{I}$. Being able to calculate them enables us to adopt the following result for logic programs.

Lemma 4 (Lemma 3 in [48]). *Let $\mathcal{I}$ be a well-serializable inconsistency measure and $X \subseteq \mathcal{P}$. Let $i \in \mathbb{N}$ and $\mathcal{C} \in \{\Sigma_i^p, \Pi_i^p\}$. Then,*

- $\mathrm{UPPER}_{\mathcal{I}}^X$ *is $\mathcal{C}$-complete iff $\mathrm{LOWER}_{\mathcal{I}}^X$ is co-$\mathcal{C}$-complete.*

- *if* UPPER$_\mathcal{I}^X$ *or* LOWER$_\mathcal{I}^X$ *is in* $\mathcal{C}$, *then* EXACT$_\mathcal{I}^X$ *is in* D_i^p.

It is straightforward to see that our proposed measures are indeed well-serializable, cf. Lemma 2.

Proposition 13. *For* $\mathcal{I} \in \{\mathcal{I}_{01}, \mathcal{I}_\pm, \mathcal{I}_+, \mathcal{I}_-, \mathcal{I}_{sd}, \mathcal{I}_\#\}$, $\mathcal{I}$ *is well-serializable.*

In the remainder of this section, we will give complexity results for the four problems for our measures on different classes of logic programs. Due to the insights above we will focus on the problem UPPER$_\mathcal{I}^X$ which allows us to easily derive complexity bounds for the other classes.

6.1 The General Case

We first turn to the complexity of our measures on general disjunctive logic programs. Recall that $\mathcal{P}$ denotes the set of all disjunctive logic programs. As already mentioned, checking the consistency of a logic program in the general case is Σ_2^p-complete.

Theorem 2 ([18]). *Deciding whether a disjunctive logic program* $P \in \mathcal{P}$ *is consistent is* Σ_2^p-*complete.*

As inconsistency measures are supposed to generalize the concept of inconsistency by obeying the *Consistency* postulate, the above result already provides a lower bound for the complexity of UPPER$_\mathcal{I}^\mathcal{P}$.

Proposition 14. *Let* $\mathcal{I}$ *be an inconsistency measure that satisfies* Consistency. *Then* UPPER$_\mathcal{I}^\mathcal{P}$ *is* Σ_2^p-*hard.*

Proof. Due to *Consistency*, checking whether $\mathcal{I}(P) \leq 0$ corresponds to checking whether the program is consistent, which is Σ_2^p-hard as stated in Theorem 2. □

However, for every measure $\mathcal{I}$ we proposed in this paper, we obtain membership in Σ_2^p for UPPER$_\mathcal{I}^\mathcal{P}$ via guess-and-check algorithms, yielding completeness for Σ_2^p.

Theorem 3. *For* $\mathcal{I} \in \{\mathcal{I}_{01}, \mathcal{I}_\pm, \mathcal{I}_+, \mathcal{I}_-, \mathcal{I}_{sd}, \mathcal{I}_\#\}$, UPPER$_\mathcal{I}^\mathcal{P}$ *is* Σ_2^p-*complete.*

Proof. Hardness follows from Proposition 14. Completeness for UPPER$_{\mathcal{I}_{01}}^\mathcal{P}$ is also obvious. Let (P, x) be an instance of UPPER$_\mathcal{I}^\mathcal{P}$.

1. UPPER$_{\mathcal{I}_\pm}^\mathcal{P} \in \Sigma_2^p$: We guess sets M, A of literals and a set $D \subseteq P$ of rules and verify in polynomial time that M is a model of $((P \cup A) \setminus D)^M$. Using an NP oracle we verify that there is no proper subset $M' \subseteq M$ that is also a model of $((P \cup A) \setminus D)^M$. Checking whether $|A| + |D| \leq x$ is also clearly in P. Hence, deciding UPPER$_{\mathcal{I}_\pm}^\mathcal{P}$ is in Σ_2^p.

2. UPPER$_{\mathcal{I}_+}^{\mathcal{P}} \in \Sigma_2^p$: analogous to above.

3. UPPER$_{\mathcal{I}_-}^{\mathcal{P}} \in \Sigma_2^p$: analogous to above.

4. UPPER$_{\mathcal{I}_{sd}}^{\mathcal{P}} \in \Sigma_2^p$: Similarly, guess a set M of literals and a set M' (a potential answer set of P^M). Checking that M' is a minimal model of P^M is in coNP, i.e., can be done using an NP oracle. Calculating $d_{sd}(M, M')$ is in P.

5. UPPER$_{\mathcal{I}_\#}^{\mathcal{P}} \in \Sigma_2^p$: Guess a set M and verify that $M \in \text{Ans}_{Con}(P^M)$. Counting the amount of complementary literals in M is in P.

□

Combining the above result with Lemma 3 and Lemma 4 we obtain the following picture on the computational complexity for general programs.

Corollary 3. *For* $\mathcal{I} \in \{\mathcal{I}_{01}, \mathcal{I}_\pm, \mathcal{I}_+, \mathcal{I}_-, \mathcal{I}_{sd}, \mathcal{I}_\#\}$, LOWER$_\mathcal{I}^\mathcal{P}$ *is* Π_2^p-*complete*, EXACT$_\mathcal{I}^\mathcal{P}$ *is in* D_2^p *and* VALUE$_\mathcal{I}^\mathcal{P}$ *is in* $\mathsf{FP}^{\Sigma_2^p[\log n]}$.

6.2 Disjunction-free Programs

We continue our investigation by considering only disjunction-free logic programs, i. e., programs with rules of the form

$$l_0 \leftarrow l_1, \ldots, l_m, \text{not } l_{m+1}, \ldots, \text{not } l_n.$$

Let $\mathcal{P}^{\mathcal{\#}} \subseteq \mathcal{P}$ be the set of all disjunction-free logic programs. For this class of programs we already have the following result from [18] pertaining to the complexity of deciding only consistency.

Theorem 4 ([18]). *Deciding whether an extended logic program P without disjunction is consistent is* **NP**-*complete*.

As before we can utilize this result to obtain a lower bound on the complexity of the problem UPPER$_\mathcal{I}^{\mathcal{P}^{\mathcal{\#}}}$.

Proposition 15. *Let $\mathcal{I}$ be an inconsistency measure that satisfies* Consistency. *Then* UPPER$_\mathcal{I}^{\mathcal{P}^{\mathcal{\#}}}$ *is* **NP**-*hard*.

We obtain the following theorem which is similar to our results above.

Theorem 5. *For* $\mathcal{I} \in \{\mathcal{I}_{01}, \mathcal{I}_\pm, \mathcal{I}_+, \mathcal{I}_-, \mathcal{I}_{sd}, \mathcal{I}_\#\}$, UPPER$_\mathcal{I}^{\mathcal{P}^{\mathcal{\#}}}$ *is* **NP**-*complete*.

Proof. For all measures, hardness follows from Proposition 15. Completeness for UPPER$_{\mathcal{I}_{01}}^{\mathcal{P}^{\mathcal{\#}}}$ is also obvious. Let (P, x) be an instance of UPPER$_\mathcal{I}^{\mathcal{P}^{\mathcal{\#}}}$.

1. $\text{UPPER}_{\mathcal{I}_\pm}^{\mathcal{P}^{\forall}} \in \text{NP}$: We guess sets M, A of literals and a set $D \subseteq P$ of rules and verify in polynomial time that M is the unique model of $((P \cup A) \setminus D)^M$. Checking whether $|A| + |D| \leq x$ is also clearly in P. Hence, $\text{UPPER}_{\mathcal{I}_\pm}^{\mathcal{P}^{\forall}}$ is in NP.

2. $\text{UPPER}_{\mathcal{I}_+}^{\mathcal{P}^{\forall}} \in \text{NP}$: analogous to above.

3. $\text{UPPER}_{\mathcal{I}_-}^{\mathcal{P}^{\forall}} \in \text{NP}$: analogous to above.

4. $\text{UPPER}_{\mathcal{I}_{sd}}^{\mathcal{P}^{\forall}} \in \text{NP}$: We can guess a set M of literals and determine the unique model M' of P^M in non-deterministic polynomial time. Computing $d_{sd}(M, M')$ can also be done in polynomial time.

5. $\text{UPPER}_{\mathcal{I}_\#}^{\mathcal{P}^{\forall}} \in \text{NP}$: We can guess a set M of literals and check whether M is indeed the unique model of P^M in non-deterministic polynomial time. Determining the number of contradictory literals in M can also be done in polynomial time.

□

Corollary 4. *For* $\mathcal{I} \in \{\mathcal{I}_{01}, \mathcal{I}_\pm, \mathcal{I}_+, \mathcal{I}_-, \mathcal{I}_{sd}, \mathcal{I}_\#\}$, $\text{LOWER}_{\mathcal{I}}^{\mathcal{P}^{\forall}}$ *is* coNP-*complete,* $\text{EXACT}_{\mathcal{I}}^{\mathcal{P}^{\forall}}$ *is in* D_1^p *and* $\text{VALUE}_{\mathcal{I}}^{\mathcal{P}^{\forall}}$ *is in* $\text{FP}^{\text{NP}[\log n]}$.

6.3 Stratified Programs

We now consider *stratified programs* as another special case, see e. g. [2].

Definition 23. Let P be a disjunction-free program consisting of rules of the form

$$l_0 \leftarrow l_1, \ldots, l_m, \text{not } l_{m+1}, \ldots, \text{not } l_n. \qquad (7)$$

We call P *stratified* if there is a mapping $\|.\| : \mathcal{L}(P) \to \mathbb{N}$ such that for each rule $r \in P$ of the form (7) the following holds:

1. $\|l_i\| \leq \|l_0\|$ for each $1 \leq i \leq m$,

2. $\|l_j\| < \|l_0\|$ for each $m+1 \leq j \leq n$.

Let $\mathcal{P}^{\|.\|}$ be the set of all stratified programs. A stratified program has a unique answer set which can be computed in polynomial time [2, 14] which gives them many advantages in practical applications. A simple corollary of this is the following observation.

Corollary 5 (implied by [2]). *Deciding whether a stratified logic program P is consistent is in* P.

The results so far and the above observation may lead to the conjecture that the decision problem $\text{UPPER}_{\mathcal{I}}^{\mathcal{P}^{\|\cdot\|}}$ for our measures can also be solved in polynomial time. At least for the measures $\mathcal{I}_{01}$ and $\mathcal{I}_\#$, this conjecture is true.

Theorem 6. $\text{UPPER}_{\mathcal{I}_{01}}^{\mathcal{P}^{\|\cdot\|}} \in P$ and $\text{UPPER}_{\mathcal{I}_\#}^{\mathcal{P}^{\|\cdot\|}} \in P$.

Proof. We can calculate the unique answer set M of P and count the number of complementary literals in M in polynomial time. □

Corollary 6. $\text{LOWER}_{\mathcal{I}_{01}}^{\mathcal{P}^{\|\cdot\|}} \in P$, $\text{EXACT}_{\mathcal{I}_{01}}^{\mathcal{P}^{\|\cdot\|}} \in P$, $\text{VALUE}_{\mathcal{I}_{01}}^{\mathcal{P}^{\|\cdot\|}} \in FP$, $\text{LOWER}_{\mathcal{I}_\#}^{\mathcal{P}^{\|\cdot\|}} \in P$, $\text{EXACT}_{\mathcal{I}_\#}^{\mathcal{P}^{\|\cdot\|}} \in P$, $\text{VALUE}_{\mathcal{I}_\#}^{\mathcal{P}^{\|\cdot\|}} \in FP$

Unfortunately, the decision problem $\text{UPPER}_{\mathcal{I}}^{\mathcal{P}^{\|\cdot\|}}$ for the other measures remains NP-complete.

Theorem 7. $\text{UPPER}_{\mathcal{I}_\pm}^{\mathcal{P}^{\|\cdot\|}}$ and $\text{UPPER}_{\mathcal{I}_-}^{\mathcal{P}^{\|\cdot\|}}$ are **NP**-complete.

Proof. Membership in **NP** is given through Theorem 5 and the fact that every stratified program is also a disjunction-free program. Hardness follows from Theorems 10 and 11 (see the next section) and the fact that every classical disjunction free program is also a stratified program. □

Theorem 8. $\text{UPPER}_{\mathcal{I}_+}^{\mathcal{P}^{\|\cdot\|}}$ and $\text{UPPER}_{\mathcal{I}_{sd}}^{\mathcal{P}^{\|\cdot\|}}$ are **NP**-complete.

Proof. We show **NP**-hardness by reduction of the Set Cover Problem, which is known to be **NP**-complete [31]. Given a set of integers $\mathcal{U} = \{1, ..., m\}$, a set of subsets $\mathcal{S} = \{S_1, ..., S_t\}$, $S_j \subseteq \mathcal{U}$ for each j, and an integer k, the Set Cover Problem asks whether there are at most k sets in $\mathcal{S}$ that cover $\mathcal{U}$.

Given an instance $\mathcal{U} = \{1, ..., m\}$, $\mathcal{S} = \{S_1, ..., S_t\}$, $S_j \subseteq \mathcal{U}$ for each j of the Set Cover Problem and an integer k, we construct the following program P:

- For each number $i \in \mathcal{U}$, we construct two rules r_i and $\neg r_i$ having literals p_i and $\neg p_i$ as head, respectively.

- For each set $S_j \in \mathcal{S}$, we let an atom a_j appear in $neg(r_i)$ and $neg(\neg r_i)$ if and only $i \in S_j$.

Example 15. For $\mathcal{U} = \{1, 2, 3\}$ and $\mathcal{S} = \{S_1, S_2, S_3\} = \{\{1, 2\}, \{2, 3\}, \{1, 3\}\}$, we obtain the following program:

$$p_1 \leftarrow \text{not } a_1, \text{not } a_3.$$
$$\neg p_1 \leftarrow \text{not } a_1, \text{not } a_3.$$
$$p_2 \leftarrow \text{not } a_1, \text{not } a_2.$$
$$\neg p_2 \leftarrow \text{not } a_1, \text{not } a_2.$$
$$p_3 \leftarrow \text{not } a_2, \text{not } a_3.$$
$$\neg p_3 \leftarrow \text{not } a_2, \text{not } a_3.$$

Clearly, this program is stratified. Furthermore, if A is a set of facts such that $P \cup A$ is consistent, then we see as above that A corresponds to a cover of $\mathcal{U}$. We obtain hardness for $\mathcal{I}_+$.

Similarly, if P^M is consistent, then M corresponds to a cover of $\mathcal{U}$. Furthermore P^M will always have $\emptyset$ as the minimal model. So, such a set M with $d_{sd}(M, \text{Ans}_{Con}(P^M)) \leq k$ corresponds to a cover with at most k sets. Hence, we obtain hardness for $\text{UPPER}_{\mathcal{I}_{sd}}^{\mathcal{P}^{\|\cdot\|}}$. □

Corollary 7. *For* $\mathcal{I} \in \{\mathcal{I}_\pm, \mathcal{I}_+, \mathcal{I}_-, \mathcal{I}_{sd}\}$, $\text{LOWER}_{\mathcal{I}}^{\mathcal{P}^{\|\cdot\|}}$ *is* coNP-*complete,* $\text{EXACT}_{\mathcal{I}}^{\mathcal{P}^{\|\cdot\|}}$ *is in* D_1^p *and* $\text{VALUE}_{\mathcal{I}}^{\mathcal{P}^{\|\cdot\|}}$ *is in* FP$^{\text{NP}[\log n]}$.

6.4 Classical Disjunction-free Programs

Finally, we investigate the computational complexity of our measures on the easiest class of logic programs, namely classical disjunction-free programs P consisting of rules of the form

$$l_0 \leftarrow l_1, \ldots, l_m. \tag{8}$$

with literals $l_0, \ldots, l_m$. Let $\mathcal{P}^{\text{not} \forall} \subseteq \mathcal{P}$ be the set of all classical disjunction-free programs.

Recall that every classical disjunction-free program has a unique answer set that can be computed in polynomial time. This also implies the following observation.

Corollary 8. *Deciding whether a classical disjunction-free program is consistent is in* P.

As every classical disjunction-free program is also a stratified program, i.e. $\mathcal{P}^{\text{not}\forall} \subseteq \mathcal{P}^{\|\cdot\|}$, Theorem 6 and Corollary 6 already give us the following results.

Corollary 9. $\text{UPPER}_{\mathcal{I}_{01}}^{\mathcal{P}^{\text{not}\forall}} \in $ P, $\text{UPPER}_{\mathcal{I}_\#}^{\mathcal{P}^{\text{not}\forall}} \in $ P, $\text{LOWER}_{\mathcal{I}_{01}}^{\mathcal{P}^{\text{not}\forall}} \in $ P, $\text{EXACT}_{\mathcal{I}_{01}}^{\mathcal{P}^{\text{not}\forall}} \in $ P, $\text{VALUE}_{\mathcal{I}_{01}}^{\mathcal{P}^{\text{not}\forall}} \in $ FP, $\text{LOWER}_{\mathcal{I}_\#}^{\mathcal{P}^{\text{not}\forall}} \in $ P, $\text{EXACT}_{\mathcal{I}_\#}^{\mathcal{P}^{\text{not}\forall}} \in $ P, $\text{VALUE}_{\mathcal{I}_\#}^{\mathcal{P}^{\text{not}\forall}} \in $ FP.

Also the problems related to the measures $\mathcal{I}_+$ and $\mathcal{I}_{sd}$ turn out to be feasible for classical disjunction-free programs.

Theorem 9. *For* $\mathcal{I} \in \{\mathcal{I}_+, \mathcal{I}_{sd}\}$, $\text{UPPER}_{\mathcal{I}}^{\mathcal{P}^{\text{not}\forall}} \in $ P, $\text{LOWER}_{\mathcal{I}}^{\mathcal{P}^{\text{not}\forall}} \in $ P, $\text{EXACT}_{\mathcal{I}}^{\mathcal{P}^{\text{not}\forall}} \in $ P, $\text{VALUE}_{\mathcal{I}}^{\mathcal{P}^{\text{not}\forall}} \in $ FP.

Proof. Note that if a classical program P is inconsistent there is no A such that $P \cup A$ becomes consistent. Therefore, for every $P \in \mathcal{P}^{\text{not}\forall}$ we have either $\mathcal{I}_+(P) = 0$ or $\mathcal{I}_+(P) = \infty$. We can decide $\mathcal{I}_+(P) \leq 0$ in polynomial time due to Corollary 8. Furthermore, note that for a classical program P we have $P = P^M$ for every set of literals M. It follows that for every $P \in \mathcal{P}^{\text{not}\forall}$ we have either $\mathcal{I}_{sd}(P) = 0$ or $\mathcal{I}_{sd}(P) = \infty$ as well.

The remaining cases follow from these observations. □

Unfortunately, for measures $\mathcal{I}_\pm$ and $\mathcal{I}_-$, all problems remain infeasible even for the class $\mathcal{P}^{not \not\vee}$.

Theorem 10. UPPER$_{\mathcal{I}_\pm}^{\mathcal{P}^{not \not\vee}}$ is NP-complete.

Proof. Membership is clear since this is a special case of Theorem 5. Again, we show NP-hardness using the Set Cover Problem [31]. Thus, we let $\mathcal{U} = \{1, ..., m\}$ be a set, $\mathcal{S} = \{S_1, ..., S_t\}$ be a set of subsets of $\mathcal{U}$ and k an integer. Recall that the Set Cover Problem asks whether there are at most k sets in $\mathcal{S}$ that cover $\mathcal{U}$.

Suppose we were given an integer k and could solve UPPER$_{\mathcal{I}_\pm}^{\mathcal{P}^{not \not\vee}}$. We construct the following classical disjunction-free program P:

- For each number $i \in \mathcal{U}$, we construct two rules r_i and $\neg r_i$ having literals p_i and $\neg p_i$ as head, respectively.

- For each set $S_j \in \mathcal{S}$, we consider an atom a_j, construct a fact "a_j." and let a_j appear in $pos(r_i)$ and $pos(\neg r_i)$ if and only $i \in S_j$.

Example 16. For $\mathcal{U} = \{1, 2, 3\}$ and $\mathcal{S} = \{S_1, S_2, S_3\} = \{\{1, 2\}, \{2, 3\}, \{1, 3\}\}$, we obtain the following classical disjunction-free program:

$$a_1. \quad a_2. \quad a_3.$$
$$p_1 \leftarrow a_1, a_3.$$
$$\neg p_1 \leftarrow a_1, a_3.$$
$$p_2 \leftarrow a_1, a_2.$$
$$\neg p_2 \leftarrow a_1, a_2.$$
$$p_3 \leftarrow a_2, a_3.$$
$$\neg p_3 \leftarrow a_2, a_3.$$

If D is a subset of the facts $\{a_1., ..., a_t.\}$ such that $P \backslash D$ is consistent, then each pair "r_i" and "$\neg r_i$" of rules is not applicable anymore. Thus, D corresponds to a cover of $\mathcal{U}$. It is only left to show that we can assume the set D of deleted rules to not contain any of the "r_i" and "$\neg r_i$". However, this is clear since deleting one of them resolves at most one conflict and can be mimicked by removing the corresponding fact from an atom which appears in the rules. (This also corresponds to adding one set S_j that covers at least one additional integer in $\mathcal{U}$ which is trivial at any point.) Moreover, the set A of added rules can be assumed to be empty since we constructed a classical program. □

We obtain the same result with the same proof for $\mathcal{I}_-$.

Theorem 11. UPPER$_{\mathcal{I}_-}^{\mathcal{P}^{not \not\vee}}$ is NP-complete.

Corollary 10. For $\mathcal{I} \in \{\mathcal{I}_{\pm}, \mathcal{I}_{-}\}$, $\text{LOWER}_{\mathcal{I}}^{\mathcal{P}^{\varnothing bt \not\forall}}$ is **coNP**-complete, $\text{EXACT}_{\mathcal{I}}^{\mathcal{P}^{\varnothing bt \not\forall}}$ is in D_1^p and $\text{VALUE}_{\mathcal{I}}^{\mathcal{P}^{\varnothing bt \not\forall}}$ is in $FP^{NP[\log n]}$.

Table 2 summarizes the results of this section.

		$\mathcal{Q} = \mathcal{P}$	$\mathcal{Q} = \mathcal{P}^{\not\forall}$	$\mathcal{Q} = \mathcal{P}^{\|\cdot\|}$	$\mathcal{Q} = \mathcal{P}^{\varnothing bt \not\forall}$
$\mathcal{I}_{\pm}$	$\text{UPPER}_{\mathcal{I}_{\pm}}^{\mathcal{Q}}$	Σ_2^p-c	NP-c	NP-c	NP-c
	$\text{LOWER}_{\mathcal{I}_{\pm}}^{\mathcal{Q}}$	Π_2^p-c	coNP-c	coNP-c	coNP-c
	$\text{EXACT}_{\mathcal{I}_{\pm}}^{\mathcal{Q}}$	D_2^p	D_1^p	D_1^p	D_1^p
	$\text{VALUE}_{\mathcal{I}_{\pm}}^{\mathcal{Q}}$	$FP^{\Sigma_2^p[\log n]}$	$FP^{NP[\log n]}$	$FP^{NP[\log n]}$	$FP^{NP[\log n]}$
$\mathcal{I}_{-}$	$\text{UPPER}_{\mathcal{I}_{-}}^{\mathcal{Q}}$	Σ_2^p-c	NP-c	NP-c	NP-c
	$\text{LOWER}_{\mathcal{I}_{-}}^{\mathcal{Q}}$	Π_2^p-c	coNP-c	coNP-c	coNP-c
	$\text{EXACT}_{\mathcal{I}_{-}}^{\mathcal{Q}}$	D_2^p	D_1^p	D_1^p	D_1^p
	$\text{VALUE}_{\mathcal{I}_{-}}^{\mathcal{Q}}$	$FP^{\Sigma_2^p[\log n]}$	$FP^{NP[\log n]}$	$FP^{NP[\log n]}$	$FP^{NP[\log n]}$
$\mathcal{I}_{+}$	$\text{UPPER}_{\mathcal{I}_{+}}^{\mathcal{Q}}$	Σ_2^p-c	NP-c	NP-c	P
	$\text{LOWER}_{\mathcal{I}_{+}}^{\mathcal{Q}}$	Π_2^p-c	coNP-c	coNP-c	P
	$\text{EXACT}_{\mathcal{I}_{+}}^{\mathcal{Q}}$	D_2^p	D_1^p	D_1^p	P
	$\text{VALUE}_{\mathcal{I}_{+}}^{\mathcal{Q}}$	$FP^{\Sigma_2^p[\log n]}$	$FP^{NP[\log n]}$	$FP^{NP[\log n]}$	FP
$\mathcal{I}_{sd}$	$\text{UPPER}_{\mathcal{I}_{sd}}^{\mathcal{Q}}$	Σ_2^p-c	NP-c	NP-c	P
	$\text{LOWER}_{\mathcal{I}_{sd}}^{\mathcal{Q}}$	Π_2^p-c	coNP-c	coNP-c	P
	$\text{EXACT}_{\mathcal{I}_{sd}}^{\mathcal{Q}}$	D_2^p	D_1^p	D_1^p	P
	$\text{VALUE}_{\mathcal{I}_{sd}}^{\mathcal{Q}}$	$FP^{\Sigma_2^p[\log n]}$	$FP^{NP[\log n]}$	$FP^{NP[\log n]}$	FP
$\mathcal{I}_{\#}$	$\text{UPPER}_{\mathcal{I}_{\#}}^{\mathcal{Q}}$	Σ_2^p-c	NP-c	P	P
	$\text{LOWER}_{\mathcal{I}_{\#}}^{\mathcal{Q}}$	Π_2^p-c	coNP-c	P	P
	$\text{EXACT}_{\mathcal{I}_{\#}}^{\mathcal{Q}}$	D_2^p	D_1^p	P	P
	$\text{VALUE}_{\mathcal{I}_{\#}}^{\mathcal{Q}}$	$FP^{\Sigma_2^p[\log n]}$	$FP^{NP[\log n]}$	FP	FP
$\mathcal{I}_{01}$	$\text{UPPER}_{\mathcal{I}_{01}}^{\mathcal{Q}}$	Σ_2^p-c	NP-c	P	P
	$\text{LOWER}_{\mathcal{I}_{01}}^{\mathcal{Q}}$	Π_2^p-c	coNP-c	P	P
	$\text{EXACT}_{\mathcal{I}_{01}}^{\mathcal{Q}}$	D_2^p	D_1^p	P	P
	$\text{VALUE}_{\mathcal{I}_{01}}^{\mathcal{Q}}$	$FP^{\Sigma_2^p[\log n]}$	$FP^{NP[\log n]}$	FP	FP

Table 2: Computational complexity of various problems related to our inconsistency measures

7 Summary and Discussion

In this paper, we addressed the challenge of measuring inconsistency in ASP by critically reviewing the propositional framework of inconsistency measurement and taking non-monotonicity into account. We developed novel rationality postulates and measures that are more apt for analyzing inconsistency in ASP than existing approaches. Intuitively, some of our measures take the effort needed to restore the consistency of programs into account ($\mathcal{I}_\pm$, $\mathcal{I}_+$, $\mathcal{I}_-$), and our results show that it does not matter whether this is done on the level of the original program or on the level of the reduct. Others measure inconsistency in terms of the quality of the produced output, e. g. $\mathcal{I}_\#$ which considers the minimal number of inconsistencies in an answer set. We showed that our new measures comply with many of our rationality postulates and illustrated their usage.

To the best of our knowledge, measuring inconsistency in extended logic programs under the answer set semantics has not been addressed before. The closest related works are by Madrid and Ojeda-Aciego, see e. g. [37, 38], who address inconsistencies in residuated logic programs under fuzzy answer set semantics. In their setting, rules such as (1) are augmented with fuzzy values in [0, 1] (or some arbitrary lattice) and inconsistency is measured by considering minimal changes in the values to restore the existence of fuzzy stable models. However, Madrid and Ojeda-Aciego do not discuss the propositional case and rationality postulates.

Inconsistency measures have numerous applications, for instance in belief revision where the degree of inconsistency of new information may provide useful guidance as to whether the new information should be accepted or not, or in belief merging where the degree of inconsistency may be used to decide whether the views of a particular agent should be taken into account. The overview paper [28] contains an analysis of these and various other potential applications. The results of this paper pave the way for similar applications in the context of ASP, which has become a popular language for declarative problem solving.

In future work, we would like to extend our analysis to more general classes of logic programs, e. g., programs with choice rules, weight constraints and aggregates. For an overview on these extensions see [20]. It would also be interesting to see whether our measures, or similar ones, can be applied to other non-monotonic formalisms, like default logic [41] or autoepistemic logic [39]. In another recent paper [11], we explored the issues of minimal inconsistent sets in general non-monotonic formalisms in more depth. There, we also hinted to possible applications for inconsistency measurement which is also part of ongoing work.

Acknowledgements We would like to thank the reviewers of the present chapter and of [50] for useful comments which also helped to improve the current chapter. This work has been partially funded by the DFG Research Training Group 1763 and DFG project BR 1817/7-2.

References

[1] M. Ammoura, Y. Salhi, B. Oukacha, and B. Raddaoui. On an MCS-based Inconsistency Measure. *International Journal of Approximate Reasoning*, 2016.

[2] K.R. Apt, H.A. Blair, and A. Walker. Towards a Theory of Declarative Knowledge. In *Foundations of Deductive Databases and Logic Programming*, pages 89–148. Morgan Kaufmann, 1988.

[3] P. Baroni, M. Caminada, and M. Giacomin. An Introduction to Argumentation Semantics. *The Knowledge Engineering Review*, 26(4):365–410, 2011.

[4] S. Benferhat, D. Dubois, and H. Prade. A Local Approach to Reasoning under Inconsisteny in Stratified Knowledge bases. In *Proceedings of the Third European Conference on Symbolic and Quantitative Approaches to Reasoning and Uncertainty (ECSQARU'95)*, pages 36–43, 1995.

[5] Ph. Besnard. Revisiting Postulates for Inconsistency Measures. In *Proceedings of the 14th European Conference on Logics in Artificial Intelligence (JELIA'14)*, pages 383–396, 2014.

[6] Ph. Besnard. Forgetting-based Inconsistency Measure. In *Proceedings of the 10th International Conference on Scalable Uncertainty Management (SUM'16)*, pages 331–337, 2016.

[7] Jean-Yves Béziau, W. Carnielli, and D. Gabbay, editors. *Handbook of Paraconsistency*. College Publications, London, 2007.

[8] H.A. Blair and V.S. Subrahmanian. Paraconsistent Logic Programming. *Theoretical Computer Science*, 68(2):135–154, 1989.

[9] R.J. Brachman and H.J. Levesque. *Knowledge Representation and Reasoning*. The Morgan Kaufmann Series in Artificial Intelligence. Morgan Kaufmann Publishers, 2004.

[10] G. Brewka, T. Eiter, and M. Truszczynski. Answer Set Programming at a Glance. *Commun. ACM*, 54(12):92–103, 2011.

[11] G. Brewka, M. Thimm, and M. Ulbricht. Strong Inconsistency in Nonmonotonic Reasoning. In *Proceedings of the 26th International Joint Conference on Artificial Intelligence (IJCAI'17)*, August 2017.

[12] L. Cholvy and A. Hunter. Information Fusion in Logic: A Brief Overview. In *Qualitative and Quantitative Practical Reasoning (ECSQARU'97/FAPR'97)*, volume 1244 of *Lecture Notes in Computer Science*, pages 86–95. Springer, 1997.

[13] S. Costantini. On the Existence of Stable Models of Non-stratified Logic Programs. *Theory and Practice of Logic Programming*, 6(1-2):169–212, January 2006.

[14] E. Dantsin, T. Eiter, G. Gottlob, and A. Voronkov. Complexity and Expressive Power of Logic Programming. *ACM Comput. Surv.*, 33(3):374–425, 2001.

[15] J.P. Delgrande, T. Schaub, H. Tompits, and S. Woltran. Belief Revision of Logic Programs under Answer Set Semantics. In *Proceedings of the 11th International Conference on Principles of Knowledge Representation and Reasoning*, pages 411–421, 2008.

[16] D. Dubois, J. Lang, and H. Prade. Inconsistency in Possibilistic Knowledge bases: To Live with it or not Live with it. In L.A. Zadeh and J. Kacprzyk, editors, *Fuzzy Logic for the Management of Uncertainty*, pages 335–351. Wiley, New York, 1992.

[17] D. Dubois and H. Prade. A Possibilistic Analysis of Inconsistency. In *Scalable Uncertainty Management - 9th International Conference, SUM 2015, Québec City, QC, Canada, September 16-18, 2015. Proceedings*, pages 347–353, 2015.

[18] T. Eiter and G. Gottlob. On the Computational Cost of Disjunctive Logic Programming: Propositional Case. *Annals of Mathematics and Artificial Intelligence*, 15(3-4):289–323, 1995.

[19] D. Gabbay, L. Giordano, A. Martelli, and N. Olivetti. Hypothetical Updates, Priority and Inconsistency in a Logic Programming Language. In *Logic Programming and Nonmonotonic Reasoning, Third International Conference, LPNMR'95, Lexington, KY, USA, June 26-28, 1995, Proceedings*, pages 203–216, 1995.

[20] M. Gebser and T. Schaub. Modeling and Language Extensions. *AI Magazine*, 37(3):33–44, 2016.

[21] M. Gelfond and N. Leone. Logic Programming and Knowledge Representation – The A-Prolog Perspective. *Artificial Intelligence*, 138(1–2):3–38, 2002.

[22] M. Gelfond and V. Lifschitz. Classical Negation in Logic Programs and Disjunctive Databases. *New Generation Comput.*, 9(3/4):365–386, 1991.

[23] J. Grant. Classifications for Inconsistent Theories. *Notre Dame Journal of Formal Logic*, 19(3):435–444, 1978.

[24] J. Grant and A. Hunter. Measuring Inconsistency in Knowledgebases. *Journal of Intelligent Information Systems*, 27:159–184, 2006.

[25] J. Grant and A. Hunter. Measuring the Good and the Bad in Inconsistent Information. In *IJCAI 2011, Proceedings of the 22nd International Joint Conference on Artificial Intelligence, Barcelona, Catalonia, Spain, July 16-22, 2011*, pages 2632–2637, 2011.

[26] J. Grant and A. Hunter. Analysing Inconsistent Information using Distance-based Measures. *International Journal of Approximate Reasoning*, In press, 2016.

[27] Sven Ove Hansson. *A Textbook of Belief Dynamics*. Kluwer Academic Publishers, Norwell, MA, USA, 2001.

[28] A. Hunter and S. Konieczny. Approaches to Measuring Inconsistent Information. In *Inconsistency Tolerance*, volume 3300 of *Lecture Notes in Computer Science*, pages 189–234. Springer International Publishing, 2004.

[29] A. Hunter and S. Konieczny. Measuring Inconsistency through Minimal Inconsistent Sets. In *Proceedings of the Eleventh International Conference on Principles of Knowledge Representation and Reasoning (KR'2008)*, pages 358–366. AAAI Press, 2008.

[30] S. Jabbour, Y. Ma, B. Raddaoui, L. Sais, and Y. Salhi. A MIS Partition based Framework for Measuring Inconsistency. In *Proceedings of the 15th International Conference on Principles of Knowledge Representation and Reasoning (KR'16)*, pages 84–93, 2016.

[31] R.M. Karp. Reducibility among Combinatorial Problems. In *Proceedings of a symposium on the Complexity of Computer Computations, held March 20-22, 1972, at the IBM Thomas J. Watson Research Center, Yorktown Heights, New York.*, pages 85–103, 1972.

[32] S. Konieczny, J. Lang, and P. Marquis. Reasoning under Inconsistency: The Forgotten Connective. In *Proceedings of IJCAI-2005*, pages 484–489, 2005.

[33] S. Konieczny and R. Pino Perez. Logic based Merging. *Journal of Philosophical Logic*, 40:239–270, 2011.

[34] J. Lang and P. Marquis. Reasoning under Inconsistency: A Forgetting-based Approach. *Artificial Intelligence*, 174(12–13):799–823, 2010.

[35] V. Lifschitz, D. Pearce, and A. Valverde. Strongly Equivalent Logic Programs. *ACM Transactions on Computational Logic*, 2(4):526–541, 2001.

[36] V. Lifschitz and H. Turner. Splitting a Logic Program. In *Logic Programming, Proceedings of the Eleventh International Conference on Logic Programming, Santa Marherita Ligure, Italy, June 13-18, 1994*, pages 23–37, 1994.

[37] N. Madrid and M. Ojeda-Aciego. Measuring Instability in Normal Residuated Logic Programs: Adding Information. In *FUZZ-IEEE 2010, IEEE International Conference on Fuzzy Systems, Barcelona, Spain, 18-23 July, 2010, Proceedings*, pages 1–7, 2010.

[38] N. Madrid and M. Ojeda-Aciego. Measuring Inconsistency in Fuzzy Answer Set Semantics. *IEEE Transactions on Fuzzy Systems*, 19(4):605–622, 2011.

[39] R.C. Moore. Autoepistemic Logic Revisited. *Artif. Intell.*, 59(1-2):27–30, 1993.

[40] K. Mu. Responsibility for Inconsistency. *International Journal of Approximate Reasoning*, 61:43–60, 2015.

[41] R. Reiter. A Logic for Default Reasoning. *Artif. Intell.*, 13(1-2):81–132, 1980.

[42] R. Reiter. A theory of Diagnosis from First Principles. *Artificial Intelligence*, 32(1):57–95, 1987.

[43] C. Schulz, K. Satoh, and F. Toni. Characterising and Explaining Inconsistency in Logic Programs. In *Logic Programming and Nonmonotonic Reasoning: 13th International Conference, LPNMR 2015, Lexington, KY, USA, September 27-30, 2015. Proceedings*, pages 467–479. Springer International Publishing, Cham, 2015.

[44] M. Thimm. Inconsistency Measures for Probabilistic Logics. *Artificial Intelligence*, 197:1–24, 2013.

[45] M. Thimm. On the Compliance of Rationality Postulates for Inconsistency Measures: A more or less complete Picture. *Künstliche Intelligenz*, 2016.

[46] M. Thimm. On the Expressivity of Inconsistency Measures. *Artificial Intelligence*, 234:120–151, 2016.

[47] M. Thimm. Stream-based Inconsistency Measurement. *International Journal of Approximate Reasoning*, 68:68–87, 2016.

[48] M. Thimm and J.P. Wallner. Some Complexity Results on Inconsistency Measurement. In *Proceedings of the 15th International Conference on Principles of Knowledge Representation and Reasoning (KR'16)*, pages 114–124, 2016.

[49] M. Truszczynski. Strong and Uniform Equivalence of Nonmonotonic Theories - An Algebraic Approach. *Annals of Mathematics and Artificial Intelligence*, 48(3-4):245-265, 2006.

[50] M. Ulbricht, M. Thimm, and G. Brewka. Measuring Inconsistency in Answer Set Programs. In *Logics in Artificial Intelligence - 15th European Conference, JELIA 2016, Larnaca, Cyprus, November 9-11, 2016, Proceedings*, pages 577-583, 2016.

Inconsistency Measures in General Fuzzy Logic Programming

Nicolás Madrid and Manuel Ojeda-Aciego

Dept Matemática Aplicada
Universidad de Málaga, Spain
{nicolas.madrid, aciego}@uma.es

Abstract

Fuzzy logic has shown to be a suitable framework to handle contradictions in which, unsurprisingly, the notion of inconsistency can be defined in different ways. This chapter analyses the notion of inconsistency in general residuated logic programming under the answer-set semantics, shows that inconsistency can be somehow decomposed into instability and incoherence and, finally, shows that each of these notions can be associated with some natural measures of inconsistency. Finally, we focus on measures of inconsistency in the particular framework of fuzzy logic programming.

1 Introduction

For many years, inconsistency has been considered as an undesirable feature to be completely ignored in our logic theories. However, inconsistency arises naturally in databases and, in many cases, seems to be unavoidable. For example, assume that a theft occurred in a classroom and we construct a knowledge-base containing the students' declarations; it is highly probable that we will obtain inconsistencies.

A typical reaction when somebody obtains an inconsistent knowledge-base is to reject it. Nevertheless, to reject the whole bulk of information provided by a knowledge-base is not a good choice, since we might be rejecting some pieces of correct and useful information. In the example, if all the students agree on the exact time when the theft occurred, this piece of data should be considered true in spite of the existence of contradictory information elsewhere.

Another way to deal with inconsistent knowledge-bases is to try to repair them. But it is worth taking into account that, in some cases, inconsistencies can provide us with useful information. Thus, it might be useful to develop some mechanism to tolerate inconsistent information instead of getting rid of it. For instance, if some students provide inconsistent declarations, then it is possible that one of them was involved in the pilfering; in this respect, a convenient *degree of inconsistency* of the declaration of a given student may be related to his/her being guilty. In this case, it would be desirable to have some means of measuring the amount of inconsistent information contained in a given logic theory.

Considering *"inconsistency-tolerant approaches"* is not new, since the family of paraconsistent logics, introduced more than 30 years ago, allows us to handle inconsistent information efficiently. Among the different approaches in the literature, we can find those related to the *consistency restoring* [2, 31] and *inconsistent information measuring* [5, 14, 36].

The existence of a number of approaches based on Belnap's lattice [3, 29] suggests that many-valued and/or fuzzy logic [4] might be an ideal framework to provide a graded notion of inconsistency [28, 33]. Actually, working in a fuzzy framework enables the possibility of restoring consistency by slightly modifying the information inferred of a given propositional symbol (there need not be either completely false or completely true).

In this chapter, we deal with measuring the amount of inconsistent information in general residuated logic programs under the fuzzy answer-set semantics, which is a convenient extension of the answer-set semantics into the fuzzy setting. The framework of answer set programming has important links with description logics and, hence, with the semantic web, as stated in [6, 12, 18, 19]. Originally, answer sets were intended to deal with two negations: one default negation and one explicit (or strong) negation. The former is related to the so-called *closed world assumption* and, for a given predicate P, it allows to extract negative knowledge about P from the absence of positive information about it. Under this semantics, a predicate is always either true or false by default. On the other hand, the explicit negation is related to the so-called *open world assumption* and a given predicate P is (explicitly) false if and only if there is information in the knowledge base supporting its falseness. In this respect, a predicate P can be either true or false or unknown or inconsistent. This fact relates the semantics of the explicit negation to many-valued logics, specifically Belnap's lattice (see [8] for extensions to more general bilattices). The use of these two types of negation is advocated in many contexts of interest, in particular in [35] their use is justified in relation to web rules.

The use of two different negations involves two different types of inconsistency. Hence, the inconsistency in fuzzy logic programming (under the fuzzy answer sets semantics) is introduced as being composed of two different levels: *"lack of stable models"* (called instability) and *"incoherent stable models"* (called incoherence). The former occurs when a set of incompatible rules appears in the logic program, whereas the latter occurs when the existing models assign *contradictory* values to p and to its explicit negation $\sim p$. Accordingly, measures of inconsistency are defined in terms of the reasons which generate inconsistency [21, 23, 24]: in the case of incoherence, the measure is intended to represent the excess of information contained in the models; in the case of instability, the measure computes the minimal amount of information which has to be either removed or added to the program so that consistency is restored.

2 Preliminary Definitions

In this section we describe the syntax of the general residuated logic programs and the fuzzy answer set semantics [20, 25]. The relation between conjunction and implication in fuzzy logic is usually represented by means of a residuated lattice.

Definition 1. *A residuated lattice is a triple* $\mathcal{L} = ((L, \leq), *, \leftarrow)$ *such that:*

1. $(L, \leq)$ *is a bounded lattice with largest element 1 and least element 0.*
2. $(L, *, 1)$ *is a commutative monoid unit element 1.*
3. $*$ *and* $\leftarrow$ *form an adjoint pair, i.e:*

$$z \leq (x \leftarrow y) \text{ iff } y * z \leq x \quad \text{for all } x, y, z \in L.$$

Hereafter, all the residuated lattices will be assumed to be *complete* (as lattices). In fuzzy logic, the complete lattice L is usually seen as the underlying set of truth-values on which the formulas are interpreted, the operator $*$ is seen as a generalized conjunction and the operator $\leftarrow$ as an implication. The most common residuated lattices on $[0, 1]$ are given by:

- Gödel residuated lattice $(([0,1], \leq), *_G, \leftarrow_G, 0, 1)$

$$x *_G y = \min\{x, y\} \qquad y \leftarrow_G x = \begin{cases} 1 & \text{if } x \leq y \\ y & \text{otherwise.} \end{cases}$$

- Product residuated lattice $(([0,1], \leq), *_P, \leftarrow_P, 0, 1)$

$$x *_P y = x \cdot y \qquad y \leftarrow_P x = \begin{cases} 1 & \text{if } x \leq y \\ \frac{y}{x} & \text{otherwise.} \end{cases}$$

- Łukasiewicz residuated lattice $(([0,1], \leq), *_\mathbf{L}, \leftarrow_\mathbf{L}, 0, 1)$

$$x *_\mathbf{L} y = \max\{x + y - 1, 0\} \qquad y \leftarrow_\mathbf{L} x = \min\{1 - x + y, 1\}$$

Hereafter, we will consider an enriched residuated lattice $(L, *, \leftarrow, \sim, \neg)$ with two negation operators. These two negations model the notions of explicit negation $\sim$ and default negation $\neg$ often used in logic programming [1, 8, 9, 27]. Both kinds of negation are modelled by a negation operator over L, this is, any decreasing mapping $n\colon L \to L$ satisfying $n(0) = 1$ and $n(1) = 0$.

The syntax of general residuated logic programs assumes a set Π of propositional symbols. If $p \in \Pi$, then both p and $\sim p$ are called *literals*. We will denote arbitrary literals with the symbol ℓ (possibly subscripted), and the set of all literals as *Lit*.

Definition 2. *Given an enriched residuated lattice* $(L, *, \leftarrow, \sim, \neg)$, *a general residuated logic program* $\mathbb{P}$ *is a set of weighted rules of the form*

$$\langle \ell \leftarrow \ell_1 * \cdots * \ell_m * \neg \ell_{m+1} * \cdots * \neg \ell_n; \quad \vartheta \rangle$$

where ϑ *is an element of* L *and* $\ell, \ell_1, \ldots, \ell_n$ *are literals.*

Note that, it need not be a relation between the negations $\sim$ and $\neg$ in the enriched residuated lattice and the negation n naturally definable in a residuated lattice by $n(p) = 0 \leftarrow p$.

The general pattern for rules in our programs is $\langle \ell \leftarrow \mathcal{B}; \vartheta \rangle$, where $\mathcal{B}$ is the *body* of the rule. Rules whose body is empty are called *facts*.

Note that the use of weights in the rules of general residuated logic programs allows us to define an ordering between programs as follows: let $\mathbb{P}_1$ and $\mathbb{P}_2$ be two general residuated logic programs, we say that $\mathbb{P}_1 \leq \mathbb{P}_2$ if for any rule $\langle \ell \leftarrow \mathcal{B}; \vartheta_1 \rangle \in \mathbb{P}_1$ there exists a rule $\langle \ell \leftarrow \mathcal{B}; \vartheta_2 \rangle \in \mathbb{P}_2$ such that $\vartheta_1 \leq \vartheta_2$.

Once we have fixed the syntax of our logic programs, we can describe the semantics. The notion of interpretation is given as follows:

Definition 3. *A fuzzy L-interpretation is a mapping* $I\colon Lit \to L$; *that is, an L-fuzzy subset of literals.*

It is worth to remark that the domain of I is the set of all literals, particularly, p and $\sim p$ are assigned truth-values independently, as if they were different propositional symbols.

In addition, note that the domain of fuzzy L-interpretations can be extended inductively from the set of literals to any formula constructed with the connectives $\leftarrow, *$ and $\neg$. In other words, given an interpretation $I\colon Lit \to L$, then

$$I\left(\ell \leftarrow \ell_1 * \cdots * \ell_m * \neg \ell_{m+1} * \cdots * \neg \ell_n\right)$$

is defined as

$$I(\ell) \leftarrow I(\ell_1) * \cdots * I(\ell_m) * \neg I(\ell_{m+1}) * \cdots * \neg I(\ell_n)$$

However, note that our approach is not compositional with respect to the connective $\sim$. That is, the truth-degree of $\sim p$ does not depend on the degree of truth of p. Thus, the truth values of p and $\sim p$ are, a priori, independent. This non-compositional approach is based on the well-known *open world assumption* semantics [7, 30, 32]. In the classical case, one has to take into account the interaction between opposite literals in order to reject inconsistent models. Therefore, in the fuzzy case we need to impose a condition between opposite literals in order to provide a criteria to either accept or reject L-interpretations.

Obviously, the information provided by two opposite literals might be discordant but not necessarily incompatible. There are several fuzzy frameworks that assign, compositionally, a truth-value to $\sim p$ from the truth value of p by using a negation operator, and such truth-values for two opposite literals are not considered inconsistent; for instance, in Łukasiewicz logic, it is admissible to assign the truth-value 0.5 to two opposite literals p and $\sim p$ [11]. Therefore, in a non-compositional framework, there are no reasons to reject L-interpretations assigning nonzero truth-values to opposite literals. So, we must determine when the information contained in an L-interpretation represents a contradiction. In order to deal with this matter, the notion of *coherence* was introduced in [20] as a suitable generalization of consistence in the residuated framework. Such a notion depends on a negation operator defined on the underlying lattice of truth-values.

Definition 4. *A fuzzy L-interpretation I over Lit is coherent if the inequality $I(\sim p) \leq \sim I(p)$ holds for every propositional symbol p.*

Note that we overload notation by using the same symbol ($\sim$) to refer to both the syntactic connective and the negation operator related to the explicit negation.

Although we will focus on the notions of coherence and inconsistency in the next section, at this point, it is worth to remark three features of *coherence* to show why it can be seen as a suitable generalization of *consistency*: firstly, coherence-checkers are easy to implement since they only depend on the negation operator (whereas other definitions use both a t-norm and a negation); secondly, it allows to handle missing information since the L-interpretation I such that $I(\ell) = 0$ for all $\ell \in Lit$ is always coherent; thirdly, our notion of coherence coincides with consistency in the classical framework (it is not difficult to check this).

The following definition introduces the notion of satisfiability of an L-interpretation.

Definition 5. *Let $I \colon Lit \to L$ be an L-interpretation. Then I satisfies a rule $\langle \ell \leftarrow \mathcal{B}; \ \vartheta \rangle$ if and only if*
$$I(\mathcal{B}) * \vartheta \leq I(\ell)$$
or, equivalently, if and only if
$$\vartheta \leq I(\ell \leftarrow \mathcal{B}).$$
Moreover we say that I is a model of $\mathbb{P}$ if it satisfies all rules in $\mathbb{P}$.

Note that we can extend the order in the residuated lattice to the set of L-interpretations. Specifically, given two L-interpretations I and J we write $I \leq J$ if $I(\ell) \leq J(\ell)$ for all $\ell \in Lit$.

Perhaps the most attractive feature of logic programming appears in the case of negation free logic programs, namely: every residuated logic program without negation has a least model. However, it is well-known that such a feature gets lost when we take into account default negation. This fact motivates the following classification of general residuated logic programs according to the kinds of negation appearing in it. A general residuated logic program $\mathbb{P}$ is said to be:

- *positive* if it does not contain negation operators.
- *normal* if it does not contain explicit negation.
- *extended* if it does not contain default negation.

Similarly to crisp logic programming, every extended residuated logic program has a least model and, moreover, this least model can be reached by means of the immediate consequence operator, which is generalized to extended logic programs as follows.

Definition 6. *Let $\mathbb{P}$ be an extended residuated logic program. The immediate consequence operator maps every L-interpretation I to the L-interpretation $T_\mathbb{P}(I)$, defined for all $\ell \in Lit$ as*

$$T_\mathbb{P}(I)(\ell) = \sup\{I(\mathcal{B}) * \theta \mid \langle \ell \leftarrow \mathcal{B}; \theta \rangle \in \mathbb{P}\}\,.$$

For general residuated logic programs, we adapt the approach given in [9, 10].

Definition 7. *Let $\mathbb{P}$ be a general residuated logic program and let I be an L-interpretation. The reduct of $\mathbb{P}$ wrt the interpretation I is the extended residuated logic program $\mathbb{P}_I$ obtained by substituting each rule in $\mathbb{P}$ of the form*

$$\langle \ell \leftarrow \ell_1 * \cdots * \ell_m * \neg \ell_{m+1} * \cdots * \neg \ell_n;\quad \vartheta \rangle$$

by the rule[1]

$$\langle \ell \leftarrow \ell_1 * \cdots * \ell_m;\quad \neg I(\ell_{m+1}) * \cdots * \neg I(\ell_n) * \vartheta \rangle$$

Notice that the reduct $\mathbb{P}_I$ is, by definition, an extended program, since it does not contain default negation. Therefore, we can consider always the least model of $\mathbb{P}_I$. Finally, the notion of fuzzy answer set of a general residuated logic program is given as a generalization of the one given originally in [9, 10] for crisp logic programming. Motivational aspects about this reduct and the definition of answer sets can be found in [17, 16].

---
[1] Note again the overloaded use of the negation symbol, as a syntactic function in the formulas and as the algebraic negation in the truth-values.

Definition 8. *Let* $\mathbb{P}$ *be a general residuated logic program and let* I *be a fuzzy L-interpretation. Then:*

- I *is said to be a* stable model *of* $\mathbb{P}$ *iff* I *is the least model of* $\mathbb{P}_I$.

- I *is said to be a* fuzzy answer set *of* $\mathbb{P}$ *iff* I *is a coherent stable model of* $\mathbb{P}$.

Theorem 1 ([20, 25]). *Let* $\mathbb{P}$ *be a general residuated logic program and let* I *be a fuzzy L-interpretation. If* I *is a stable model of* $\mathbb{P}$ *then,* I *is a minimal model of* $\mathbb{P}$.

3 Inconsistency of General Residuated Logic Programs

The fuzzy answer set semantics described in the previous section allows us to provide a meaning to every general residuated logic program. Under such a semantics, a natural definition of inconsistency arises: a general residuated logic program is inconsistent if and only if it has no fuzzy answer sets. Then, from a technical point of view, inconsistency of a residuated logic program $\mathbb{P}$ can be due to one of the following two reasons:

- **Incoherence**: Every stable model is incoherent.

- **Instability**: There are no stable models.

This technical distinction allows to classify inconsistent logic programs into two classes, namely, the class of *incoherent* and the class of *unstable* programs. Note that, although the above distinction between types of inconsistency can be also done in crisp logic programming as well, it is in the fuzzy framework where instability can be avoided for any finite general residuated logic program by the choice of a convenient underlying residuated lattice. This is given by the following theorem:

Theorem 2 ([26]). *Let* $\mathcal{L} = ([0,1], \leq, *, \leftarrow, \sim, \neg)$ *be an enriched residuated lattice. If* $*$ *and* $\neg$ *are continuous operators, then for every finite general program* $\mathbb{P}$ *there exists at least a stable model.*

Note that the use of a convenient residuated lattice guarantees, by the theorem above, the existence of stable models for any general program. It is important to recall that most connectives in fuzzy logic are defined on the unit interval $[0, 1]$. Moreover, most t-norms used currently in fuzzy logic are continuous (Gödel, Łukasiewicz, product, ...), therefore the theorem establishes that in the most used fuzzy frameworks, the existence of fuzzy stable models is always guaranteed.

Example 1. Let us consider the lattice $L = [0, 1]$ and the following simple normal residuated logic program $\mathbb{P}$:

$$\langle p \leftarrow \neg p ; \quad 1 \rangle$$

Obviously the existence of stable models for the program above is not always guaranteed. For instance, if we consider the Gödel residuated lattice with the negation

$$\neg x = 0 \leftarrow_G x = \begin{cases} 0 & \text{if } x = 1 \\ 1 & \text{otherwise} \end{cases}$$

the program $\mathbb{P}$ has no stable models. However, if we use the standard negation $\neg x = 1-x$ to model the default negation instead of the previous one, the program turns out to be consistent with the only stable model given by $M = \{(p, 0.5)\}$. Note that in this latter case, both connectives, $*$ and $\neg$, are continuous.

Example 2. Let us consider the following general residuated logic program $\mathbb{P}$ with truth-values in $[0, 1]$:

$$\langle p \leftarrow \neg q; \quad 0.8 \rangle$$
$$\langle q \leftarrow \neg(\sim p); \quad 0.7 \rangle$$
$$\langle \sim p \leftarrow \neg p; \quad 0.9 \rangle$$

As in the previous example, the existence of stable models for the program above is not always guaranteed. To avoid the cause of instability, we consider the product residuated lattice with $\neg x = 1 - x$; i.e., continuous connectives. Under such an assumption, the only interpretation satisfying $I = \text{lfp}(T_{\mathbb{P}_I})$ is

$$I = \{(p, 0.49); (\sim p, 0.45); (q, 0.38)\}$$

In this case, the program would be consistent if and only if I is coherent. Note that this last assertion depends on the negation operator considered to model $\sim$. In this way, if $\sim x = \neg x = 1 - x$, then I is coherent (since $I(\sim p) = 0.45 \leq \sim I(p) = 1 - 0.49 = 0.51$). However, if we would have considered

$$\neg x = \begin{cases} 0 & \text{if } x = 1 \\ 1 & \text{otherwise} \end{cases}$$

the program would have turned out to be incoherent, since $I(\sim p) = 0.45 \not\leq \sim I(p) = 0$.

4 Measures of Instability

As stated in the previous section, instability can be avoided by the choice of continuous operators in the residuated lattice. However, in some cases operators cannot be chosen to be continuous and, in such a case, the only way to recover the stability of a general residuated program is by modifying its rules; that is, by modifying the information of the logic program. In this section, we introduce two measures of instability which focus on measuring the amount of information we have to add and/or remove in order to recover stability. Contrary to crisp logic programming, where the only possible way to add or remove information is by including or removing rules respectively, in residuated logic programming we can proceed by increasing or reducing weights of rules. Such a procedure provides a family of degrees for each rule, since the greater the increasing (resp. decreasing) of the weight, the greater is the information included (resp. removed). In order to measure the information included or removed, we define the following measure of information that represents the degree of inherent information for each element in the residuated lattice.

Definition 9. *Let L be a lattice. An* information measure *is an operator $m\colon L \to \mathbb{R}^+$ such that:*

- $m(x) = 0$ *if and only if $x = 0$*
- m *is monotonic*
- $m(\sup(x,y)) \geq m(x) + m(y) - m(\inf(x,y))$ *for all $x, y \in L$*

In what follows, we propose two different measures of instability, one to measure the amount of information we have to reduce to reach the stability and another to measure the amount of information we have to add to reach the stability. Consider a general residuated logic program $\mathbb{P}$. Then, $\overline{\mathbb{P}}$ is defined as

$$\overline{\mathbb{P}} = \mathbb{P} \cup \{\langle \ell_i \leftarrow\ ; 0\rangle \mid \ell_i \in Lit_\mathbb{P} \text{ and } \langle \ell_i \leftarrow\ ; \vartheta\rangle \notin \mathbb{P}\}$$

Note that $\overline{\mathbb{P}}$ only includes in $\mathbb{P}$ facts with weight 0. As a result $\overline{\mathbb{P}}$ is semantically equivalent to $\mathbb{P}$ and it neither includes nor removes information. The reason of working on $\overline{\mathbb{P}}$ instead of on $\mathbb{P}$ is simply technical, since it allows us to add information for literals that do not appear in the head of any rule in $\mathbb{P}$. Let us consider now a set $X = \{\langle r_i; \vartheta_i\rangle\}_i$ of rules in $\mathbb{P}$; a set $\overline{X} = \{\langle \overline{r_i}; \overline{\vartheta_i}\rangle\}_i$ of rules in $\overline{\mathbb{P}}$, and a set of values $\{\varphi_i\}_i$ in L. Then, we can define the following two new general residuated logic programs:

- $\Theta_\mathbb{P}(X, \{\varphi\}_i) = (\mathbb{P} \smallsetminus X) \cup \{\langle r_i; t(\vartheta_i, \varphi_i)\rangle\}_i$
- $\Omega_\mathbb{P}(\overline{X}, \{\varphi_i\}_i) = (\overline{\mathbb{P}} \smallsetminus \overline{X}) \cup \{\langle \overline{r_i}; s(\overline{\vartheta_i}, \varphi_i)\rangle\}_i$

where t and s denote a t-norm and a t-conorm respectively. Note that $\Theta_\mathbb{P}(X, \{\varphi\}_i)$ reduces the weights of every rule in X whereas $\Omega_\mathbb{P}(\overline{X}, \{\varphi_i\}_i)$ increases the weight of every rule in $\overline{X}$. Now, we can define the following measures of instability associated with subsets of rules.

Definition 10. Let $((L, \leq), *, \leftarrow, \neg, \sim)$ be an enriched residuated lattice and let m be a measure of information on L. Given a general residuated logic program $\mathbb{P}$, a set of rules $X = \{\langle r_i, \vartheta_i \rangle\}_i \subseteq \mathbb{P}$ and a set $\overline{X} = \{\langle \overline{r_i}, \overline{\vartheta_i} \rangle\}_i \subseteq \overline{\mathbb{P}}$ we define the measures of instability associated with X and $\overline{X}$ by removing and adding information as:

$$\text{INST}_{\mathbb{P}}^-(X) = \inf_{\{\varphi_i\}_i} \left\{ \sum_{i \in \mathbb{I}} (m(1) - m(\varphi_i)) \mid \Theta_{\mathbb{P}}(X, \{\varphi_i\}_i) \text{ is stable} \right\}$$

and

$$\text{INST}_{\mathbb{P}}^+(\overline{X}) = \inf_{\{\varphi_i\}_i} \left\{ \sum_{i \in \mathbb{I}} m(\varphi_i) \mid \Omega_{\overline{\mathbb{P}}}(\overline{X}, \{\varphi_i\}_i) \text{ is stable} \right\},$$

respectively.

It is worth to note that, on the one hand, the consideration of a subset of rules X to be modified, instead of taking the whole program $\mathbb{P}$, has to do with the particular application one has in mind; for instance, *conflict mediation* in which each party in the negotiation provides different sets of consistent rules that cannot be changed, or some form of *belief revision* in which our program can be divided into a set of reliable consistent rules and others that are not so. On the other hand, the underlying interpretation of both measures above are based on the amount of information changed in order to recover stability. In [25], the interested reader can find a procedure to compute the stable version of a given unstable program which, in addition, can be used to directly compute the measures $\text{INST}_{\mathbb{P}}^- X)$ and $\text{INST}_{\mathbb{P}}^+$ defined above.

The first result associated with these two measures of stability is related to monotonicity.

Proposition 1 ([25]). *Let $\mathbb{P}$ be a general residuated logic program and let $X \subseteq Y$ be two sets of rules of $\mathbb{P}$ (resp. of $\overline{\mathbb{P}}$). Then*

$$\text{INST}_{\mathbb{P}}^-(X) \geq \text{INST}_{\mathbb{P}}^-(Y) \qquad (\text{resp. } \text{INST}_{\mathbb{P}}^+(X) \geq \text{INST}_{\mathbb{P}}^+(Y)).$$

Finally, the following proposition states the extreme cases of measure of instability 0.

Proposition 2 ([25]). *Let $\mathbb{P}$ be a general residuated logic program and X a set of rules of $\mathbb{P}$:*

- *If $\mathbb{P}$ is stable then $\text{INST}_{\mathbb{P}}^-(X) = \text{INST}_{\mathbb{P}}^+(X) = 0$ for all X.*

- *If $\text{INST}_{\mathbb{P}}^-(X) = 0$, then for all $\varepsilon > 0$ there exists a set $\{\varphi_i\}_i \subseteq L$ such that $\Theta_{\mathbb{P}}(X, \{\varphi_i\}_i)$ is stable and $\sum_{i \in \mathbb{I}} (m(\top) - m(\varphi_i)) < \varepsilon$.*

- *If $\text{INST}_{\mathbb{P}}^+(X) = 0$, then for all $\varepsilon > 0$ there exists a set $\{\varphi_i\}_i$ of values in L such that $\Omega_{\overline{\mathbb{P}}}(X, \{\varphi_i\}_i)$ is stable and $\sum_{i \in \mathbb{I}} m(\varphi_i) < \varepsilon$.*

5 Measures of Incoherence

As we have briefly described previously, *coherence* establishes in which cases the information assigned to opposite literals by an L-interpretation is admissible. In the first part of this section, we motivate the notion of incoherence as a convenient generalization of inconsistency; then, we show that incoherence involves an inconsistency related to an excess of information inferred by a residuated logic program; later, we introduce some measures of incoherence for L-interpretations which are, finally, used for defining measures of incoherence on residuated logic programs.

5.1 Coherence

Underlying the notion of inconsistency there are many different ideas: conflicting inference, the inference of a contradictory formula, lack of models, etc. Whereas in classical logic, most such ideas lead to the same notion, in fuzzy logic each of those lead to a different notion of inconsistency; for instance, the notion of α-inconsistency [11, 15] is based on the inference of a contradictory formula whereas the consistency over bilattice structures [13, 8] is mainly based on conflicting inference and excess of information. Roughly speaking, *"incoherence models a conflict with the inference of $\sim p$ by excess"*. Let us clarify this assertion. In fuzzy logic, a negation connective $\sim$ is identified with a negation operator $n \colon L \to L$. So, if a propositional symbol p has a truth-value θ, then the negated propositional symbol $\sim p$ has a truth value $n(\theta)$. However, although this rule is also applied in classical logic, this kind of inference is not used in logic programming for explicit negation. The reason is because explicit negation models the open-world assumption and default negation models the close-world assumption. Let us work temporarily on a crisp logic programming environment to explain better the idea generalized by coherence. Note that in an empty program the least model is the empty interpretation, and it states that neither p nor $\sim p$ is true for all $p \in \mathbb{P}$. So it is clear that explicit negation can contradict the usual inference for negation, but not every interpretation contradicting such an inference is rejected. For instance, the empty interpretation above is considered a consistent model. So the only case that is rejected is when the inference is contradicted by inferring that both p and $\sim p$ are true for some $p \in \mathbb{P}$; i.e. by excess.

It is clear that an incoherent L-interpretation contradicts the inference of $\sim p$ by the usual inference rule. Is it given by an excess of information? Certainly. In order to show this, it is convenient, firstly, to take a look to a result that relates coherence with the ordering among L-interpretations. Note that the complete lattice structure of $(L, \leq)$ induces a complete lattice structure in the set of L-interpretations by considering a point-wise ordering; i.e. let I and J be two L-interpretation, then $J \leq I$ if and only $J(\ell) \leq I(\ell)$ for all literals

$\ell \in Lit$.

Proposition 3 ([22]). *Let I and J be two L-interpretation such that $J \leq I$. If I is coherent, then J is coherent as well.*

The result above states that, given an incoherent L-interpretation I, the only possibility to obtain a coherent L-interpretation from I is by reducing the degrees of truth assigned by I to literals; in other words, by removing information. Therefore, incoherence represents a conflict by excess of information inferred by the program.

Note that, in order to recover the coherence of a program, we just have to choose a suitable explicit negation operator. Moreover, the L-interpretations assign a truth-value to $\sim p$ independently from the truth value of p and the chosen negation operator. Therefore, it makes sense to talk about the coherence of an L-interpretation according to two different negation operators. Note that, as in the case of the set of L-interpretations, the complete lattice structure of $(L, \leq)$ induces a complete lattice structure in the set of negation operators defined over $(L, \leq)$ (denoted hereafter by $\mathfrak{N}_L$); i.e. let n_1 and n_2 be two negation operators, then $n_1 \leq n_2$ if and only if $n_1(x) \leq n_2(x)$ for all $x \in L$. Additionally, we say that $n_1 < n_2$ if and only if $n_1 \leq n_2$ and there is $x \in L$ such that $n_1(x) \neq n_2(x)$.

Proposition 4 ([22]). *Let n_1 and n_2 be two negation operators such that $n_1 \leq n_2$. Then any coherent L-interpretation w.r.t. n_1 is coherent w.r.t. n_2 as well.*

As a consequence of the previous result, in some sense, we can consider the negation operator used to define coherence as a *level* up to which we admit negative information in an L-interpretation. From the ordering defined above, it is straightforward to determine the expression of the greatest and the least elements in $\mathfrak{N}_L$, which are, respectively:

$$n_\top = \begin{cases} 1 & \text{if } x \neq 1 \\ 0 & \text{if } x = 1 \end{cases} \quad \text{and} \quad n_\bot = \begin{cases} 1 & \text{if } x = 0 \\ 0 & \text{if } x \neq 0 \end{cases}$$

Besides, as a consequence of Proposition 4, both negation operators determine the weakest and the strongest restriction imposed by coherence, respectively.

Corollary 1 ([22]). *Let I be an L-interpretation*

- *If I is coherent w.r.t. $n_\bot$, then I is coherent w.r.t. any negation operator.*

- *If I is coherent w.r.t. some negation operator, then I is coherent w.r.t. $n_\top$.*

Thus, if we consider $n_\bot$ as the negation operator associated with coherence we are assuming the strongest restriction, whereas if we consider $n_\top$ we are assuming the weakest one. Note that not every L-interpretation has to be coherent even w.r.t. the least restrictive $n_\top$, since there exist L-interpretations which are non-coherent w.r.t. any negation operator, as the following example shows:

Example 3. *Consider the $[0,1]$-interpretation I, with domain just one propositional symbol p, given by $I(p) = I(\sim p) = 1$. Then, I is not coherent w.r.t. any negation operator n since $1 = I(\sim p) > n(I(p)) = n(1) = 0$.*

Note that incoherence is intrinsic to interpretations, so we can measure the incoherence of arbitrary L-interpretations and, subsequently, measure the incoherence of a given stable residuated logic program by measuring the incoherence in each stable model. This will be done in the following sections.

5.2 Measures of Incoherence on L-interpretations

The first reasonable approach to measure incoherence is to consider the *ratio of incoherent propositional symbols*.

Definition 11. *Let $I: \Pi \to L$ be an L-interpretation and let $\sim: L \to L$ be a negation operator.*

- *A propositional symbol $p \in \Pi$ is incoherent w.r.t. the negation operator $\sim$ if $I(\sim p) \not\leq \sim I(p)$.*

- *The set of incoherent propositional symbols of I w.r.t. $\sim$ is denoted by $\mathit{Incoh}(I)$.*

- *The ratio of incoherent propositional symbols is given by:*

$$\mathcal{I}\mathit{ncohRatio}(I) = \frac{|\mathit{Incoh}(I)|}{|\Pi_\mathbb{P}|}$$

The following result shows some properties of $\mathcal{I}\mathit{ncohRatio}$.

Proposition 5. *[[21]] Let I and J be two L-interpretations with finite domain (i.e., a finite set of propositional symbols) and let $\sim: L \to L$ be a negation operator. Then:*

- *if $I \leq J$ then, $\mathcal{I}\mathit{ncohRatio}(I) \leq \mathcal{I}\mathit{ncohRatio}(J)$.*

- *$\mathcal{I}\mathit{ncohRatio}(I) = 0$ if and only if I is coherent.*

- *$\mathcal{I}\mathit{ncohRatio}(I) \neq 0$ if and only if I is incoherent.*

Note that the ideas underlying the previous measure are somehow crisp. The following approaches provide measures of incoherence characteristic of a fuzzy setting. By Proposition 4 we can consider negation operators as indexes of incoherence. Hence, we can choose a set of negation operators $\mathcal{N}$ and measure the incoherence of an L-interpretation by checking the coherence w.r.t. each of them. For computational and technical reasons, we will consider a finite set $\mathcal{N} \subseteq \mathfrak{N}_L$ and require that $n_\bot \in \mathcal{N}$ and $n_\top \notin \mathcal{N}$.

Definition 12. *Let $I: \Pi \to L$ be an L-interpretation and let $\mathcal{N} \subseteq \mathfrak{N}_L$ be a finite subset of negations operators such that $n_\bot \in \mathcal{N}$ and $n_\top \notin \mathcal{N}$. The $\mathcal{N}$-index of incoherence of I is defined by*

$$\mathcal{N}\text{-}index(I) = \inf\{n \in \mathcal{N} \mid I \text{ is coherent w.r.t. } n\}$$

Note that the $\mathcal{N}$-index of inclusion is well defined because $\mathfrak{N}_L$ is a complete lattice. Moreover, note that if an L-interpretation I is not coherent w.r.t. any negation operator in $\mathcal{N}$, then

$$\mathcal{N}\text{-}index(I) = \inf\{\varnothing\} = n_\top$$

By Proposition 4 we have that the greater the $\mathcal{N}$-index of incoherence, the lesser the number of negations $n \in \mathcal{N}$ that fulfill the coherence and, hence, the greater the incoherence as well. The following result shows some convenient properties of the $\mathcal{N}$-index of incoherence in order to be considered as a representative index to represent the incoherence in an L-interpretation.

Proposition 6. *Let I and J be two L-interpretations and let $\mathcal{N} \subseteq \mathfrak{N}_L$ be a finite subset of negation operators such that $n_\bot \in \mathcal{N}$ and $n_\top \notin \mathcal{N}$.*

- *The mapping $\mathcal{N}$-index is increasing w.r.t. the order of L-interpretations.*
- *$\mathcal{N}\text{-}index(I) = n_\bot$ if and only if I is coherent w.r.t. every $n \in \mathcal{N}$.*
- *$\mathcal{N}\text{-}index(I) = n_\top$ if and only if I is incoherent w.r.t. every $n \in \mathcal{N}$.*

Proof. The result follows from properties of infimum and Corollary 1. □

Proposition 7. *Let I be an L-interpretation and let $\mathcal{N}_1 \subseteq \mathfrak{N}_L$ and $\mathcal{N}_2 \subseteq \mathfrak{N}_L$ be two finite subsets of negation operators such that $n_\bot \in \mathcal{N}_1 \cap \mathcal{N}_2$ and $n_\top \notin \mathcal{N}_1 \cup \mathcal{N}_2$ and $\mathcal{N}_1 \subseteq \mathcal{N}_2$. Then*

$$\mathcal{N}_1\text{-}index(I) \geq \mathcal{N}_2\text{-}index(I)$$

Proof. The result follows from properties of infimum and Proposition 4. □

The goal of the following example is to clarify the main idea underlying the notion of $\mathcal{N}$-index of incoherence.

Example 4. *On the set of $[0,1]$-interpretations, let us consider the set of negation indexes given by*

$$\mathcal{N} = \{n_\perp\} \cup \{n_i(x) = 1 - x^i\}_{i=1,2,3,4}$$

Note that the five negations in $\mathcal{N}$ form a chain under the ordering of $\mathfrak{N}_{[0,1]}$; i.e., $n_\perp < n_1 < n_2 < n_3 < n_4$. As a result, the $\mathcal{N}$-index of incoherence of a $[0,1]$-interpretation represents a grade of incoherence with respect to such a chain. For instance, let us consider the $[0,1]$-interpretation I on the set of propositional symbols $\Pi = \{p,q,r\}$ given by:

$$I(p) = 0.5 \qquad I(q) = 0.9 \qquad I(r) = 0.2$$
$$I(\sim p) = 0.6 \qquad I(\sim q) = 0.2 \qquad I(\sim r) = 0.5$$

It is easy to check that I is not coherent w.r.t. $n_\perp(x), n_1(x)$ or $n_2(x)$, but it is w.r.t. $n_3(x)$ and $n_4(x)$. Since $n_3(x) \leq n_4(x)$, then the $\mathcal{N}$-index of incoherence of I is n_3.

On the other hand, the $[0,1]$-interpretation J given by:

$$J(p) = 0.3 \qquad J(q) = 0.8 \qquad J(r) = 0.2$$
$$J(\sim p) = 0.6 \qquad J(\sim q) = 0.1 \qquad J(\sim r) = 0.5$$

is not coherent w.r.t. $n_\perp(x)$ but it is w.r.t. the other four. As a result, the $\mathcal{N}$-index of incoherence of J is n_1. Since $n_1 < n_3$, we can assert that the $\mathcal{N}$-index of incoherence of I is greater than the $\mathcal{N}$-index of incoherence of J. Finally, consider the $[0,1]$-interpretation G given by:

$$G(p) = 1 \qquad G(q) = 0.8 \qquad G(r) = 0.2$$
$$G(\sim p) = 1 \qquad G(\sim q) = 0.9 \qquad G(\sim r) = 0.1$$

which is not coherent with respect to any negation in $\mathcal{N}$. Then, the $\mathcal{N}$-index of coherence of G is $n_\top$ since $n_\top$ is the greatest element of $\mathfrak{N}_{[0,1]}$. Moreover, since $n_1 < n_3 < n_\top$, then G is more incoherent than I and J.

In the case of working in the unit interval, we can define the following natural measure of incoherence from the $\mathcal{N}$-index of incoherence.

Definition 13. *Let $I: \Pi \to [0,1]$ be an $[0,1]$-interpretation and let $\mathcal{N} \subseteq \mathfrak{N}_{[0,1]}$ be a finite subset of negation operators such that $n_\perp \in \mathcal{N}$ and $n_\top \notin \mathcal{N}$. The $\mathcal{N}$-measure of incoherence of I is defined by*

$$\mathcal{N}\text{-}\mathcal{I}ncoh(I) = \int_0^1 \mathcal{N}\text{-}index(I)(x)\,dx$$

Some properties of the previous measure:

Proposition 8. *Let I and J be two $[0,1]$-interpretations and let $\mathcal{N} \subseteq \mathfrak{N}_{[0,1]}$ be a finite subset of negation operators such that $n_\perp \in \mathcal{N}$. Then:*

- *If $I \leq J$ then, $\mathcal{N}\text{-}\mathcal{I}ncoh(I) \leq \mathcal{N}\text{-}\mathcal{I}ncoh(J)$.*
- *$\mathcal{N}\text{-}\mathcal{I}ncoh(I) = 0$ if and only if I is coherent w.r.t. every $n \in \mathcal{N}$.*
- *$\mathcal{N}\text{-}\mathcal{I}ncoh(I) = 1$ if and only if I is incoherent w.r.t. every $n \in \mathcal{N}$.*

Proof. The result follows from properties of integrals and Proposition 6. □

Example 5. *Let us continue with Example 4. The $\mathcal{N}$-measures of incoherence of I, J and G are :*

$$\mathcal{N}\text{-}\mathcal{I}ncoh(I) = \int_0^1 n_3(x)\,dx = \int_0^1 (1-x^3)\,dx = 0.75$$

$$\mathcal{N}\text{-}\mathcal{I}ncoh(J) = \int_0^1 n_1(x)\,dx = \int_0^1 (1-x)\,dx = 0.5$$

$$\mathcal{N}\text{-}\mathcal{I}ncoh(G) = \int_0^1 n_\top(x)\,dx = 1$$

5.3 Measures of Incoherence on Stable General Residuated Logic Programs

We have already mentioned that the inconsistence of general residuated logic programs can be split into two hierarchical features: instability and incoherence. On the one hand, if a general residuated logic program is stable, the only reason which can cause inconsistency is incoherence. On the other hand, if a general residuated logic program is not stable, coherence plays no role. Therefore, in this section we assume that $\mathbb{P}$ is a stable general residuated logic program and then, we make use of the set of stable models of $\mathbb{P}$ to extend the measures of coherence presented in the previous section for L-interpretations.

Definition 14. *Let $\mathbb{P}$ be a stable general residuated logic program. We define the measure of incoherence of $\mathbb{P}$ based on the ratio of incoherent propositional symbols by:*

$$\mathcal{I}ncohRatio(\mathbb{P}) = \inf\{\mathcal{I}ncohRatio(I) \mid I \text{ is a stable model of } \mathbb{P}\}$$

Definition 15. *Let $\mathbb{P}$ be a stable general residuated logic program and let $\mathcal{N} \subseteq \mathfrak{N}_L$ be a finite subset of negation operators such that $n_\perp \in \mathcal{N}$ and $n_\top \notin \mathcal{N}$. The $\mathcal{N}$-index of incoherence of $\mathbb{P}$ is defined by*

$$\mathcal{N}\text{-}index(\mathbb{P}) = \inf\{\mathcal{N}\text{-}index(I) \mid I \text{ is a stable model of } \mathbb{P}\}$$

Moreover, if $L = [0,1]$ then, the $\mathcal{N}$-measure of incoherence of $\mathbb{P}$ is

$$\mathcal{N}\text{-}\mathcal{I}ncoh(\mathbb{P}) = \inf\{\mathcal{N}\text{-}\mathcal{I}ncoh(I) \mid I \text{ is a stable model of } \mathbb{P}\}$$

Because of the non-monotonic character of general residuated logic programs, the previous measures are not monotonic. The following example shows this feature.

Example 6. *Let us consider the general residuated logic program $\mathbb{P}$ on the residuated lattice $(([0,1), \leq), *_L, \leftarrow_L, \neg, \sim)$ with $\neg(x) = 1 - x$ and $\sim(x) = n_\perp$ given by*

$$\langle p \leftarrow \neg p * \sim q; \quad 1 \rangle$$
$$\langle \sim p \leftarrow \neg \sim p; \quad 1 \rangle$$
$$\langle \sim q \leftarrow \neg q; \quad 1 \rangle$$

The only stable model of $\mathbb{P}$ is $M = \{(p, 0.5), (\sim p, 0.5), (q, 0), (\sim q, 1)\}$, and is an incoherent model with measures $\mathcal{I}ncoh\mathcal{R}atio(\mathbb{P}) = 0.5$ and $\mathcal{N}\text{-}\mathcal{I}ncoh(\mathbb{P}) = 0.5$. However, if we add the following fact to the previous general residuated logic program:

$$\langle q \leftarrow; \quad 1 \rangle,$$

then the only stable model turns out to be

$$M = \{(p, 0), (\sim p, 0.5), (q, 1), (\sim q, 0)\}$$

which is coherent and, hence, M is a fuzzy answer set and $\mathcal{I}ncoh\mathcal{R}atio(\mathbb{P}) = \mathcal{N}\text{-}\mathcal{I}ncoh(\mathbb{P}) = 0$. In other words, we have shown that

$$\mathcal{I}ncoh\mathcal{R}atio(\mathbb{P} \cup \{\langle q \leftarrow; \quad 1 \rangle\}) < \mathcal{I}ncoh\mathcal{R}atio(\mathbb{P})$$

and

$$\mathcal{N}\text{-}\mathcal{I}ncoh(\mathbb{P} \cup \{\langle q \leftarrow; \quad 1 \rangle\}) < \mathcal{N}\text{-}\mathcal{I}ncoh(\mathbb{P})$$

Those measures turn out to be monotonic when we restrict them to extended residuated logic programs, as the following results shows.

Theorem 3. *Let $\mathbb{P}$ and $\mathbb{Q}$ be two finite stable extended residuated logic programs defined on an enriched residuated lattice $(([0,1], \leq), *, \leftarrow, \neg, \sim)$ such that $\mathbb{P} \subseteq \mathbb{Q}$, and let $\mathcal{N} \subseteq \mathfrak{N}_{[0,1]}$ be a finite subset of negations operators such that $n_\perp \in \mathcal{N}$. Then:*

- $\mathcal{N}\text{-}\mathcal{I}ncoh(\mathbb{P}) \leq \mathcal{N}\text{-}\mathcal{I}ncoh(\mathbb{Q})$.

- $\mathcal{I}ncoh\mathcal{R}atio(\mathbb{P}) \leq \mathcal{I}ncoh\mathcal{R}atio(\mathbb{Q})$.

Proof. It is not difficult to prove that the least fix point of the immediate consequence operator (Definition 6) is monotonic also with the ordering between extended logic programs. As a result, the least model of $\mathbb{P}$ is less than $\mathbb{Q}$. Then, both inequalities of the theorem follows as consequences of Proposition 8 and Proposition 5, respectively. □

The extreme cases, i.e., a measure of incoherence 0 and 1, are characterized by the following theorem. Note that in this case we do not need to restrict the theorem to extended residuated logic programs.

Theorem 4. *Let $\mathbb{P}$ and $\mathbb{Q}$ be two finite stable general residuated logic programs defined on a residuated lattice with negations $(([0,1], \leq), *, \leftarrow, \neg, \sim)$ and let $\mathcal{N} \subseteq \mathfrak{N}_{[0,1]}$ be a finite subset of negation operators such that $n_\perp \in \mathcal{N}$ and $\sim \, \in \mathcal{N}$. Then:*

- *$\mathbb{P}$ is consistent if and only if $\mathcal{N}$-index$(\mathbb{P}) \leq \sim$.*
- *If $\mathcal{N}$-$\mathcal{I}ncoh(\mathbb{P}) = 0$, then $\mathbb{P}$ is consistent.*
- *If $\mathbb{P}$ is consistent then, $\mathcal{N}$-$\mathcal{I}ncoh(\mathbb{P}) \leq \int_0^1 \sim(x)\, dx$*
- *If $\sim \, \neq n_\top$ then, $\mathcal{N}$-$\mathcal{I}ncoh(\mathbb{P}) = 1$ if and only if $\mathbb{P}$ is inconsistent.*
- *$\mathbb{P}$ is consistent if and only if $\mathcal{I}ncohRatio(\mathbb{P}) = 0$.*
- *$\mathcal{I}ncohRatio(\mathbb{P}) \neq 1$ if and only if $\mathbb{P}$ is inconsistent.*

Proof. The result follows from the definitions of *consistency*, $\mathcal{I}ncohRatio$ and $\mathcal{N}$-$\mathcal{I}ncoh(\mathbb{P})$, Propositions 4, 5 and 8, and Corollary 1. □

6 Conclusions and Future Work

In this chapter we have dealt with inconsistencies in General Residuated Logic Programming under the fuzzy answer set semantics. Such paradigm is a generalization of the classical answer set semantics of logic programming studied in [34, this volume].

We focus on the two possible sources of inconsistency: on the one hand, on the lack of stable models (termed instability) and, on the other hand, on the contradictory nature of every stable model (termed incoherence), and provide specific measures for each one. Instability is measured in terms of the minimal amount of changes the program needs in order to recover stability; these changes can be classified as incremental or decremental and, as a result, we define two measures related to each of these types. In turn, incoherence is inherently related to models and, hence, we measure the incoherence of each model of the program and, finally, combine them in order to provide a measure of incoherence of the whole program.

This approach opens up an interesting starting point for further research. Firstly, different postulates to measure inconsistency developed in the context of classical logic should be studied in the framework of fuzzy and non-monotonic logics. Secondly, from an application point of view, we expect to develop belief systems founded on these measures of inconsistency using the proviso of "the more contradictory, the less reliable" in order to choose which subset of information is more prone to be updated.

Acknowledgements This work has been partially funded by the Spanish Ministry of Science project TIN2015-70266-C2-1-P, co-funded by the European Regional Development Fund (ERDF).

References

[1] J.J. Alferes, L.M. Pereira, and T.C. Przymusinski. Strong and Explicit Negation in Non-monotonic Reasoning and Logic Programming. *Lecture Notes in Artificial Intelligence* 1126:143–163, 1996.

[2] M. Balduccini. CR-prolog as a Specification Language for Constraint Satisfaction Problems. *Lecture Notes in Artificial Intelligence* 5753:402–408, 2009.

[3] N.D. Belnap. A Useful Four-valued Logic. In G. Epstein and J. M. Dunn, editors, *Modern Uses of Multiple-Valued Logic*, pages 7–37. Reidel Publishing Company, Boston, 1977.

[4] M.E. Coniglio, F. Esteva, and L. Godo. Logics of Formal Inconsistency Arising from Systems of Fuzzy Logic. *Logic Journal of the IGPL* 22:880–904, 2014.

[5] D. Dubois, S. Konieczny, and H. Prade. Quasi-possibilistic Logic and its Measures of Information and Conflict. *Fundamenta Informaticae*, 57(2-4):101–125, 2003.

[6] T. Eiter, G. Ianni, T. Lukasiewicz, R. Schindlauer, and H. Tompits. Combining Answer Set Programming with Description Logics for the Semantic Web. *Artificial Intelligence* 172:1495–1539, 2008.

[7] A. Elçi, B. Rahnama, and S. Kamran. Defining a Strategy to Select either of Closed/Open World Assumptions on Semantic Robots. In *Proc of the IEEE Conf. on Computer Software and Applications (COMPSAC)*, pages 417–423, 2008.

[8] M. Fitting. Bilattices and the Semantics of Logic Programming. *Journal of Logic Programming*, 11:91–116, 1991.

[9] M. Gelfond and V. Lifschitz. The Stable Model Semantics for Logic Programming. In *Proc of the Intl Conf on Logic Programming (ICLP)*, pages 1070–1080, 1988.

[10] M. Gelfond and V. Lifschitz. Classical Negation in Logic Programs and Disjunctive Databases. *New Generation Computing*, 9:365–385, 1991.

[11] P. Hájek. *Metamathematics of Fuzzy Logic*. Trends in Logic. Kluwer Academic, 1998.

[12] S. Heymans, D. Van Nieuwenborgh, and D. Vermeir. Open Answer Set Programming for the Semantic Web. *Journal of Applied Logic*, 5(1):144–169, 2007.

[13] M. Ginsberg. Multivalued Logics: A Uniform Approach to Inference in Artificial Intelligence. *Computational Intelligence*, 4:265–316, 1988.

[14] J. Grant and A. Hunter. Measuring Inconsistency in Knowledgebases. *J. Intelligent Information Systems*, 27(2):159–184, 2006.

[15] J. Janssen, S. Schockaert, D. Vermeir, and M. De Cock. General Fuzzy Answer Set Programs. *Lecture Notes in Computer Science*, 5571:352–359, 2009.

[16] J. Lloyd. *Foundations of Logic Programming*. Springer Verlag, 1987.

[17] Y. Loyer and U. Straccia. Epistemic Foundation of Stable Model Semantics. *Journal of Theory and Practice of Logic Programming*, 6:355–393, 2006.

[18] T. Lukasiewicz. Fuzzy Description Logic Programs under the Answer Set Aemantics for the Semantic Web. *Fundamenta Informaticae*, 82(3):289–310, 2008.

[19] T. Lukasiewicz and U. Straccia. Tightly Integrated Fuzzy Description Logic Programs under the Answer Set Semantics for the Semantic Web. *Lectures Notes in Computer Science* 4524:289–298, 2007.

[20] N. Madrid and M. Ojeda-Aciego. Towards a Fuzzy Answer Set Semantics for Residuated Logic Programs. In *Proc of WI-IAT'08. Workshop on Fuzzy Logic in the Web*, pages 260–264, 2008.

[21] N. Madrid and M. Ojeda-Aciego. On the Measure of Incoherence in Extended Residuated Logic Programs. In *IEEE Intl Conf on Fuzzy Systems (FUZZ-IEEE'09)*, pages 598–603, 2009.

[22] N. Madrid and M. Ojeda-Aciego. On Coherence and Consistence in Fuzzy Answer Set Semantics for Residuated Logic Programs. *Lecture Notes in Computer Science* 5571:60–67, 2009.

[23] N. Madrid and M. Ojeda-Aciego. Measuring Instability in Normal Residuated Logic Programs: Adding Information. In *IEEE Intl Conf on Fuzzy Systems (FUZZ-IEEE'10)*, pages 2244–2250, 2010.

[24] N. Madrid and M. Ojeda-Aciego. Measuring Instability in Normal Residuated Logic Programs: Discarding Information. *Communications in Computer and Information Science*, 80:128–137, 2010.

[25] N. Madrid and M. Ojeda-Aciego. Measuring Inconsistency in Fuzzy Answer Set Semantics. *IEEE Transactions on Fuzzy Systems*, 19(4):605–622, 2011.

[26] N. Madrid and M. Ojeda-Aciego. On the Existence and Unicity of Stable Models in Normal Residuated Logic Programs. *Int. J. Computer Mathematics* 89(3): 310–324, 2012.

[27] D. Pearce. Reasoning with negative information, II: Hard negation, strong negation and logic programs. *Lectures Notes in Computer Science* 619: 63–79, 1992.

[28] D. Picado-Muiño. Measuring and repairing inconsistency in knowledge bases with graded truth. *Fuzzy Sets and Systems* 197:108–122, 2012.

[29] G. Priest. *Paraconsistent Logic*, volume 6 of *Handbook of philosophical logic*, pages 287–393. Springer, 2002.

[30] R. Rosati. On the Finite Controllability of Conjunctive Query Answering in Databases under Open-world Assumption. *Journal of Computer and System Sciences* 77(3):572–594, 2011.

[31] T.C. Son and C. Sakama. Negotiation using Logic Programming with Consistency Restoring Rules. In *Proc of the Intl Joint Conf on Artificial Intelligence (IJCAI)*, pg 930–935, 2009.

[32] U. Straccia and Y. Loyer. Any-world Assumptions in Logic Programming. *Theorical Computer Science*, 342:351–381, 2005.

[33] M. Thimm. Measuring Inconsistency with Many-valued Logics. *Int. J. Approximate Reasoning* 86:1–23, 2017.

[34] M. Ulbricht, M. Thimm, and G. Brewka. Inconsistency Measures for Disjunctive Logic Programs. This volume, 2017.

[35] G. Wagner. Web Rules Need Two Kinds of Negation. *Lecture Notes in Computer Science*, 2901:33–50, 2003.

[36] P. Wong and Ph. Besnard. Paraconsistent Reasoning as an Analytic Tool. *Logic Journal of the IGPL* 9:217–229, 2001.

Inconsistency Measures in Hybrid Logics

Diana Costa, Manuel A. Martins

CIDMA - Center for R&D in Mathematics and Applications
Dept. Mathematics, University of Aveiro, Portugal
{dianafcosta,martins}@ua.pt

Abstract

We start by discussing the way paraconsistency has been coupled with other logical systems; we address particularly Quasi-classical logic, modal systems where paraconsistency has been introduced, and dynamic logic with belnapian truth values.

Then, following Grant and Hunter's approach to Quasi-classical Logic, we introduce a paraconsistent version of hybrid multimodal logic at the level of propositional variables and discuss two measures of inconsistency. We examine the role of quantifiers and we consider the idea of allowing inconsistent information in other aspects of hybrid logic, namely in nominals and in the accessibility relations.

1 Introduction

The digital era in which we live has offered us a large, and getting bigger by the minute, amount of information. However, lots of data does not always mean perfect data; in fact, our everyday lives are constantly prone to contradictions, as different sources of information are not always in tune. Being able to reason in these situations is crucial, and that is the main cause why paraconsistency has drawn so much attention in the last years.

A paraconsistent logic allows inconsistencies without trivialization, which means that we can still reach sensible conclusions even though something is, somehow and somewhere, wrong.

Several variants of paraconsistent logic have been proposed, often to meet different aims or target specific applications. Research has been driven not only by theoretical interest, but also by genuine problems in different scientific domains, namely Computer Science, Medicine and Robotics. Thus researchers have not only developed completely new logics, as Da Costa did with his C-systems [9], but also adapted several pre-existing classical logics by attaching them with a paraconsistent reasoning, as is the case of Quasi-classical logic [2],

Four-valued modal logic [23] and Quasi-hybrid logic [8], amongst many others. These can be viewed as the combination of logical systems, which has been an important research topic in Logic. Notably, the combination of features of a logical system "on top" of another one has been considered in several contexts, and abstract notions of combination, as well as concrete combinations, have been proposed. Such is the case of Fibring [13], Temporalization [12], Hybridization [22] and Dynamization [20]. Moving towards paraconsistency, Souza, Costa-Leite and Dias even proposed a systematic procedure to construct a paraconsistent version of a given logic in [11]. Their method is developed in a framework of category theory and the process is obtained via an endofunctor on the category CON of consequence structures, which they called paraconsistentization functor. The special case of paraconsistentization of classical logic is discussed in that paper.

1.1 An Overview of Paraconsistency in Logical Systems

So far, a large number of formal techniques to invalidate the *Principle of Explosion* – which states that from contradictory premises any formula can be derived – have been developed. The main reason to reject this principle is that we should be able to isolate inconsistencies and move on with our reasoning. Special examples of combining paraconsistency with modal logics and some of its extensions will be discussed in the next sections, although not thoroughly.

Quasi-Classical Logic is a gateway to deal with inconsistencies in classical logic; it was introduced by Besnard and Hunter in [2] and afterwards Grant and Hunter used it to find the four-valued models of a knowledgebase [14]. The idea explored in the latter is to decouple the link between a proposition and its negation. For that, two valuations are required: one for dealing with the positive propositions and another one for dealing with the negative propositions. The satisfaction is defined using these two valuations, plus a feature called focus, which asserts the validity of the disjunctive syllogism in Quasi-Classical Logic. Since it is possible to use the method of Robinson diagrams, models are defined as sets of literals. The measure of inconsistency is introduced as the ratio between the number of inconsistencies in the model, and the total number of inconsistencies that the model could have. Measuring inconsistency is the key to an effective management of information, namely for comparing between different models *w.r.t.* contradictions, allowing us to choose the ones with less conflicts.

In [21], João Marcos discusses the connection between paraconsistency and modal logics. He shows that Jaśkowski's **D2** (a paraconsistent extension of the positive fragment of classical logic) is not a modal logic as it fails the replacement law; and, surprisingly enough, he states that "not only is it possible to start from a (non-degenerate) normal modal logic and define operators that represent a paraconsistent negation and a consistency connective, it is also possible to do it *the other way around"*.

More recently, Rivieccio et al. in [23] studied a family of modal expansions of the Belnap–Dunn four-valued logic where Kripke models are such that the accessible relation and the valuation are four-valued functions. They also introduced Hilbert-style calculi for the least modal logic over the four-element Belnap lattice.

In [5], Troben Braüner presented axiomatic systems for a hybridized version of the constructive and paraconsistent logic **N4**. We should point out that this is a different approach from ours, and that he did not discuss the notion of measure of inconsistency neither explored the syntactic representation of models as we did in [8].

Also quite novel is the work of Sedlár who introduced BPDL, a logic which combines dynamic logic with the basic four-valued modal logic BK (see [24]). Some technical results about BPDL were presented, namely a decidability proof and a sound and weakly complete axiomatisation for it.

2 Paraconsistency in Hybrid Logic

In this section we study paraconsistency in Hybrid Multimodal logic in an approach inspired by the work of Grant and Hunter [14, 15]. In Quasi-hybrid multimodal logic, and analogously to the assumption in [14], where it is assumed that all formulas are in *Prenex Conjunctive Normal Form*, we will assume that all formulas are in *Negation Normal Form*, since it does not lead to loss of generality.

We define the concepts of hybrid multimodal bistructure, which we will refer to simply as *bistructure*, decoupled and strong satisfaction and model for a set of formulas. We use the paraconsistent diagram of a bistructure in order to represent a bistructure by the set of quasi-hybrid literals that are true in it. Afterwards, we consider minimal QH models and we present some examples with illustrations.

We propose a measure of inconsistency so that we can make comparisons between models and check which one is more *reliable*. Then we consider that propositional variables do not all have the same relevance and introduce a weighted measure of inconsistency. The idea of having different *kinds* of propositional variables is depicted in [1], where Arruda's paraconsistent logic V1 assumes that there are propositions that behave classically and others that behave paraconsistently. We conclude this section with an applied example in the field of health care.

Let us make a quick remark to say that we will merely introduce the semantics for this paraconsistent logic. However, going one step further one could also explore some of its proof-theoretical aspects. We were able to develop a complete tableau system for Quasi-hybrid logic, by combining the tableau system for Quasi-classical and Hybrid logics, in [7].

2.1 Quasi-Hybrid Multimodal Logic

Hybrid logics constitute a simple formalism for working with relational structures (or multigraphs) with the ability to name worlds (or states), to assert equalities and describe accessibility relations between them.

This is obtained by adding to modal logic a new class of atomic formulas, called *nominals*, and using a new operator, @, called the *satisfaction operator*. A nominal is true at exactly one state: the one it names. The satisfaction operator allows us to move to the state it indicates and evaluate formulas there.

Moreover, hybrid logics are strictly more expressive than their modal fragment. And although stronger than modal logic, the basic hybrid logic does not increase the complexity of the problem of determining whether a formula is valid or not. Actually, it remains a decidable system.

This section will extend the work in [8] to a multimodal setting.

2.1.1 Formal Representation

The basic hybrid language introduces nominals and the satisfaction operator into the modal logic. It might seem a simple extension, but it carries great power in terms of expressivity.

Without adding complexity, we can also introduce modalities which can be viewed as modes of truth. Amongst these are temporal modalities, deontic modalities, epistemic modalities, etc. Thus we get that the syntactic structure of hybrid multimodal logic is given by:

Definition 2.1. *Let* $L_\pi = \langle \text{Prop}, \text{Nom}, \text{Mod} \rangle$ *be a hybrid multimodal similarity type where* Prop *is a set of propositional symbols,* Nom *is a set disjoint from* Prop *and* Mod *is a set of modality labels. We use* p, q, r, *etc. to refer to the elements in* Prop. *The elements in* Nom *are called nominals and we typically write them as* i, j, k, *etc. Modalities are usually represented by* π, π', *etc. The well-formed formulas over* L_π, $\text{Form}_@(L_\pi)$, *are defined by the following grammar:*

$$WFF := i \mid p \mid \bot \mid \top \mid \neg \varphi \mid \varphi \vee \psi \mid \varphi \wedge \psi \mid \langle \pi \rangle \varphi \mid [\pi] \varphi \mid @_i \varphi$$
where $i \in \text{Nom}, p \in \text{Prop}, \pi \in \text{Mod}$.

For any nominal i, any formula φ, $@_i \varphi$ is called a *satisfaction statement*.

In order to generalize the approach in [14] to the hybrid multimodal case, we have to consider formulas in *negation normal form* (*i.e.*, formulas in which the negation symbol occurs immediately before propositional symbols and/or nominals).

We define the notion of negation normal form for hybrid multimodal logic and we establish an analogous result to the one in [4] for classical logic that states that any modal formula is logically equivalent to one in the negation normal form; we would also like to point out that the same result was presented without proof in [10] for the modal case. Therefore we can restrict our attention to formulas in negation normal form.

Definition 2.2. *Let $L_\pi = \langle \text{Prop}, \text{Nom}, \text{Mod} \rangle$ be a hybrid multimodal similarity type. The* negation normal form *of a formula, for short* NNF, *is defined just as in classical logic: a formula is said to be in* NNF *if negation only appears directly before propositional variables and/or nominals. The set of NNF formulas over L_π, $\text{Form}_{\text{NNF}(@)}(L_\pi)$, is recursively defined as follows:*

For $p \in \text{Prop}$, $i \in \text{Nom}$, $\pi \in \text{Mod}$,

1. $\bot, \top$ are in NNF;

2. p, i, $\neg p$, $\neg i$ are in NNF;

3. If φ, ψ are formulas in NNF, then $\varphi \vee \psi$, $\varphi \wedge \psi$ are in NNF;

4. If φ is in NNF, then $[\pi]\varphi$, $\langle\pi\rangle\varphi$ are in NNF;

5. If φ is in NNF, then $@_i\varphi$ is in NNF.

The next proposition states that we can consider only formulas in negation normal form.

Proposition 2.3. *Every formula $\varphi \in \text{Form}_@(L_\pi)$ is logically equivalent to a formula $\varphi^* \in \text{Form}_{\text{NNF}(@)}(L_\pi)$.*

Proof. The proof can be found in [8] for the case with a single modality. □

It follows from the proof that a recursive procedure that puts formulas in negation normal form can be formulated. Formally, $nnf : \text{Form}_@(L_\pi) \to \text{Form}_{\text{NNF}(@)}(L_\pi)$ is defined as follows:

1. $nnf(l) \stackrel{def}{=} l$, if l is of the form $@_i p$, $@_i \neg p$, $@_i j$, $@_i \neg j, @_i \langle \pi \rangle j, @_i [\pi] \neg j$ for $i, j \in \text{Nom}, p \in \text{Prop}, \pi \in \text{Mod}$

2. $nnf(\psi_1 \vee \psi_2) \stackrel{def}{=} nnf(\psi_1) \vee nnf(\psi_2);$

3. $nnf(\psi_1 \wedge \psi_2) \stackrel{def}{=} nnf(\psi_1) \wedge nnf(\psi_2);$

4. $nnf(\neg(\psi_1 \vee \psi_2)) \stackrel{def}{=} nnf(\neg\psi_1) \wedge nnf(\neg\psi_2);$

5. $nnf(\neg(\psi_1 \wedge \psi_2)) \stackrel{def}{=} nnf(\neg\psi_1) \vee nnf(\neg\psi_2)$;

6. $nnf([\pi]\psi) \stackrel{def}{=} [\pi]nnf(\psi)$;

7. $nnf(\neg[\pi]\psi) \stackrel{def}{=} \langle\pi\rangle nnf(\neg\psi)$;

8. $nnf(\langle\pi\rangle\psi) \stackrel{def}{=} \langle\pi\rangle nnf(\psi)$;

9. $nnf(\neg\langle\pi\rangle\psi) \stackrel{def}{=} [\pi]nnf(\neg\psi)$;

10. $nnf(\neg\neg\psi) \stackrel{def}{=} nnf(\psi)$;

11. $nnf(@_i\psi) \stackrel{def}{=} @_i nnf(\psi)$;

12. $nnf(\neg @_i\psi) \stackrel{def}{=} @_i nnf(\neg\psi)$.

Thus, without loss of generality, we will assume that all formulas are in negation normal form, *i.e.*, given a hybrid multimodal similarity type $L_\pi = \langle \text{Prop}, \text{Nom}, \text{Mod} \rangle$, the set of formulas is $\text{Form}_{\text{NNF}(@)}(L_\pi)$.

Definition 2.4. *Let θ be a formula in* NNF. *We define the complementation operation $\sim$ from $\sim \theta := nnf(\neg\theta)$.*

The $\sim$ operator is not part of the object hybrid similarity type but it makes some definitions clearer.

A *hybrid structure* for a hybrid multimodal similarity type L_π is a tuple $(W, (R_\pi)_{\pi \in \text{Mod}}, N, V)$, were W is a non-empty set called the *domain* whose elements are called *states* or *worlds*; each R_π is a binary relation such that $R_\pi \subseteq W \times W$ and is called the π-*accessibility relation*; $N : \text{Nom} \to W$ is a function called *hybrid nomination* that assigns nominals to elements in W such that for any nominal i, $N(i)$ is the element of W named by i and V is a *hybrid valuation*, which means that V is a function with domain Prop and range $Pow(W)$ such that $V(p)$ tells us at which states (if any) each propositional symbol is true. However, in order to accommodate contradictions in a model, we will use two valuations for propositions: V^+ and V^-.

Definition 2.5. *A hybrid multimodal bistructure is a tuple $(W, (R_\pi)_{\pi \in \text{Mod}}, N, V^+, V^-)$, where $(W, (R_\pi)_{\pi \in \text{Mod}}, N, V^+)$ and $(W, (R_\pi)_{\pi \in \text{Mod}}, N, V^-)$ are hybrid multimodal structures.*

The map V^+ is interpreted as the acceptance of a propositional symbol, and V^- as the rejection. This is formalized in the definition of decoupled satisfaction.

Definition 2.6. *For a bistructure* $E = (W, (R_\pi)_{\pi \in \text{Mod}}, N, V^+, V^-)$ *we define a satisfiability relation* $\models_d$ *called* decoupled satisfaction *at* $w \in W$ *for propositional symbols and nominals as follows:*

1. $E, w \models_d p$ *iff* $w \in V^+(p)$;
2. $E, w \models_d i$ *iff* $w = N(i)$;
3. $E, w \models_d \neg p$ *iff* $w \in V^-(p)$;
4. $E, w \models_d \neg i$ *iff* $w \neq N(i)$.

Classically, if a propositional symbol is satisfied in a structure at a certain world, its negation cannot be satisfied in that structure at the same world. In this setting, we break this link as we allow both a propositional symbol and its negation to be simultaneously satisfiable.

This decoupling gives us the basis for a semantics for paraconsistent reasoning.

Definition 2.7. *We define* strong satisfaction, $\models_s$, *in a bistructure* E *at a world* $w \in W$ *as follows:*

1. $E, w \models_s \top$ *always;*
2. $E, w \models_s \bot$ *never;*
3. $E, w \models_s p$ *iff* $E, w \models_d p$;
4. $E, w \models_s \neg p$ *iff* $E, w \models_d \neg p$;
5. $E, w \models_s i$ *iff* $E, w \models_d i$;
6. $E, w \models_s \neg i$ *iff* $E, w \models_d \neg i$;
7. $E, w \models_s \theta_1 \vee \theta_2$ *iff* $[E, w \models_s \theta_1$ *or* $E, w \models_s \theta_2]$ *and* $[E, w \models_s \sim \theta_1$ *implies* $E, w \models_s \theta_2]$ *and* $[E, w \models_s \sim \theta_2$ *implies* $E, w \models_s \theta_1]$;
8. $E, w \models_s \theta_1 \wedge \theta_2$ *iff* $E, w \models_s \theta_1$ *and* $E, w \models_s \theta_2$;
9. $E, w \models_s \langle \pi \rangle \theta$ *iff* $\exists w'(w R_\pi w'$ *and* $E, w' \models_s \theta)$;
10. $E, w \models_s [\pi] \theta$ *iff* $\forall w'(w R_\pi w'$ *implies* $E, w' \models_s \theta)$;
11. $E, w \models_s @_i \theta$ *iff* $E, w' \models_s \theta$ *where* $w' = N(i)$.

One effortlessly sees that disjunction is stronger than usual, and that is because we keep the notion of disjunctive syllogism. It follows that some principles must be abandoned, for example disjunction introduction fails.

We define *strong satisfaction in a bistructure* as follows:

$$E \models_s \theta \text{ iff for all } w \in W, \ E, w \models_s \theta.$$

2.1.2 Quasi-Hybrid Multimodal Models

Let Δ be a set of formulas in $\text{Form}_@(L_\pi)$. We say that a bistructure E is a *quasi-hybrid multimodal model* of Δ iff for all $\theta \in \Delta$, $E \models_s \theta$.

To make the text easier to follow, we will assume that N maps nominals to themselves; hence W will always contain all the nominals in L_π. This also means that all nominals are mapped to distinct elements, *i.e.*, N is an inclusion map. Hence, for a given hybrid similarity type $L_\pi = \langle \text{Prop}, \text{Nom}, \text{Mod} \rangle$ and a domain W of a bistructure we must have $\text{Nom} \subseteq W$.

It is known that hybrid logic can specify Robinson diagrams [3]. Recall that a Robinson diagram of a structure in hybrid logic is the set of all (closed) literals over the extended hybrid similarity type that contains one nominal for each world, which are true in the structure (see [17] for diagrams in first-order logic). Following the assumption that N is injective, in order to define diagrams for this paraconsistent case we do not need the hybrid literals regarding equality between nominals, *i.e.*, $@_i j$, $@_i \neg j$. Therefore, in this context, we define the notion of atom and literal as follows:

Definition 2.8. *For a hybrid multimodal similarity type $L_\pi = \langle \text{Prop}, \text{Nom}, \text{Mod} \rangle$,*

1. *QH atoms over L_π:*
   $\text{QHAt}(L_\pi) = \{@_i p,\ @_i \langle \pi \rangle j \mid i, j \in \text{Nom}, p \in \text{Prop}, \pi \in \text{Mod}\}$;

2. *QH literals over L_π:*
   $\text{QHLit}(L_\pi) = \{@_i p,\ @_i \neg p,\ @_i \langle \pi \rangle j,\ @_i [\pi] \neg j \mid i, j \in \text{Nom}, p \in \text{Prop}, \pi \in \text{Mod}\}$;

In order to build the paraconsistent diagram, we add new nominals for the elements of W which are not named yet, and we denote this expanded similarity type by $L_\pi(W)$, *i.e.*, $L_\pi(W) = \langle \text{Prop}, W, \text{Mod} \rangle$ (recall that $\text{Nom} \subseteq W$). As in the standard case, $E(W)$ denotes the natural expansion of the bistructure E to the hybrid multimodal similarity type $L_\pi(W)$, by taking N to be the identity for all the nominals in $L_\pi(W)$. Moreover, we will assume that Prop, Nom and Mod are finite sets for any multimodal hybrid similarity type $L_\pi = \langle \text{Prop}, \text{Nom}, \text{Mod} \rangle$, as is the domain W of any bistructure.

Definition 2.9. *Let $L_\pi = \langle \text{Prop}, \text{Nom}, \text{Mod} \rangle$ be a hybrid similarity type, and $E = (W, (R_\pi)_{\pi \in \text{Mod}}, N, V^+, V^-)$ a bistructure over L_π. The elementary paraconsistent diagram of E, denoted by $Pdiag(E)$, is the set of quasi-hybrid literals over $L_\pi(W)$ that hold in $E(W)$, i.e.,*

$$Pdiag(E) = \{\alpha \in \text{QHLit}(L_\pi(W)) \mid E(W) \models_s \alpha\}$$

Based on this definition, we may state that two distinct bistructures over L_π with the same domain W and the same set of nominals N induce two distinct paraconsistent diagrams (over $L_\pi(W)$).

Thus the paraconsistent diagram $Pdiag(E)$ defines the bistructure E. Therefore, in the sequel, we will represent a (finite) bistructure $E = (W, (R_\pi)_{\pi \in \text{Mod}}, N, V^+, V^-)$ by its (finite) paraconsistent diagram $Pdiag(E)$. This syntactic representation will play an important role throughout this chapter.

Let $L_\pi = \langle \text{Prop}, \text{Nom}, \text{Mod} \rangle$ be a hybrid multimodal similarity type, $\Delta \subseteq \text{Form}_{\text{NNF}(@)}(L_\pi)$ and W a finite set. We write $\text{QH}(L_\pi, \Delta, W)$ to denote the set of representations (i.e., paraconsistent diagrams) of bistructures that are models of Δ with domain W. Recall that the domain and the hybrid similarity type are considered to be finite. This implies that the bistructures are finite and consequently the representations of QH models are also finite. This fact is relevant in the next section when discussing the measure of inconsistency of a model.

The syntactic representations of models will be denoted by $\mathcal{M}$, $\mathcal{M}_1$, etc. Let $\mathcal{M}$ be the representation of E with domain W. For $w \in W$, we write $\mathcal{M}, w \models_s \varphi$ if $E, w \models_s \varphi$. Analogously, we write $\mathcal{M} \models_s \varphi$ if $E \models_s \varphi$.

2.1.3 Construction of Syntactic QH Models

In order to make it easier to construct QH models as sets of quasi-hybrid literals, we will prove some properties about the satisfaction operator, and we will introduce a very important construction which will transform a formula in negation normal form into a quasi-equivalent positive boolean combination of QH-literals.

Definition 2.10. *A formula* $\varphi^* \in \text{Form}_{\text{NNF}(@)}(L_\pi)$ *is said to be quasi-equivalent to a formula* $\varphi \in \text{Form}_{\text{NNF}(@)}(L_\pi)$, *denoted* $\varphi \equiv_q \varphi^*$, *iff for all hybrid bistructures* $E = (W, (R_\pi)_{\pi \in \text{Mod}}, N, V^+, V^-)$ *and any* $w \in W$,

$$E, w \models_s \varphi \Leftrightarrow E, w \models_s \varphi^*.$$

We now present some properties of the satisfaction operator in quasi-hybrid logic:

Lemma 2.11. *Let* $L_\pi = \langle \text{Prop}, \text{Nom}, \text{Mod} \rangle$ *be a hybrid multimodal similarity type, and* $\varphi, \psi \in \text{Form}_{\text{NNF}(@)}(L_\pi)$ *be hybrid formulas in negation normal form. Then,*

1. $@_i(\varphi \vee \psi) \equiv_q @_i\varphi \vee @_i\psi$;

2. $@_i(\varphi \wedge \psi) \equiv_q @_i\varphi \wedge @_i\psi$;

3. $@_i@_j\varphi \equiv_q @_j\varphi$;

4. $\neg @_i\varphi \equiv_q @_i\neg\varphi$.

Proof. (1) Let E be an arbitrary bistructure, and w an arbitrary world in E:

$$\begin{aligned}
E, w \models_s @_i(\varphi \vee \psi) &\Leftrightarrow E, w' \models_s \varphi \vee \psi, w' = N(i) \\
&\Leftrightarrow [E, w' \models_s \varphi \text{ or } E, w' \models_s \psi] \\
&\quad \text{and } [E, w' \models_s nnf(\neg\varphi) \text{ implies } E, w' \models_s \psi] \\
&\quad \text{and } [E, w' \models_s nnf(\neg\psi) \text{ implies } E, w' \models_s \varphi], w' = N(i) \\
&\Leftrightarrow [E, w \models_s @_i\varphi \text{ or } E, w \models_s @_i\psi] \\
&\quad \text{and } [E, w \models_s @_i(nnf(\neg\varphi)) \text{ implies } E, w \models_s @_i\psi] \\
&\quad \text{and } [E, w \models_s @_i(nnf(\neg\psi)) \text{ implies } E, w \models_s @_i\varphi] \\
&\Leftrightarrow [E, w \models_s @_i\varphi \text{ or } E, w \models_s @_i\psi] \\
&\quad \text{and } [E, w \models_s nnf(\neg(@_i\varphi)) \text{ implies } E, w \models_s @_i\psi] \\
&\quad \text{and } [E, w \models_s nnf(\neg(@_i\psi)) \text{ implies } E, w \models_s @_i\varphi] \\
&\Leftrightarrow E, w \models_s @_i\varphi \vee @_i\psi
\end{aligned}$$

Items (2), (3) and (4) are proved similarly. $\square$

The distributive law does not hold as shown in the following example.

Example 1. *Let $L_\pi = \langle \{p, q, r\}, \{i\}, \{\pi\} \rangle$ be a hybrid multimodal similarity type. Consider the bistructure E, with domain $W = \{i\}$, $R_\pi = \emptyset$ and the valuation V defined by $V^+(p) = \emptyset, V^-(p) = \{i\}, V^+(q) = V^-(q) = \{i\}, V^+(r) = \{i\}, V^-(r) = \emptyset$.*

The formula $@_i p \vee (@_i q \wedge @_i r)$ is strongly satisfied in E. The proof of $E \models_s @_i p \vee (@_i q \wedge @_i r)$ is left to the reader. Observe that $\sim (@_i q \wedge @_i r) = @_i \neg q \vee @_i \neg r$, thus disjunctive syllogism is invoked twice.

However, the formula $(@_i p \vee @_i q) \wedge (@_i p \vee @_i r)$ is not strongly satisfied in E. This shows that in Quasi-Hybrid Multimodal Logic the distributive law does not hold.

For the representation of bistructures using quasi-hybrid literals, consider that $L_\pi = \langle \text{Prop}, \text{Nom}, \text{Mod} \rangle$ is a hybrid multimodal similarity type, and $\mathcal{M}$ is the representation of a finite bistructure over L_π with domain W.

We start by noticing that, with a finite domain $W = \{i_1, i_2, \ldots, i_n\}$, for any formula $\varphi \in \text{Form}_{\text{NNF}(@)} L_\pi$, we have

$$\mathcal{M} \models_s \varphi \Leftrightarrow \mathcal{M} \models_s @_{i_1}\varphi \wedge @_{i_2}\varphi \wedge \cdots \wedge @_{i_n}\varphi.$$

Since the conjunction of positive boolean combinations of quasi-hybrid literals remains a positive boolean combination of quasi-hybrid literals, we define a procedure that transforms any formula $@_{i_*}\varphi$ into a quasi-equivalent positive boolean combination of quasi-hybrid literals, PBCL for short, in the following way:

- if $\varphi = p$, $@_{i_*} p$ is a PBCL;
- if $\varphi = \neg p$, $@_{i_*} \neg p$ is a PBCL;

- if $\varphi = i$, $@_{i_*}i$ is quasi-equivalent to $\begin{cases} \top, & \text{if } i = i_* \\ \bot, & \text{if } i \neq i_* \end{cases}$ which is a PBCL;

- if $\varphi = \neg i$, $@_{i_*}\neg i$ is quasi-equivalent to $\begin{cases} \top, & \text{if } i \neq i_* \\ \bot, & \text{if } i = i_* \end{cases}$ which is a PBCL;

For the induction step, suppose that $@_{i_*}\phi, @_{i_*}\psi$ are quasi-equivalent to PBCL formulas. Then,

- if $\varphi = \phi \vee \psi$, $@_{i_*}\phi \vee \psi$ is quasi-equivalent to $@_{i_*}\phi \vee @_{i_*}\psi$ which by the inductive hypothesis is quasi-equivalent to a PBCL;

- if $\varphi = \phi \wedge \psi$, $@_{i_*}\phi \wedge \psi$ is quasi-equivalent to $@_{i_*}\phi \wedge @_{i_*}\psi$ which by the inductive hypothesis is quasi-equivalent to a PBCL;

- if $\varphi = [\pi]\phi$, $@_{i_*}[\pi]\phi$ is quasi-equivalent to $(@_{i_*}[\pi]\neg i_1 \vee @_{i_1}\phi) \wedge (@_{i_*}[\pi]\neg i_2 \vee @_{i_2}\phi) \wedge \cdots \wedge (@_{i_*}[\pi]\neg i_n \vee @_{i_n}\phi)$ which by the inductive hypothesis is quasi-equivalent to a PBCL;

- if $\varphi = \langle\pi\rangle\phi$, $@_{i_*}\langle\pi\rangle\phi$ is quasi-equivalent to $(@_{i_*}\langle\pi\rangle i_1 \wedge @_{i_1}\phi) \vee (@_{i_*}\langle\pi\rangle i_2 \wedge @_{i_2}\phi) \vee \cdots \vee (@_{i_*}\langle\pi\rangle i_n \wedge @_{i_n}\phi)$ which by the inductive hypothesis is quasi-equivalent to a PBCL;

- if $\varphi = @_{i_k}\phi$, $@_{i_*}@_{i_k}\phi$ is quasi-equivalent to $@_{i_k}\phi$ which by the inductive hypothesis is quasi-equivalent to a PBCL.

As we have already pointed out, we can syntactically represent bistructures as paraconsistent diagrams and thus we want to build models as sets of quasi-hybrid literals. So, for a given set Δ, it is easier to decompose each formula into quasi-hybrid literals using this procedure, and then construct the model upon those literals.

It is simple to prove that the procedure works and produces quasi-equivalent formulas. Formally,

Theorem 2.12. *Let $L_\pi = \langle \text{Prop}, \text{Nom}, \text{Mod} \rangle$ be a hybrid multimodal similarity type and $\mathcal{M}$ a representation of a finite bistructure over L_π with domain W. For any formula $\varphi \in \text{Form}_{\text{NNF}(@)}(L_\pi)$,*

$$\mathcal{M} \models_s \varphi \Leftrightarrow \mathcal{M} \models_s \overline{\varphi}$$

for $\overline{\varphi}$ a PBCL which is the result of applying the previous procedure to φ.

2.1.4 Minimal QH Models

The next definition will be the basis to prove that we can deal with only minimal models.

Definition 2.13. *Let L_π be a hybrid multimodal similarity type. For a set K of QH models over the same domain W, the set of (strongly) satisfied literals in K is the set $SLit(K)$ defined as follows:*

$$\mathrm{SLit}(K) = \{\alpha \in \mathrm{QHLit}(L_\pi(W)) \mid \forall \mathcal{M} \in K, \ \mathcal{M} \models_s \alpha\}$$

Observe that if $K' \subseteq K$ then $\mathrm{SLit}(K) \subseteq \mathrm{SLit}(K')$.

Since different hybrid similarity types contain different sets of formulas, it will be useful to have L_π as a parameter when we discuss concepts about models. Also, we will assume that Δ is a set of formulas of L_π. The domain is also important and we consider it as a parameter.

Minimal models are those where each formula is absolutely necessary to keep it a *model*, according to the following definition:

Definition 2.14. *Let L_π be a hybrid multimodal similarity type, and consider a set $\Delta \subseteq \mathrm{Form}_{\mathrm{NNF}(@)}(L_\pi)$ and a non-empty domain W. The set of minimal QH models of Δ with domain W is the set $\mathrm{MQH}(L_\pi, \Delta, W)$, defined as:*

$$\mathrm{MQH}(L_\pi, \Delta, W) = \{\mathcal{M} \in \mathrm{QH}(L_\pi, \Delta, W) \mid \ \text{if} \ \mathcal{M}' \subset \mathcal{M} \ \text{then} \ \mathcal{M}' \notin \mathrm{QH}(L_\pi, \Delta, W)\}$$

Clearly, every QH model contains a minimal QH model, *i.e.*, for every QH model $\mathcal{M}_1$, there is a minimal QH model $\mathcal{M}_2$ such that $\mathcal{M}_2 \subseteq \mathcal{M}_1$.

It is not difficult to see that, if a variable $p \in \mathrm{Prop}$ does not occur in Δ, then p also does not occur in any model $\mathcal{M} \in \mathrm{MQH}(L_\pi, \Delta, W)$.

Our interest in using $\mathrm{MQH}(L_\pi, \Delta, W)$ rather than $\mathrm{QH}(L_\pi, \Delta, W)$, for a set of formulas Δ, is that the models in $\mathrm{MQH}(L_\pi, \Delta, W)$ do not contain irrelevant information for analysing inconsistency, and we do not lose any useful information, according to the next theorem:

Theorem 2.15. *Let L_π be a hybrid multimodal similarity type, $\Delta \subseteq \mathrm{Form}_{\mathrm{NNF}(@)}(L_\pi)$ and W a non-empty set. Then*

$$\mathrm{SLit}(\mathrm{QH}(L_\pi, \Delta, W)) = \mathrm{SLit}(\mathrm{MQH}(L_\pi, \Delta, W))$$

Proof. Since $\mathrm{MQH}(L_\pi, \Delta, W) \subseteq \mathrm{QH}(L_\pi, \Delta, W)$, by the observation made after Definition 2.13, it follows that $\mathrm{SLit}\,(\mathrm{QH}(L_\pi, \Delta, W)) \subseteq \mathrm{SLit}\,(\mathrm{MQH}(L_\pi, \Delta, W))$.

To prove the other inclusion, let $\varphi \in \mathrm{SLit}(\mathrm{MQH}(L_\pi, \Delta, W))$. So, $\mathcal{M}_i \models_s \varphi$, for all $\mathcal{M}_i \in \mathrm{MQH}(L_\pi, \Delta, W)$.

For all $\mathcal{M}'_j \in \mathrm{QH}(L_\pi, \Delta, W)$, there is a subset $\mathcal{N}_j \subseteq \mathrm{QHLit}(L_\pi(W))$ and a model $\mathcal{M}_i \in \mathrm{MQH}(L_\pi, \Delta, W)$ such that $\mathcal{M}_i \cup \mathcal{N}_j = \mathcal{M}'_j$. Since $\mathcal{M}_i \models_s \varphi$, for all $\mathcal{M}_i \in \mathrm{MQH}(L_\pi, \Delta, W)$, then $\mathcal{M}_i \cup \mathcal{N}_j \models_s \varphi$, for all $\mathcal{M}_i \in \mathrm{MQH}(L_\pi, \Delta, W)$ and any $\mathcal{N}_j \subseteq \mathrm{QHLit}(L_\pi(W))$. So, $\mathcal{M}'_j \models_s \varphi$, for all $\mathcal{M}'_j \in \mathrm{QH}(L_\pi, \Delta, W)$. Therefore, $\mathrm{SLit}(\mathrm{MQH}(L_\pi, \Delta, W)) \subseteq \mathrm{SLit}(\mathrm{QH}(L_\pi, \Delta, W))$.

Thus $\mathrm{SLit}(\mathrm{QH}(L_\pi, \Delta, W)) = \mathrm{SLit}(\mathrm{MQH}(L_\pi, \Delta, W))$. □

The previous theorem does not hold if we consider all satisfied formulas, say SForm(K), instead of only satisfied literals. We have the following counter-example:

Example 2. Let $L_\pi = (\{p, q, r\}, \{i, j\}, \{\pi\})$, $W = \{i, j\}$ and $\Delta = \{@_i p \vee @_i q, @_i \neg p \vee @_i \neg q, @_j r\}$.

The two minimal QH models of Δ are:
$\mathcal{M}_1 = \{@_i p, @_i \neg q, @_j r\}$;
$\mathcal{M}_2 = \{@_i q, @_i \neg p, @_j r\}$.

It is easy to see that:
$@_i r \vee @_j r \in \text{SForm}(\text{MQH}(L_\pi, \Delta, W))$.

However, $@_i r \vee @_j r \notin \text{SForm}(\text{QH}(L_\pi, \Delta, W))$. In fact, if we consider the model $\mathcal{M} = \{@_i p, @_i \neg q, @_j r, @_j \neg r\}$, we have that $\mathcal{M} \not\models_s @_i r \vee @_j r$ (this happens because $\mathcal{M} \models_s \neg @_j r$ implies that $\mathcal{M}$ would have to satisfy $@_i r$, which is false). Thus $\text{SForm}(\text{MQH}(L_\pi, \Delta, W)) \nsubseteq \text{SForm}(\text{QH}(L_\pi, \Delta, W))$.

Next we will present several examples that illustrate how to build models (which are sets of quasi-hybrid literals) for a set of formulas Δ, using Theorem 2.12.

It is not always the case that minimal models have the same number of contradictions, as we will see below.

Example 3. Let $L_\pi = \langle \{p\}, \{i\}, \{\pi\} \rangle$, $W = \{i\}$, and $\Delta = \{@_i[\pi]p, @_i[\pi]\neg p\}$.
There are exactly two minimal models with domain $W = \{i\}$, which are:
$\mathcal{M}_1 = \{@_i[\pi]\neg i\}$;
$\mathcal{M}_2 = \{@_i\langle\pi\rangle i, @_i p, @_i \neg p\}$.
We can verify that $\mathcal{M}_1$ has no contradictions while $\mathcal{M}_2$ has one.
The minimal models $\mathcal{M}_1$ and $\mathcal{M}_2$ are represented in Figure 1.

Example 4. Let $L_\pi = \langle \{p, q\}, \{i, j\}, \{\pi\} \rangle$, $W = \{i, j\}$, and $\Delta = \{@_i(p \vee q), @_i \neg p, @_j \langle\pi\rangle \neg q\}$.

Not all formulas in Δ are PBCLs. Using the properties of the satisfaction operator and the method described in the proof of Theorem 2.12, let us make the necessary adjustments:

- $@_i(p \vee q) \equiv_q @_i p \vee @_i q$

Since the formula $@_i \neg p$ is mandatory in every model, from $(@_i p \vee @_i q)$ it follows that $@_i q$ is mandatory too. Hence, any minimal model of Δ must contain the formulas: $@_i \neg p$ and $@_i q$.

In order to satisfy the formula $@_j \langle\pi\rangle \neg q$ in Δ, we can choose between two options: either there is a connection from j to i, or there is a connection from j to itself. Thus any minimal model must contain exactly one of these sets:

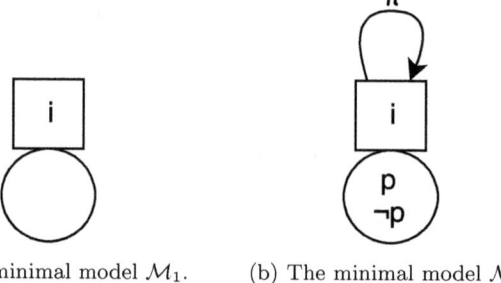

(a) The minimal model $\mathcal{M}_1$. (b) The minimal model $\mathcal{M}_2$.

Figure 1: Minimal models – Example 3.

$\Omega_1 = \{@_i\neg p, @_i q, @_j \langle \pi \rangle j, @_j \neg q, @_j [\pi] \neg i\}$
$\Omega_2 = \{@_i \neg p, @_i q, @_j \langle \pi \rangle i, @_i \neg q, @_i p, @_j [\pi] \neg j\}$

Recall that all connections and lack of them must be specified in minimal models. Therefore, we have 4 models for each Ω, depending on the connections leaving i.

We have for instance, for Ω_1:
$\mathcal{M}_1 = \{@_i \neg p, @_i q, @_j \langle \pi \rangle j, @_j \neg q, @_j [\pi] \neg i, @_i [\pi] \neg i, @_i [\pi] \neg j\}$;
$\mathcal{M}_2 = \{@_i \neg p, @_i q, @_j \langle \pi \rangle j, @_j \neg q, @_j [\pi] \neg i, @_i \langle \pi \rangle i, @_i [\pi] \neg j\}$;

And for Ω_2:
$\mathcal{M}_3 = \{@_i \neg p, @_i q, @_j \langle \pi \rangle i, @_i \neg q, @_i p, @_j [\pi] \neg j, @_i [\pi] \neg i, @_i \langle \pi \rangle j\}$
$\mathcal{M}_4 = \{@_i \neg p, @_i q, @_j \langle \pi \rangle i, @_i \neg q, @_i p, @_j [\pi] \neg j, @_i \langle \pi \rangle i, @_i \langle \pi \rangle j\}$

We verify that there are no contradictions in models $\mathcal{M}_1$ and $\mathcal{M}_2$. Models $\mathcal{M}_3$ and $\mathcal{M}_4$ have two contradictions each.

The minimal models $\mathcal{M}_1, \mathcal{M}_2, \mathcal{M}_3$ and $\mathcal{M}_4$ are represented in Figure 2.

2.2 Measure of Inconsistency

A theory may have different minimal QH models depending on the hybrid similarity type and domain considered. Now we will introduce a way to measure the inconsistency of a QH model. This measure is a ratio between 0 and 1 whose numerator is the number of contradictions in the model, and whose denominator is the total possible number of contradictions in the underlying hybrid similarity type. This is somewhat similar to the degree of inconsistency discussed by Konieczny, Lang and Marquis in [19] for Priest's paraconsistent logic LP.

The measure of inconsistency of a model is crucial in a diverse range of applications in artificial intelligence to compare knowledge bases. As supported

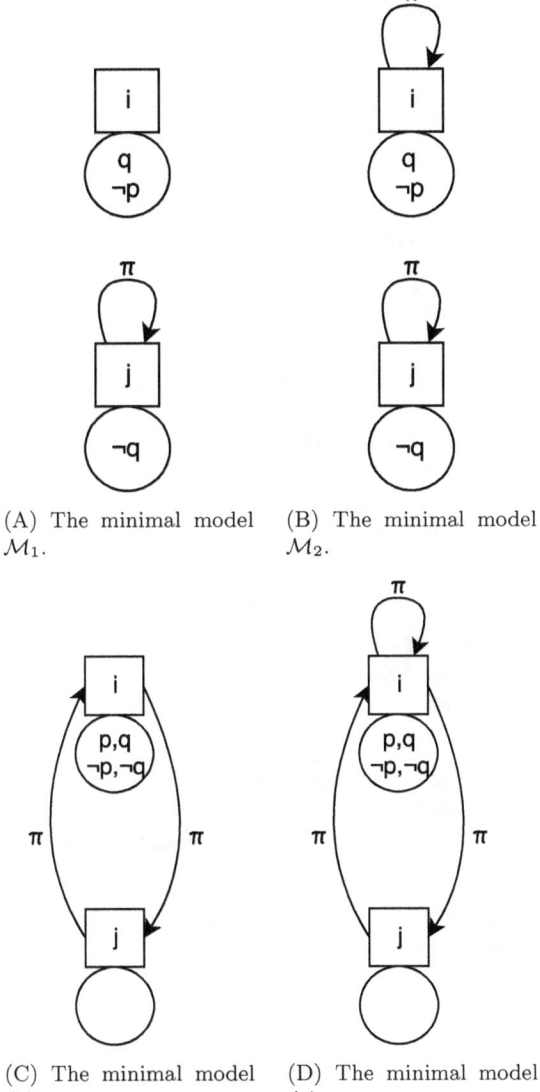

Figure 2: Minimal models – Example 4.

in [18], it may be a useful tool in analysing various information types, such as news reports, software specifications, integrity constraints and e-commerce protocols.

To make the notation in the next definition simpler, we define the set of *inconsistency atoms* over L_π and W as

$$IA(L_\pi, W) = \{@_i p \mid i \in W, p \in \text{Prop}\}$$

Definition 2.16. *For a* QH *model* $\mathcal{M}$,

$$\text{Conflictbase}(\mathcal{M}) = \{@_i p \in IA(L_\pi, W) \mid @_i p \in \mathcal{M} \ \& \ @_i \neg p \in \mathcal{M}\}$$

Our inconsistency measure is as follows:

Definition 2.17. *The measure of inconsistency for a model $\mathcal{M}$ with domain W in the context of a hybrid similarity type L_π (which must contain at least all propositional variables, modalities and nominals present in $\mathcal{M}$) is given by the* MInc *function, giving a value between 0 and 1, as follows:*

$$\text{MInc}(\mathcal{M}, L_\pi) = \frac{|\text{Conflictbase}(\mathcal{M})|}{|IA(L_\pi, W)|}$$

MInc is anti-monotonic in the following sense:

Theorem 2.18. *Let $L_{\pi,1}$ and $L_{\pi,2}$ be hybrid similarity types such that $L_{\pi,1} \subseteq L_{\pi,2}$, i.e., $\text{Prop}_1 \subseteq \text{Prop}_2$, $\text{Nom}_1 \subseteq \text{Nom}_2$, and $\text{Mod}_1 \subseteq \text{Mod}_2$. Then, for each model $\mathcal{M}$ in $L_{\pi,1}$ (which is obviously also a model in $L_{\pi,2}$), $\text{MInc}(\mathcal{M}, L_{\pi,1}) \geq \text{MInc}(\mathcal{M}, L_{\pi,2})$.*

Proof. The proof is straightforward: the number of elements in the Conflictbase of the model $\mathcal{M}$ is fixed; however, the denumerator in $\text{MInc}(\mathcal{M}, L_{\pi,2})$ is greater than or equal to the denumerator in $\text{MInc}(\mathcal{M}, L_{\pi,1})$ and thus the result follows. □

Example 5. *In this example we consider again the minimal models presented in Examples 3 and 4, and we compute their measures of inconsistency, i.e., the* MInc *function.*

1. From Example 3, $\text{Conflictbase}(\mathcal{M}_1) = \{\}$. Then,

$$\text{MInc}(\mathcal{M}_1, L_\pi) = \frac{|\text{Conflictbase}(\mathcal{M}_1)|}{|IA(L_\pi, W)|} = \frac{0}{1} = 0$$

However, $\text{Conflictbase}(\mathcal{M}_2) = \{@_i p\}$, *thus*

$$\text{MInc}(\mathcal{M}_2, L_\pi) = \frac{1}{1} = 1$$

2. From Example 4, Conflictbase($\mathcal{M}_1$) = Conflictbase($\mathcal{M}_2$) = {}. Then,

$$\mathrm{MInc}(\mathcal{M}_1, L_\pi) = \mathrm{MInc}(\mathcal{M}_2, L_\pi) = \frac{0}{4} = 0$$

On the other hand, Conflictbase($\mathcal{M}_3$) = Conflictbase($\mathcal{M}_4$) = {$@_i p, @_i q$}. So,

$$\mathrm{MInc}(\mathcal{M}_3, L_\pi) = \mathrm{MInc}(\mathcal{M}_4, L_\pi) = \frac{2}{4} = \frac{1}{2}$$

2.2.1 Weighted Measure of Inconsistency

There is also the possibility of adding weights to propositional variables, such that, for example, a contradiction regarding p is worse than a contradiction regarding q. Let us formalize this idea:

Definition 2.19. *Let $\omega : \mathrm{Prop} \to \mathbb{N}$ be a function that assigns to each propositional variable its relevance, i.e., the higher the value of $\omega(p)$ the more valuable it is to maintain consistency about p.*

And now for the weighted measure of inconsistency we get:

Definition 2.20. *The weighted measure of inconsistency for a model $\mathcal{M}$ with domain W, in the context of a hybrid multimodal similarity type L_π with associated weights given by ω, is given by the MInc_ω function, giving a value between 0 and 1, as follows:*

$$\mathrm{MInc}_\omega(\mathcal{M}, L_\pi) = \frac{\sum_{p \in \mathrm{Prop}} \omega(p) \cdot |\{i \in \mathrm{Nom} \mid @_i p \in \mathcal{M} \ \& \ @_i \neg p \in \mathcal{M}\}|}{\sum_{p \in \mathrm{Prop}} \omega(p) \cdot |\mathrm{Nom}|}$$

Let us introduce an illustrative example:

Example 6. *Let $L_\pi = \langle \{p, q\}, \{i\}, \{\} \rangle$, $W = \{i\}$, and $\Delta = \{@_i(p \wedge \neg p) \vee @_i(q \wedge \neg q))\}$.*

Clearly both $\mathcal{M}_1 = \{@_i p, @_i \neg p\}$ and $\mathcal{M}_2 = \{@_i q, @_i \neg q\}$ are minimal models for Δ.

However, if $\omega(p) = 1$ and $\omega(q) = 2$, then $\mathrm{MInc}_\omega(\mathcal{M}_1, L_\pi) = \frac{1}{3}$ and $\mathrm{MInc}_\omega(\mathcal{M}_2, L_\pi) = \frac{2}{3}$. Thus $\mathcal{M}_1$ is less inconsistent than $\mathcal{M}_2$ with respect to ω.

2.2.2 An Applied Example

An application in the health area where a measure of inconsistency would be useful has been discussed in [6], and concerns the path of a patient in the health care system, subject to contradictory diagnoses. Here we present an updated version which uses modalities.

The flow of a patient in the health care system. Figure 3 represents an *hypothetical* fragment of the clinical flow of patients in a central hospital. This is intentionally a simplified example.

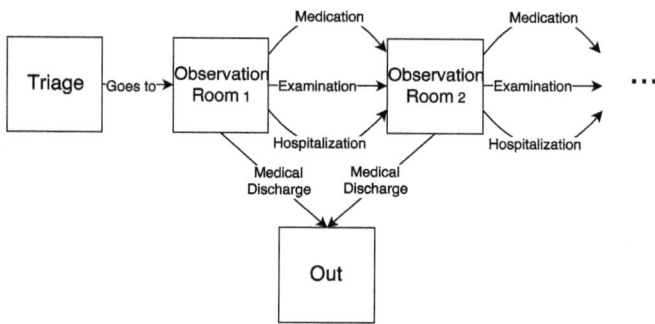

Figure 3: Patient flow in the health care delivery process.

This figure intends to represent that a patient coming into a hospital is consulted at the triage station. The next step in the care delivery process is the observation of the patient by physicians in an observation room. At this stage, several things can happen: (1) the patient may need to get some medication, after which is again observed by physicians; (2) the patient may need to take an examination, after which is again observed by physicians; (3) the patient may need to be hospitalized, after which is again observed by physicians; or (4) the patient may be discharged. Observe that a combination of processes (1) to (3) is possible, for example, the patient may get some medication *and* take an examination before being observed again. However, (4) cannot be combined with any other process.

For each patient, his path obeys the formulas in the following set:

$$\Delta = \Big\{ @_{\text{Triage}}\langle\text{Goes to}\rangle\text{Observation Room 1},$$

$$\bigwedge_{\substack{k \in \mathbb{N}: \\ \text{Patient has been in} \\ \text{Observation Room k}}} \Big(@_{\text{Observation Room k}}(\langle\text{Medication}\rangle\text{Observation Room k+1} \vee$$

$$\vee \langle\text{Examination}\rangle\text{Observation Room k+1} \vee \langle\text{Hospitalization}\rangle\text{Observation Room k+1}) \,\dot\vee$$

$$\dot\vee\, @_{\text{Observation Room k}}\langle\text{Medical Discharge}\rangle\text{Out} \Big) \Big\}$$

The symbol $\dot\vee$ stands for exclusive disjunction, *i.e.*, $\varphi \dot\vee \psi$ is true if and only if φ is true or ψ is true, but not both.

The path of a patient can be represented by a Kripke frame and the reports made at each stage are represented by a decoupled valuation. Note that the decoupling of the valuation is mandatory since very often the diagnosis is not deterministic and we have to allow inconsistencies; actually, a team of physicians may not agree on the diagnosis of a specific disease or even an exam can be inconclusive (for example a CT Screening for lung cancer may give inconclusive evidence).

The propositional variables are used to represent the data in the patient report. More specifically, a propositional variable can be seen as a health feature observed in the patient (for example fever, cancer, cough, pallor). Some propositional variables are classical, however, some others are paraconsistent. Nominals are used to name referential states (*i.e.* important moments of diagnosis). Modalities are used to label transitions in the flow, for example transitions induced by the administration of a certain medicine or by a specific examination.

Note that the patient eventually leaves the hospital (our models will thus be finite).

An Example.

Consider the following scenario: a patient appears at triage with fever and a headache. In the first observation he is diagnosed by a physician with disease X while another one disagrees and says that the patient does not have disease X, he has disease Y. The first physician says nothing about disease Y. The patient takes some medication, and is observed a second time. His fever did not go down and both physicians keep their opinions. Then the patient takes an examination. For the third time in an observation room, the patient receives his final diagnosis: he has disease X and not disease Y. He is finally discharged, and goes home with due recommendations.

From the medical pathway of this patient we can extract a paraconsistent model and we can evaluate its measure of inconsistency in order to be able to compare it with other models. (Observe that Figure 3 does not intend to graphically depict our model!)

We assume that the propositional variables for the symptoms are classical and that inconsistencies are only present in the medical diagnoses. Thus the set Prop includes ClassicalProp – propositional variables for symptoms, which behave classically – and ParaconsistentProp – propositional variables for diseases, which are prone to inconsistency. In this example, ClassicalProp = {Fever, Headache} and ParaconsistentProp = {Disease X, Disease Y}.

Our domain W is the set composed by the Triage, the Observation Rooms visited by the patient, and the "Out of the hospital"-state. In this particular case, W={*Triage, Observation Room* 1, *Observation Room* 2, *Observation Room* 3, *Out*}.

Since there are no diagnoses in the Triage and when the patient is Out of

the hospital, we do not consider those in our measure of inconsistency, which has been adapted to this example as follows:
- We regard as inconsistency atoms those of the form @$_{\text{Observation Room k}}\Upsilon$ such that Observation Room k $\in W$ and $\Upsilon \in$ ParaconsistentProp.
- The Conflictbase is clearly the set

$$\{@_{\text{Observation Room 1}}\text{Disease X}, @_{\text{Observation Room 2}}\text{Disease X}\}$$

Ergo, the measure of inconsistency for this model is $\frac{2}{6} = \frac{1}{3}$.

Further considerations: A. Transitions between states can also be labeled with modalities that might correspond to specific medications or examinations. If we had the chance to fully axiomatize the medical guideline – which unfortunately we cannot – we would have the perfect case. The axiomatization of the medical guideline would include: (1) the complete (not a fragment) clinical flow of patients in a central hospital, *i.e.*, all the possible transitions between different stages, (2) the action of specific medication in the symptoms of the patient, for example having $p \to \langle A \rangle \neg p$ meaning that a patient with problem p who takes medicine A will get cured, and furthermore, (3) the diagnosis of a disease by means of a specific examination, *i.e.*, for any disease there would be a conclusive examination — the disease is present or not.

B. Assigning weights to propositional variables in this particular problem can be useful if we consider that inconsistencies in some diagnoses are more severe than in others. It is certainly more worrisome if physicians cannot agree on the diagnosis of a cancer than in the diagnosis of a flu, for example.

3 Paraconsistency in Other Constituents of Hybrid Logic

3.1 Quasi-Hybrid Multimodal Logic with Quantifiers

This intends to be a brief introduction to a quantified version of (classical) hybrid multimodal logic.

In order to add quantification, consider the addition of a set of *world variables*, WVar, (typically written as s, t, u, etc), distinct from both nominals and propositional variables.

The result of this new machinery is a powerful hybrid logic, $\mathcal{H}(@, \forall)$, whose grammar is defined as follows:

Definition 3.1. *Let* $L_\pi = \langle \text{Prop}, \text{Nom}, \text{WVar}, \text{Mod} \rangle$ *be a hybrid multimodal similarity type where* Prop *and* Nom *are as usual the set of propositional variables and the set of nominals, and* WVar *is the set of world variables. The well-formed formulas over* L_π, Form$_{@,\forall}(L_\pi)$, *are defined by the following grammar:*

$WFF := i \mid p \mid s \mid \bot \mid \top \mid \neg\varphi \mid \varphi \vee \psi \mid \varphi \wedge \psi \mid \langle \pi \rangle \varphi \mid [\pi]\varphi \mid @_i\varphi \mid @_s\varphi \mid \forall s.\varphi \mid \exists s.\varphi$

Note that @ can make use of world variables.

The definition of hybrid structure remains the same as for basic hybrid logic. But now there is a need for a mechanism for coping with free world variables, so it is considered an *assignment* $g :$ WVar $\to W$.

If g, g' are assignments of values to variables in $\mathcal{H}$, recall that g' is an *s-variant assignment* of g iff $g'(t) = g(t)$, for all $t \in$ WVar, $t \neq s$; in such a case we write $g' \overset{s}{\sim} g$.

Satisfaction is defined in the following way:

Definition 3.2 (Satisfaction). *The local satisfaction relation $\models$ between a hybrid structure $\mathcal{H} = (W, (R_\pi)_{\pi \in \text{Mod}}, N, V)$, a state $w \in W$, an assignment g and a hybrid formula is an extension of the satisfaction relation previously introduced such that for the new formulas:*

- $\mathcal{H}, g, w \models s$ *iff* $w = g(s)$;
- $\mathcal{H}, g, w \models @_s\varphi$ *iff* $\mathcal{H}, g, w' \models \varphi$, *where* $w' = g(s)$;
- $\mathcal{H}, g, w \models \forall s.\varphi$ *iff* $\mathcal{H}, g', w \models \varphi$, *for all* $g' \overset{s}{\sim} g$;
- $\mathcal{H}, g, w \models \exists s.\varphi$ *iff* $\mathcal{H}, g', w \models \varphi$, *for some* $g' \overset{s}{\sim} g$;

Unfortunately, but not unexpectedly, if one takes a look at what happens in [14], when allowing contradictions and applying the same paraconsistent reasoning as before, we notice a non-standard behaviour of the existential quantifier due to the difference in the semantics for disjunction in quasi-hybrid logic and classical hybrid logic.

Here is an example:

Example 7. Let $L_\pi = \langle \{p, q\}, \{i, j\}, \{u\}, \{\pi\}\rangle$ and $W = \{i, j\}$.
$\Delta_1 = \{\exists u.@_u p, \neg@_i p \vee @_i q, \neg@_i q\}$
$\Delta_2 = \{@_i p \vee @_j p, \neg@_i p \vee @_i q, \neg@_i q\}$
$\mathcal{M}_1 = \{@_i q, \neg@_i q, @_i p, \neg@_i p\}$
$\mathcal{M}_2 = \{\neg@_i q, \neg@_i p, @_j p\}$
$\mathcal{M}_1$ and $\mathcal{M}_2$ are both models (with domain W) for Δ_1. However, only $\mathcal{M}_2$ is a model for Δ_2.

Nonetheless, this is not a major problem, as one could have added a second disjunction to the language, with a classical behaviour, thus creating a link between the existential quantifier and this (classical) disjunction.

We believe this could be interesting, especially when we talk about preferred models (*i.e.*, models with the lowest MInc for the same Δ) and intrinsic and extrinsic inconsistency, concepts adapted from [14] to Quasi-hybrid logic and that the reader can consult in [8].

3.2 The Role of Inconsistency in Nominals

As opposed to propositional variables, the "valuation" for nominals, which we designated by denotation, cannot be a set with multiple worlds; it must be a singleton.

If one were to consider that nominals could be naming worlds in an inconsistent manner, then we would have to drop the property just mentioned.

Furthermore, when we consider nominals in a classical approach, we know that if i denotes a world w, then in any other world $\neg i$ must be true.

When we move into a paraconsistent version and decouple N into two new nominations as we did for V, then neither N^+ is required to be a singleton nor N^- is set to be its complement on W.

This raises problems at the level of the accessibility relation as it is defined, because if $N^+(i)$ is not a singleton, how can one interpret $N^+(i)RN^+(j)$ which stands as the semantics of $@_i \Diamond j$?

However, there may be a way to achieve inconsistency at the level of nominals, certainly by making some adjustments in the accessibility relation as well as in the satisfaction of formulas envolving nominals. It is definitely a challenge for the future.

3.3 Adding Inconsistency to the Accessibility Relation

Lastly, let us explore the accessibility relation as a source of inconsistency. For simplicity, we will detail our thoughts on a unimodal version, which can easily be expanded to a multimodal one.

Such effect will be possible by decoupling the link between information about the existence and the absence of a connection between two worlds, thus considering two accessibility relations instead of one, say R^+ and R^-, and breaking the property $\Diamond \varphi \equiv \neg \Box \neg \varphi$. This will allow us to reason over knowledge bases where $@_i \Diamond \varphi$ and $@_i \neg \Diamond \varphi$ are present.

In our previous setting of quasi-hybrid multimodal logic (recall that we are considering a single modality), if φ were to represent a propositional variable p, then $@_i \neg \Diamond p$ would be equivalently considered as $@_i \Box \neg p$. This formula, in conjunction with $@_i \Diamond p$, would entail a contradiction over the propositional variable p in a world accessible from the one denoted by i.

Our goal here is to re-interpret $@_i \neg \Diamond \varphi$, as *"there is no connection from the world denoted by i to a world where φ holds"*. Analogously, $@_i \neg \Box \varphi$ would mean that *"not all connections leaving i and leading to φ exist"*.

Formulas with $\Diamond$ and $\Box$ will be associated with R^+, with semantics as defined before, and formulas with $\neg \Diamond$ and $\neg \Box$ will be associated with the relation R^-.

We can intuitively think about a measure of inconsistency for a model that allows for inconsistencies in both the propositional variables and the accessibility relation in the following way:

$$\mathrm{MInc}_R(\mathcal{M}, L_\pi, W) = \frac{|\{@_i j \in \mathcal{M} \text{ such that } iR^+j \ \& \ iR^-j\}| + |\mathrm{Conflictbase}(\mathcal{M})|}{|W|^2 + |W| \cdot |\mathrm{Prop}|}$$

Observe that we can adapt this measure to a measure with weights on the relation and propositional variables.

4 Conclusions and Further Work

This chapter started with an overview on paraconsistency in logical systems, namely with an analysis of the work on Quasi-classical logic, and a brief inspection about the literature relating modal logic, its extensions, and paraconsistency.

Then, inspired by the approach of Grant and Hunter to QC Logic in [14], we introduced a paraconsistent version of multimodal hybrid logic and showed the roles of inconsistency and measure of inconsistency at the level of propositional variables, nominals, and accessibility relations.

Although there is some work done with paraconsistency in Dynamic Logic, namely by Sedlár who introduced belnapian truth values for propositions in [24], nothing has been said about considering paraconsistency at the level of programs.

In the future, we aim to develop a new logic that combines the work by Rivieccio et al. in [23] – in which they introduce a modal logic provided with not only a belnapian valuation, but also with a belnapian accessibility relation – together with the notion of programs found in dynamic logic. Namely we would like to know what happens in the case of the composition of two Belnap-style relations, the program of choice between two of those relations, etc.

As we have already mentioned, we would like to come up with an idea on how to allow inconsistencies at the level of nominals, so that naming worlds incorrectly would be considered as possible. This would end our cycle of exploring inconsistency *(everywhere)* in hybrid logic: we have already introduced inconsistency in propositional variables, in the accessibility relation, and once we do it in nominals, all three major aspects of hybrid logic would have been dealt with.

Another topic on our to-do list is to explore the concept of distance between inconsistent models, discussed in [16] but assuming that a knowledgebase contains only consistent formulas; one could define the distance between two finite bistructures over the same domain in QH Logic as the sum of the distances in each world, adapting the notion just mentioned.

It would also be worth to investigate inconsistency measures in the construction described in [11].

Acknowledgements This work was supported in part by the Portuguese Foundation for Science and Technology (FCT) through CIDMA within project UID/MAT/04106/2013, by the ERDF – European Regional Development Fund - through the Operational Programme for Competitiveness and Internationalisation - COMPETE 2020 Programme - and by National Funds through the Portuguese funding agency FCT within project POCI-01-0145-FEDER-016692. Diana Costa also thanks the support of FCT via the PhD scholarship PD/BD/105730/2014.

References

[1] A. Arruda. On the Imaginary Logic of N. A. Vasil'ev. In *Non-Classical Logics, Model Theory and Computability: Proceedings of the 3rd Latin-American Symposium on Mathematical Logic*, volume 89 of *Studies in Logic and the Foundations of Mathematics*, pages 3–24, 1976.

[2] P. Besnard and A. Hunter. Quasi-classical Logic: Non-trivializable Classical Reasoning from Inconsistent Information. In *Proceedings of the 3rd European Conference on Symbolic and Quantitative Approaches to Reasoning and Uncertainty (ECSQARU'1995)*, volume 946 of *Lecture Notes in Computer Science*, pages 44–51, 1995.

[3] P. Blackburn. Internalizing Labelled Deduction. *Journal of Logic and Computation*, 10(1):137–168, 2000.

[4] G. Boolos, J. Burgess, and R. Jeffrey. *Computability and Logic. 5th ed.* Cambridge University Press, 5th ed. edition, 2007.

[5] T. Braüner. Axioms for Classical, Intuitionistic, and Paraconsistent Hybrid Logic. *Journal of Logic, Language and Information*, 15(3):179–194, October 2006.

[6] D. Costa and M. Martins. Inconsistencies in Health Care Knowledge. In *2014 IEEE 16th International Conference on e-Health Networking, Applications and Services (Healthcom)*, pages 37–42, October 2014.

[7] D. Costa and M. Martins. A Tableau System for Quasi-Hybrid Logic. In *Automated Reasoning. IJCAR 2016. Lecture Notes in Computer Science, vol 9706*, pages 435–451, 2016.

[8] D. Costa and M. Martins. Paraconsistency in Hybrid Logic. *Journal of Logic and Computation*, 27(6):1825–1852, 2017.

[9] N. Da Costa, D. Krause, and O. Bueno. Paraconsistent Logics and Paraconsistency. *Philosophy of logic*, 5:655–781, 2007.

[10] H. de Nivelle, R. Schmidt, and U. Hustadt. Resolution-based Methods for Modal Logics. *Logic Journal of the IGPL*, 8(3):265–292, 2000.

[11] E. de Souza, A. Costa-Leite, and D. Dias. On a Paraconsistentization Functor in the Category of Consequence Structures. *Journal of Applied Non-Classical Logics*, 26(3):240–250, 2016.

[12] M. Finger and D. Gabbay. Adding a Temporal Dimension to a Logic System. *Journal of Logic, Language and Information*, 1(3):203–233, September 1992.

[13] D. Gabbay. *Fibring Logics*. Oxford Logic Guides. Clarendon Press, 1998.

[14] J. Grant and A. Hunter. Measuring Inconsistency in Knowledgebases. *Journal of Intelligent Information Systems*, 27(2):159–184, 2006.

[15] J. Grant and A. Hunter. Analysing Inconsistent First-order Knowledgebases. *Artif. Intell.*, 172(8-9):1064–1093, 2008.

[16] J. Grant and A. Hunter. Analysing Inconsistent Information using Distance-based Measures. *International Journal of Approximate Reasoning, Special Issue on Theories of Inconsistency Measures and Their Applications*, 89(Supplement C):3 – 26, 2017.

[17] W. Hodges. *Model Theory*. Encyclopedia of Mathematics and its Applications. Cambridge University Press, 1993.

[18] A. Hunter and S. Konieczny. On the Measure of Conflicts: Shapley Inconsistency Values. *Artificial Intelligence*, 174(14):1007–1026, 2010.

[19] S. Konieczny, J. Lang, and P. Marquis. Quantifying Information and Contradiction in Propositional Logic through Epistemic Tests. In *18th International Joint Conference on Artificial Intelligence (IJCAI'03)*, pages 106–111, 2003.

[20] A. Madeira, R. Neves, and M. Martins. An Exercise on the Generation of Many-valued Dynamic Logics. *Journal of Logical and Algebraic Methods in Programming*, 85(5, Part 2):1011 – 1037, 2016.

[21] J. Marcos. Modality and Paraconsistency. In *Logica Yearbook 2004*, pages 213–222. Filosofia, Prague, 2005.

[22] M. Martins, A. Madeira, R. Diaconescu, and L. Barbosa. Hybridization of Institutions. In *Algebra and Coalgebra in Computer Science: 4th International Conference, CALCO 2011, Lecture Notes in Computer Science, vol 6859*, pages 283–297. Springer, Berlin, Heidelberg, 2011.

[23] U. Rivieccio, A. Jung, and R. Jansana. Four-valued Modal Logic: Kripke Semantics and Duality. *Journal of Logic and Computation*, 27(1):155–199, 2017.

[24] I. Sedlár. Propositional Dynamic Logic with Belnapian Truth Values. In *Proceedings of the 11th Conference on Advances in Modal Logic*, pages 503–519, 2016.

Inconsistency Measuring over Multisets of Formulas

Philippe Besnard

IRIT-CNRS
Université Paul Sabatier Toulouse, France
besnard@irit.fr

Abstract

A number of postulates for inconsistency measures have been proposed in the literature. They are all based on identifying knowledge bases with sets of logical formulas. However, there are other forms for knowledge bases, e.g., in belief merging, where the number of occurrences of a formula is taken into account. In this chapter, considering knowledge bases as multisets of formulas leads us to revisit postulates for inconsistency measures. We first show that some common restrictions in postulates become otiose, e.g., the well-known Dominance postulate gets a simpler form in the multiset version. We develop a series of results for these revised postulates.

1 Introduction

Inconsistency measures are meant to tell how much inconsistency a knowledge base carries, in the sense that *the higher the amount of inconsistency in the knowledge base, the greater the value returned by the inconsistency measure.* As an illustration, consider the assertion *"This idea is provocative but helpful"*. It is contradicted by the statement *"If it is provocative then it is not helpful"*. Alternatively, it can be contradicted by the statement *"It is not helpful and not even provocative"*. The latter expresses that both claims *"the idea is helpful"* and *"the idea is provocative"* in the initial assertion are false whereas the former only objects that either *"the idea is helpful"* is false or *"the idea is provocative"* is false. That is to say, the knowledge base

$$K_1 = \left\{ \begin{array}{c} \text{This idea is provocative and helpful} \\ \text{If it is provocative then it is not helpful} \end{array} \right\}$$

can be viewed as *less inconsistent than* the knowledge base

$$K_2 = \left\{ \begin{array}{l} \textit{This idea is provocative and helpful} \\ \textit{It is not helpful and not provocative} \end{array} \right\}.$$

Formally, for I denoting an inconsistency measure,

$$I(K_1) < I(K_2).$$

A wealth of inconsistency measures exist, see [13, 14, 17, 8, 18, 10, 23, 24, 21, 5, 9, 12, 1], mostly in the context of classical logic $\vdash$ over a language $\mathcal{L}$ where knowledge bases are finite sets of formulas of $\mathcal{L}$. Formally, an inconsistency measure I maps all finite sets of formulas of $\mathcal{L}$ to values in $\mathbf{R}^+ \cup \{\infty\}$.

However, identifying knowledge bases with finite sets of formulas lacks generality as tradition with belief merging indicates the importance of taking into account duplicates, e.g., through the notion of a profile that is (except in the case of sensitivity to syntactical representation) a multiset of formulas [11]. Moreover, the unicity of elements in sets disrupts regularities expected with assessing the amount of inconsistency. Indeed, as being more consistent can be viewed as enlarging the set of models from classical logic, being more inconsistent can be viewed as a converse progression. In terms of formulas, such a converse progression would be a step from a formula φ to a formula $\varphi \wedge \psi_1$, then to a formula $\varphi \wedge \psi_1 \wedge \psi_2$ and so on. For a knowledge base K, this would mean a progression from $K \cup \{\varphi\}$ to $K \cup \{\varphi \wedge \psi_1\}$, to $K \cup \{\varphi \wedge \psi_1 \wedge \psi_2\}$, ... but a problem arises if $\varphi \wedge \psi_1 \wedge \ldots \wedge \psi_i$ is in K for some i, because the purported chain reaches a point where the first and ith items are the same, namely K. In contrast, by taking duplicates into account, the essence of the progression is preserved (please keep in mind that an inconsistency measure ranges over a linearly ordered set!). Other reasons include minor differences in form, that make for example $\{p \wedge q, q \wedge p, \neg p\}$ to be treated differently from the situation where $p \wedge q$ occurs twice and $\neg p$ once. Thus, considering multisets of formulas seems a sensible starting point for knowledge bases when it comes to inconsistency measuring. This chapter indeed explores the case from the perspective of postulates for inconsistency measures.

The chapter is organised as follows. To begin with, Section 2 introduces working definitions so that knowledge bases can be identified with multisets of logical formulas. Section 3 gives the original notion of inconsistency postulates, for sets of formulas. Various features are presented there to be later explored in multiset guise in Section 4. More than a dozen inconsistency postulates for multisets are examined, individually and in combinations, in Section 5. Lastly, Section 6 concludes.

2 Knowledge Bases as Finite Multisets of Logical Formulas

To simplify things a lot, a fixed propositional language $\mathfrak{L}$ is assumed throughout that has infinitely many propositional variables (also, $\bot$ stands for absurdity, $\neg$ for negation, $\wedge$ for conjunction, $\vee$ for disjunction). Since a multiset can be defined as a pair (X, μ) where X is a set and $\mu : X \to \mathbf{N}$ is a function, it is now possible to identify knowledge bases over $\mathfrak{L}$ with functions $\mu : F_\mathfrak{L} \to \mathbf{N}$ where $F_\mathfrak{L}$ is the set of all formulas of $\mathfrak{L}$.

Notation. Throughout, calligraphics $\mathcal{K}, \mathcal{S}, \ldots$ are used for multisets whereas $F, K, S, \ldots$ are reserved to indicate *sets*.

For an introduction to the topic, [19] gives a brief account of multisets that deals with all the definitions in this section.

Definition 1 *A multiset of formulas $\mathcal{K}$ over $\mathfrak{L}$ is an ordered pair $(F_\mathfrak{L}, \mu)$ where $F_\mathfrak{L}$ is the set of formulas of $\mathfrak{L}$ and $\mu : F_\mathfrak{L} \to \mathbf{N}$ is a function. The value of $\mu(\varphi)$ is the multiplicity of φ in $\mathcal{K}$. The cardinality of $\mathcal{K}$ is*

$$\operatorname{card}(\mathcal{K}) \stackrel{\text{def}}{=} \sum_{\varphi \in F_\mathfrak{L}} \mu(\varphi).$$

The carrier of $\mathcal{K}$, denoted $C(\mathcal{K})$, is the largest subset of $F_\mathfrak{L}$ with no null multiplicity in $\mathcal{K}$. In symbols,

$$C(\mathcal{K}) \stackrel{\text{def}}{=} \{\varphi \in F_\mathfrak{L} \mid \mu(\varphi) \geq 1\}.$$

Notation. The delimiters for multisets are still { and }, with repetitions of elements as needed. Specifically, singletons are still written $\{\alpha\}$ with multiplicity defined as: $\mu_{\{\alpha\}}(\varphi) = 1$ for $\varphi = \alpha$ else $\mu_{\{\alpha\}}(\varphi) = 0$.

Definition 2 *The empty multiset is denoted { } with multiplicity defined as: $\mu_{\{\}}(\varphi) = 0$ for all $\varphi \in F_\mathfrak{L}$.*

Thus, the empty multiset { } can be identified with the empty subset of $F_\mathfrak{L}$. Of course, { } has no member where membership is defined as follows.

Definition 3 *Let $\mathcal{K} = (F_\mathfrak{L}, \mu_\mathcal{K})$ be a multiset of formulas of $\mathfrak{L}$. For all $\varphi \in F_\mathfrak{L}$, $\varphi \in \mathcal{K}$ iff $\mu_\mathcal{K}(\varphi) \geq 1$.*

Notation. Definition 3 examplifies the convention that, for the sake of clarity, multiplicity functions are indexed by the name of the multiset. For instance, $\mu_\mathcal{K}$ denotes the multiplicity function of $\mathcal{K}$ and $\mu_{\mathcal{K}_i}$ denotes the multiplicity function of $\mathcal{K}_i$, ...

Definition 4 *Two multisets $\mathcal{K}_1 = (F_\mathfrak{L}, \mu_{\mathcal{K}_1})$ and $\mathcal{K}_2 = (F_\mathfrak{L}, \mu_{\mathcal{K}_2})$ of formulas of $\mathfrak{L}$ are disjoint iff there exist no $\varphi \in F_\mathfrak{L}$ such that $\varphi \in \mathcal{K}_1$ and $\varphi \in \mathcal{K}_2$.*

Against various alternative definitions of a submultiset, the following notion for a subbase is in order.

Definition 5 *Let $\mathcal{K}_1 = (F_\mathfrak{L}, \mu_{\mathcal{K}_1})$ and $\mathcal{K}_2 = (F_\mathfrak{L}, \mu_{\mathcal{K}_2})$ be two multisets of formulas of $\mathfrak{L}$. $\mathcal{K}_1$ is a submultiset of $\mathcal{K}_2$, written $\mathcal{K}_1 \Subset \mathcal{K}_2$, iff $\mu_{\mathcal{K}_1} \leq \mu_{\mathcal{K}_2}$.*

This is an instance where the assumption of a fixed language $\mathfrak{L}$ helps as it implies that $\mu_{\mathcal{K}_1}$ and $\mu_{\mathcal{K}_2}$ have the same domain (otherwise, Definition 5 would not only require comparing restrictions of $\mu_{\mathcal{K}_1}$ and $\mu_{\mathcal{K}_2}$ but also require the extra condition that the preimage of $\mathbf{N}^*$ for $\mu_{\mathcal{K}_1}$ be a subset of that for $\mu_{\mathcal{K}_2}$).

Remark. $\mathcal{K}_1 = \mathcal{K}_2$ iff $\mathcal{K}_1 \Subset \mathcal{K}_2$ and $\mathcal{K}_2 \Subset \mathcal{K}_1$.

Among various ways to join two multisets into a bigger multiset, of interest here is multiset sum defined as follows.

Definition 6 *The sum of two multisets $\mathcal{K}_1 = (F_\mathfrak{L}, \mu_{\mathcal{K}_1})$ and $\mathcal{K}_2 = (F_\mathfrak{L}, \mu_{\mathcal{K}_2})$ of formulas of $\mathfrak{L}$ is defined as*

$$\mathcal{K}_1 \uplus \mathcal{K}_2 \stackrel{\text{def}}{=} (F_\mathfrak{L}, \mu_{\mathcal{K}_1} + \mu_{\mathcal{K}_2}).$$

That is, $\mu_{\mathcal{K}_1 \uplus \mathcal{K}_2} = \mu_{\mathcal{K}_1} + \mu_{\mathcal{K}_2}$.

Instead of joining, removing is also an important operation for multisets. The following definition is useful later in the chapter.

Definition 7 *For two multisets $\mathcal{K}_1 = (F_\mathfrak{L}, \mu_{\mathcal{K}_1})$ and $\mathcal{K}_2 = (F_\mathfrak{L}, \mu_{\mathcal{K}_2})$ of formulas of $\mathfrak{L}$, multiset difference is defined as*

$$\mathcal{K}_1 - \mathcal{K}_2 \stackrel{\text{def}}{=} (F_\mathfrak{L}, \mu_{\mathcal{K}_1 - \mathcal{K}_2})$$

where $\mu_{\mathcal{K}_1 - \mathcal{K}_2}(\varphi) = \max(0, \mu_{\mathcal{K}_1}(\varphi) - \mu_{\mathcal{K}_2}(\varphi))$ for all $\varphi \in F_\mathfrak{L}$.

Warning. Throughout the text, only finite multisets are considered. That is, $\mathcal{K}, \mathcal{K}', \ldots$ always refer to finite multisets of formulas of $\mathfrak{L}$.

2.1 Inference and Consistency for Multisets of Formulas

Considering multisets of logical formulas, the first question is about inference. Should classical inference not be distorted, inference would apply to carriers of multisets of formulas. That is, duplicates would play a role wrt measuring inconsistency but not wrt inference: In symbols,

$$\mathcal{K} \Vdash \varphi \text{ is defined as } C(\mathcal{K}) \vdash \varphi.$$

Hence, a knowledge base $\mathcal{K}$ is consistent, $\mathcal{K} \not\Vdash \bot$, iff the carrier of $\mathcal{K}$ is consistent, $C(\mathcal{K}) \not\vdash \bot$.

Warning. Importantly, inference is *unchanged*: it is still the case that, e.g., $\mathcal{K} \Vdash \varphi \to (\varphi \wedge \varphi)$ holds.[1]

Reminder. *Given a set K of formulas, a set $S \subseteq K$ is a minimal unsatisfiable subset of K if and only if $S \vdash \bot$ and $S \setminus \{\varphi\} \nvdash \bot$ for all $\varphi \in S$. The set of minimal unsatisfiable subsets of K is denoted $MUS(K)$. A formula φ is free for K if and only if $\varphi \notin \bigcup MUS(K \cup \{\varphi\})$.*

Generalizing to the case that a knowledge base is a multiset of formulas, what is a minimal inconsistent subbase? In fact, it is a set (thereby supporting a notion of a minimal inconsistent subset of a multiset of formulas). Indeed, although duplicates may happen to cause *more* inconsistency, duplicates are helpless wrt causing inconsistency in the first place: e.g., $\{p, p, p, \neg p, \neg p\}$ is inconsistent and so is $\{p, \neg p\}$. However, the minimal inconsistent subsets of a multiset (of formulas) form a multiset as it matters whether, and how many, duplicates are used: If $\mathcal{K} = \{p, p, \neg p\}$ then the multiset of its minimal inconsistent subsets is $\{\{p, \neg p\}, \{p, \neg p\}\}$. These ideas result in the next definition.

Definition 8 *Let $\mathcal{K} = (F_\mathfrak{L}, \mu_\mathcal{K})$ be a multiset of formulas over $\mathfrak{L}$. The multiset of the minimal inconsistent subsets of $\mathcal{K}$ is defined as*

$$MIS(\mathcal{K}) \stackrel{\text{def}}{=} (2^{F_\mathfrak{L}}, \mu_{MIS(\mathcal{K})})$$

where for all $T \subseteq F_\mathfrak{L}$

$$\mu_{MIS(\mathcal{K})}(T) = \begin{cases} \prod_{\alpha \in T} \mu_\mathcal{K}(\alpha) & \text{if } T \in MUS(C(\mathcal{K})) \\ 0 & \text{else.} \end{cases}$$

A formula φ of $\mathfrak{L}$ is free for $\mathcal{K}$ iff $\varphi \notin \bigcup MUS(C(\mathcal{K}) \cup \{\varphi\})$.

Since $MIS(\mathcal{K})$ is not a multiset of formulas of $\mathfrak{L}$ but a multiset of sets of formulas of $\mathfrak{L}$, its underlying set is not $F_\mathfrak{L}$ but $2^{F_\mathfrak{L}}$, namely the powerset of the set of all formulas of $\mathfrak{L}$. Moreover, please notice the difference with a multiset of subbases of $\mathcal{K}$, which would have $\mathbf{N}^{F_\mathfrak{L}}$ as its underlying set.

When $\mathcal{K}$ is actually a set, the notion of the collection of minimal inconsistent subsets of $\mathcal{K}$ reduces to the classical case because it then happens both that $MUS(\mathcal{K})$ is defined and $MIS(\mathcal{K}) = MUS(\mathcal{K})$.

It can be checked that a formula φ is free for $\mathcal{K}$ iff no $\mathcal{K}' \Subset \mathcal{K}$ is such that $\mathcal{K}' \nVdash \bot$ and $\mathcal{K}' \uplus \{\varphi\} \Vdash \bot$. Clearly, Definition 8 implies that a formula φ is free for $\mathcal{K}$ iff φ is free for $C(\mathcal{K})$. When $\mathcal{K}$ is actually a set, the notion of a free formula accordingly reduces to the classical case.

---

[1] Using $\to$ to denote material implication, defined as usual from $\neg$ and $\vee$.

3 Postulates for Inconsistency Measures over Sets of Formulas

In the literature, an inconsistency measure is a function I that maps every finite set K of logical formulas to a value in $\mathbb{R}^+ \cup \{\infty\}$. Of course, not all such functions can be regarded as an inconsistency measure. A number of conditions have been proposed, called postulates for inconsistency measures [6, 7, 15, 16, 20, 2, 22, 9]. Here are four most famous postulates, due to Hunter and Konieczny [6].

- $I(K) = 0$ iff $K \not\vdash \bot$ (Consistency Null)
- $I(K \cup K') \geq I(K)$ (Monotony)
- If $\alpha \vdash \beta$ and $\alpha \not\vdash \bot$ then $I(K \cup \{\alpha\}) \geq I(K \cup \{\beta\})$ (Dominance)
- If α is free for K then $I(K \cup \{\alpha\}) = I(K)$ (Free Formula Independence)

Since only finite sets of formulas are considered, (Monotony) is to be replaced throughout by

- $I(K \cup \{\alpha\}) \geq I(K)$ (Unit Monotony)

The postulates for inconsistency measures that have been proposed in the literature cover a large spectrum. In particular, it seems relevant to mention that some of them are devoted to a connective, for instance,

- $I(K \cup \{\alpha \wedge \beta\}) \geq I(K \cup \{\alpha\})$ (Conjunction Dominance)

Such postulates are called here "set"-postulates in order to stress the fact that they are given for sets of formulas, in contrast to postulates for multisets of formulas as developed from Section 4 on.

3.1 Form of Postulates

In the literature, postulates for inconsistency measures are of a similar format. As an evidence for this, apart from the single postulate (Almost Consistency), all others in the collection of 18 existing postulates examined in [22] fall into one of four categories (where $=$ can be replaced by $\leq$ or $\geq$ or $<$ or $>$ or $\neq$) as follows:

(1) If ... then $I(K) = X$ where X can be $I(K \cup \{\alpha\})$ or a value in $\mathbb{R}^+ \cup \{\infty\}$

(2) If ... then $I(K \cup \{\alpha\}) = I(K \cup \{\beta\})$

(3) If ... then $I(K) = I(K')$

(4) If ... then $I(K \cup K') = I(K) + I(K')$

Conforming to (1) are 8 postulates out of 18 from [22] and 3 more from [2]. More figures witnessing that many postulates are of the form (1)-(4) are given in the table below:

Format of postulates	postulates from [22]	from [2]
(1)	8	3
(2)	2	5
(3)	5	2
(4)	2	

To be more illustrative, (Consistency Null) is in category (1) as split into its if part and contrapositive of its only if part:
- If $K \not\vdash \bot$ then $I(K) = 0$
- If $K \vdash \bot$ then $I(K) > 0$

It may happen that the condition replaced with omission marks in (1)-(4) is automatically fulfilled, as (Unit Monotony) illustrates:
- $I(K \cup \{\alpha\}) \geq I(K)$

In addition to (Dominance), another postulate from the collection of 18 in [22] belongs to category (2), namely (Irrelevance of Syntax), which can be turned into (Unit Irrelevance of Syntax) by taking adavantage of the fact that only finite sets of formulas are considered for K throughout the chapter:
- If $\alpha \equiv \beta$ [2] then $I(K \cup \{\alpha\}) = I(K \cup \{\beta\})$ (Unit Irrelevance of Syntax)

Conditions can be quite diverse, and sometimes, for the sake of readability, rewritten using restricted quantifiers *"For $Y \in Z$, if ... then ..."*, as in the following example of category (3)
- For $K, K' \in MUS(K'')$, if $\text{card}(K) = \text{card}(K')$ then $I(K) = I(K')$
    (Equal Conflict)

Out of the 18 existing postulates collected in [22], (Almost Consistency) is indeed the only one that fails to belong to any of the categories (1)-(4) above:
- For a sequence of MUSes $K_1, K_2, \ldots$ if $\lim_{i \to \infty} \text{card}(K_i) = \infty$ then $\lim_{i \to \infty} I(K_i) = 0$
    (Almost Consistency)

The reason for the above exposition about the form of postulates is that the next four pages rely on the form of postulates to pinpoint some properties that turn to be of interest for multiset versions of postulates for inconsistency measures, both in Section 4 and Section 5.

---
[2] $\alpha \equiv \beta$ is shorthand for $\alpha \vdash \beta$ and $\beta \vdash \alpha$.

which has already been proposed in [1] and [4] under the equivalent (see footnote 3) formulation

- For $\alpha \notin K$, if $\alpha \vdash \beta$ and $\alpha \not\vdash \bot$ then $\boldsymbol{I}(K \cup \{\alpha\}) \geq \boldsymbol{I}(K \cup \{\beta\})$.

3.2.2 Equivalently using Set Union or Set Difference

A number of "set"-postulates can be formulated either by using set union or by resorting to set difference. A frequent form (where $\mathfrak{C}[\alpha, \beta]$ stands for a condition on α and β, *with no occurrence of K*) is

$$\text{If } \mathfrak{C}[\alpha, \beta] \text{ then } \boldsymbol{I}(K \cup \{\alpha\}) \geq \boldsymbol{I}(K \cup \{\beta\}).$$

As just mentioned, a proviso $\alpha \notin K$ must be introduced in order to avoid problems with the case that α is in K but β is not (since $\boldsymbol{I}(K) \geq \boldsymbol{I}(K \cup \{\beta\})$ would be required). Then, the form of interest is

$$\text{For } \alpha \notin K, \text{ if } \mathfrak{C}[\alpha, \beta] \text{ then } \boldsymbol{I}(K \cup \{\alpha\}) \geq \boldsymbol{I}(K \cup \{\beta\}) \tag{\dag}$$

for which a substitute using set difference would be

$$\text{For } \alpha \in K, \text{ if } \mathfrak{C}[\alpha, \beta] \text{ then } \boldsymbol{I}(K \setminus \{\beta\}) \geq \boldsymbol{I}(K \setminus \{\alpha\}) \tag{\dag\dag}$$

In (†)–(††), if β does not occur in the condition $\mathfrak{C}[\alpha, \beta]$ then $K \cup \{\beta\}$ and $K \setminus \{\beta\}$ must be replaced by K.

Theorem 1 *In the presence of (Unit Monotony), (†) and (††) are equivalent.*

Proof First, $\alpha \neq \beta$ is assumed since $\alpha = \beta$ makes (†) and (††) tautological. [(†) → (††)] Let $\alpha \in K$. Assume $\beta \in K$. Apply (†) wrt $K \setminus \{\alpha, \beta\}$, that is, $\mathfrak{C}[\alpha, \beta]$ implies $\boldsymbol{I}((K \setminus \{\alpha, \beta\}) \cup \{\alpha\}) \geq \boldsymbol{I}((K \setminus \{\alpha, \beta\}) \cup \{\beta\})$. It ensues from $\alpha \neq \beta$ that $\mathfrak{C}[\alpha, \beta]$ implies $\boldsymbol{I}(K \setminus \{\beta\}) \geq \boldsymbol{I}(K \setminus \{\alpha\})$. Assume $\beta \notin K$ instead. Trivially, $K \setminus \{\beta\} = K$. Thus, the instance of (††) to be proved is: For $\alpha \in K$, if $\mathfrak{C}[\alpha, \beta]$ then $\boldsymbol{I}(K) \geq \boldsymbol{I}(K \setminus \{\alpha\})$. This is a fairly simple consequence of an instance of (Unit Monotony), that is, $\boldsymbol{I}(K' \cup \{\alpha\}) \geq \boldsymbol{I}(K')$ for $K' = K \setminus \{\alpha\}$. [(††) → (†)] Let $\alpha \notin K$ and assume $\beta \notin K$. Apply (††) wrt $K \cup \{\alpha, \beta\}$, i.e., $\mathfrak{C}[\alpha, \beta]$ implies $\boldsymbol{I}((K \cup \{\alpha, \beta\}) \setminus \{\beta\}) \geq \boldsymbol{I}((K \cup \{\alpha, \beta\}) \setminus \{\alpha\})$. Accordingly, $\mathfrak{C}[\alpha, \beta]$ implies $\boldsymbol{I}(K \cup \{\alpha\}) \geq \boldsymbol{I}(K \cup \{\beta\})$ because $\alpha \neq \beta$. The alternative assumption is $\beta \in K$. Trivially, $K \cup \{\beta\} = K$. Thus, the instance of (†) to be proved is: For $\alpha \notin K$, if $\mathfrak{C}[\alpha, \beta]$ then $\boldsymbol{I}(K \cup \{\alpha\}) \geq \boldsymbol{I}(K)$. This is an easy consequence of (Unit Monotony). ∎

Clearly, there is no need to deal with $\leq$ since it suffices to use the converse [4] $\mathfrak{C}^{-1}$ of $\mathfrak{C}$ thereby turning

$$\text{If } \mathfrak{C}[\alpha,\beta] \text{ then } \boldsymbol{I}(K \cup \{\alpha\}) \leq \boldsymbol{I}(K \cup \{\beta\})$$

into

$$\text{If } \mathfrak{C}^{-1}[\alpha,\beta] \text{ then } \boldsymbol{I}(K \cup \{\alpha\}) \geq \boldsymbol{I}(K \cup \{\beta\}).$$

As to postulates with $=$ instead of $\geq$, the discussion in [3] recommends introducing the twofold proviso $\alpha \notin K$ and $\beta \notin K$. Then, the following result (*again for $\mathfrak{C}[\alpha,\beta]$ independent of K*) arises.

Theorem 2 *If $?_=$ is uniformously replaced by $\geq$ or $\leq$ or $=$ in both ($\ddagger$) and ($\ddagger\ddagger$) below, then*

$$\text{For } \alpha \notin K, \beta \notin K, \text{ if } \mathfrak{C}[\alpha,\beta] \text{ then } \boldsymbol{I}(K \cup \{\alpha\}) ?_= \boldsymbol{I}(K \cup \{\beta\}) \qquad (\ddagger)$$

and

$$\text{For } \alpha \in K, \beta \in K, \text{ if } \mathfrak{C}[\alpha,\beta] \text{ then } \boldsymbol{I}(K \setminus \{\beta\}) ?_= \boldsymbol{I}(K \setminus \{\alpha\}) \qquad (\ddagger\ddagger)$$

are equivalent.

Proof First, $\alpha \neq \beta$ is assumed since $\alpha = \beta$ makes ($\ddagger$) and ($\ddagger\ddagger$) tautological. [($\ddagger$) $\to$ ($\ddagger\ddagger$)] Let $\alpha \in K$ and $\beta \in K$. Apply ($\ddagger$) wrt $K \setminus \{\alpha,\beta\}$, which means that $\mathfrak{C}[\alpha,\beta]$ implies $\boldsymbol{I}((K \setminus \{\alpha,\beta\}) \cup \{\alpha\}) ?_= \boldsymbol{I}((K \setminus \{\alpha,\beta\}) \cup \{\beta\})$. By $\alpha \neq \beta$, it ensues that $\mathfrak{C}[\alpha,\beta]$ implies $\boldsymbol{I}(K \setminus \{\beta\}) ?_= \boldsymbol{I}(K \setminus \{\alpha\})$. [($\ddagger\ddagger$) $\to$ ($\ddagger$)] Let $\alpha \notin K$ and $\beta \notin K$. Apply ($\ddagger\ddagger$) wrt $K \cup \{\alpha,\beta\}$, which directly means that $\mathfrak{C}[\alpha,\beta]$ implies $\boldsymbol{I}((K \cup \{\alpha,\beta\}) \setminus \{\beta\}) ?_= \boldsymbol{I}((K \cup \{\alpha,\beta\}) \setminus \{\alpha\})$. Then, $\mathfrak{C}[\alpha,\beta]$ implies $\boldsymbol{I}(K \cup \{\alpha\}) ?_= \boldsymbol{I}(K \cup \{\beta\})$ because $\alpha \neq \beta$. ∎

Interestingly, some postulates out of the scope of Theorem 1 (similarly, Theorem 2) still enjoy the equivalence because the condition they rely upon, although depending on K, can be appropriately arranged. Next is such an example about a postulate introduced in [16].

Lemma 1 *The following two postulates are equivalent.*

- *If α is free for K then $\boldsymbol{I}(K) \geq \boldsymbol{I}(K \setminus \{\alpha\})$* (*Free Formula Dilution*)
- *If α is free for K then $\boldsymbol{I}(K \cup \{\alpha\}) \geq \boldsymbol{I}(K)$*

Proof For clarity, consider

- If α is free for K then $\boldsymbol{I}(K) \geq \boldsymbol{I}(K \setminus \{\alpha\})$ (i)
- If β is free for K then $\boldsymbol{I}(K \cup \{\beta\}) \geq \boldsymbol{I}(K)$ (ii)

---

[4] Since $\mathfrak{C}[\alpha,\beta]$ ascribes true or false to ordered pairs (α,β), it is a binary relation and the converse is a well-known notion: here, $\mathfrak{C}^{-1}[\alpha,\beta]$ holds iff $\mathfrak{C}[\beta,\alpha]$ holds.

As to (i) → (ii), assume β is free for K. If $\beta \in K$ then $\boldsymbol{I}(K \cup \{\beta\}) \geq \boldsymbol{I}(K)$ trivially holds. Else, $\beta \notin K$. That β is free for K gives $\beta \notin \bigcup MUS(K \cup \{\beta\})$. Therefore, β is free for $K \cup \{\beta\}$. Applying (i), $\boldsymbol{I}(K \cup \{\beta\}) \geq \boldsymbol{I}((K \cup \{\beta\}) \setminus \{\beta\})$. However, $(K \cup \{\beta\}) \setminus \{\beta\}$ is K since $\beta \notin K$. Therefore, $\boldsymbol{I}(K \cup \{\beta\}) \geq \boldsymbol{I}(K)$. As to (ii) → (i), assume α is free for K. If $\alpha \notin K$ then $\boldsymbol{I}(K) \geq \boldsymbol{I}(K \setminus \{\alpha\})$ trivially holds. Otherwise, $\alpha \in K$. Thus, $K = (K \setminus \{\alpha\}) \cup \{\alpha\}$. Since α is free for K, it is the case that $\alpha \notin \bigcup MUS(K \cup \{\alpha\})$ which also means that α is free for $K \setminus \{\alpha\}$. By (ii), $\boldsymbol{I}((K \setminus \{\alpha\}) \cup \{\alpha\}) \geq \boldsymbol{I}(K \setminus \{\alpha\})$ ensues. Again, $K = (K \setminus \{\alpha\}) \cup \{\alpha\}$ hence it follows that $\boldsymbol{I}(K) \geq \boldsymbol{I}(K \setminus \{\alpha\})$. ∎

3.2.3 Subsumption between "Set"-postulates

It is known (see [2] or [22] for instance) that dependencies exist between various "set"-postulates. In the prospect of studying how such dependencies may persist or break when passing from sets to multisets, it is presumably instructive to start with a few cases of "set"-postulates subsumed by other "set"-postulates. The first relationship of interest here is about two postulates expressing that the amount of inconsistency cannot change through tautologies:

- If $\vdash \alpha$ then $\boldsymbol{I}(K \cup \{\alpha\}) = \boldsymbol{I}(K)$ (Tautology Independence)
- If $\vdash \alpha$ then $\boldsymbol{I}(K \cup \{\alpha \wedge \beta\}) = \boldsymbol{I}(K \cup \{\beta\})$ (⊤-conjunct Independence)

Lemma 2 *Assuming (Consistency Null), (Unit Monotony), and (Conjunction Dominance), (⊤-conjunct Independence) entails (Tautology Independence).*

Proof Let $\vdash \alpha$. Assume $K \neq \emptyset$. There must exist $\beta \in K$, and, accordingly, $\boldsymbol{I}(K) = \boldsymbol{I}(K \cup \{\beta\})$ trivially holds. Applying (⊤-conjunct Independence) gives $\boldsymbol{I}(K \cup \{\alpha \wedge \beta\}) = \boldsymbol{I}(K \cup \{\beta\})$. Therefore, $\boldsymbol{I}(K \cup \{\alpha \wedge \beta\}) = \boldsymbol{I}(K)$. By virtue of (Conjunction Dominance), $\boldsymbol{I}(K \cup \{\alpha\}) \leq \boldsymbol{I}(K \cup \{\alpha \wedge \beta\})$. It follows that $\boldsymbol{I}(K \cup \{\alpha\}) \leq \boldsymbol{I}(K)$. The converse holds by (Unit Monotony). As a result, $\boldsymbol{I}(K \cup \{\alpha\}) = \boldsymbol{I}(K)$ holds for $K \neq \emptyset$. The remaining case $K = \emptyset$ is settled by (Consistency Null) because $\boldsymbol{I}(\emptyset) = 0 = \boldsymbol{I}(\{\alpha\})$ whenever $\vdash \alpha$. ∎

Another case of subsumption with an interesting outcome in the multiset framework is as follows.

Lemma 3 *In the presence of (Unit Monotony), (Dominance) is equivalent to the following postulate.*
- For $\alpha \in K$, if $\alpha \not\vdash \bot$ and $\alpha \vdash \beta$ then $\boldsymbol{I}(K \cup \{\beta\}) = \boldsymbol{I}(K)$ (A_1)

Proof Let α and β be such that $\alpha \not\vdash \bot$ and $\alpha \vdash \beta$. Assume (A_1). Trivially, $\alpha \in K \cup \{\alpha\}$ hence (A_1) gives $\boldsymbol{I}(K \cup \{\alpha\} \cup \{\beta\}) = \boldsymbol{I}(K \cup \{\alpha\})$. According to (Unit Monotony), $\boldsymbol{I}(K \cup \{\alpha\} \cup \{\beta\}) \geq \boldsymbol{I}(K \cup \{\beta\})$. By transitivity, it follows that $\boldsymbol{I}(K \cup \{\alpha\}) \geq \boldsymbol{I}(K \cup \{\beta\})$. Conversely, assume (Dominance). Consider K such that $\alpha \in K$. (Dominance) gives $\boldsymbol{I}(K \cup \{\alpha\}) \geq \boldsymbol{I}(K \cup \{\beta\})$. Accordingly, $\boldsymbol{I}(K) \geq \boldsymbol{I}(K \cup \{\beta\})$ since $\boldsymbol{I}(K \cup \{\alpha\}) = \boldsymbol{I}(K)$ in view of $\alpha \in K$. The converse, i.e., $\boldsymbol{I}(K \cup \{\beta\}) \geq \boldsymbol{I}(K)$ holds by (Unit Monotony). ∎

4 Inconsistency Measures over Multisets of Formulas

This section is devoted to the topics of Sections 3.2.1-3.2.2 now investigated in the multiset case. That is, the following questions are studied. What about guarding multiset postulates? What about expressing a multiset postulate with multiset sum or multiset difference? In between, a counter-example is given that fails the same property whether in multiset version or in set version.

Here, an inconsistency measure is a function I that maps every (finite) $\mathcal{K}$ to a value in $\mathbf{R}^+ \cup \{\infty\}$. Again, not all such functions can be viewed as an inconsistency measure and postulates are in order, e.g., the following versions of two postulates listed above:

- $I(\mathcal{K}) = 0$ iff $\mathcal{K} \not\Vdash \bot$ (Consistency Null)^M
- $I(\mathcal{K} \uplus \{\alpha\}) \geq I(\mathcal{K})$ (Unit Monotony)^M

Notation. As just exemplified, the multiset counterpart of a "set"-postulate is given the same name with a final superscript "M" in order to easily distinguish it from the original version.

More generally, in what way does a multiset version of a postulate differ from the original "set"-postulate? The answer is certainly based on what items occurring in "set"-postulates can be changed to? In fact, the list is not so long:

$$
\begin{array}{rcl}
K & \longrightarrow & \mathcal{K} \\
\cup & \longrightarrow & \uplus \\
\setminus & \longrightarrow & - \\
\subseteq & \longrightarrow & \Subset \\
\vdash & \longrightarrow & \Vdash \\
MUS & \longrightarrow & MIS
\end{array}
$$

4.1 Multiset Postulates do not Need to be Guarded

These substitutions are the only way a defect of a "set"-postulate can disappear in the extended setting. Here is an illustration. For a formula α belonging to a set K, $K \cup \{\alpha\} = K$. In sharp contrast, α belonging to a multiset $\mathcal{K}$ fails to entail $\mathcal{K} \uplus \{\alpha\} = \mathcal{K}$. If a "set"-postulate goes astray by requiring, say, $I(K \cup \{\alpha\}) \geq I(K')$ for some K', this does not mean that requiring $I(\mathcal{K} \uplus \{\alpha\}) \geq I(\mathcal{K}')$ is mistaken: A relationship between K' and $K \cup \{\alpha\}$ (possibly due to $K \cup \{\alpha\} = K$) that makes $I(K \cup \{\alpha\}) \geq I(K')$ wrong need not hold between $\mathcal{K}'$ and $\mathcal{K} \uplus \{\alpha\}$.

Example 2 *(Adapted from [16]) Consider $K = \{p, \neg q, p \wedge q\}$. Take $\alpha = p \wedge q$ and $\beta = \neg p \vee q$. So, $\alpha \not\vdash \bot$ and $\alpha \vdash \beta$. Since K is a set, (Dominance) requires* here $I(K) \geq I(K \cup \{\beta\})$, i.e.,

$$I\left(\left\{\begin{array}{c} p, \\ \neg q, \\ p \wedge q \end{array}\right\}\right) \geq I\left(\left\{\begin{array}{c} p, \\ \neg q, \\ p \wedge q, \\ \neg p \vee q \end{array}\right\}\right)$$

As to multisets, the fact that α is a member of $\mathcal{K}$ does not mean that $\mathcal{K}$ supplemented with α can be identified with $\mathcal{K}$. Stated otherwise, the multiset version of (Dominance) would only require here $I(\mathcal{K} \uplus \{\alpha\}) \geq I(\mathcal{K} \uplus \{\beta\})$, i.e.,

$$I\left(\left\{\begin{array}{c} p, \\ \neg q, \\ p \wedge q, \\ p \wedge q \end{array}\right\}\right) \geq I\left(\left\{\begin{array}{c} p, \\ \neg q, \\ p \wedge q, \\ \neg p \vee q \end{array}\right\}\right)$$

In general, multiset postulates do not need guarded variants. The reason is that the proviso $\alpha \notin \mathcal{K}$ (possibly supplemented with $\beta \notin \mathcal{K}$, too) actually occurs to preclude $K = K \cup \{\alpha\}$ (similarly, to preclude $K = K \cup \{\beta\}$). However, it is always the case that $\mathcal{K} \neq \mathcal{K} \uplus \{\alpha\}$ (similarly, $\mathcal{K} \neq \mathcal{K} \uplus \{\beta\}$) hence the need for such a proviso vanishes. This is established by the next result.

Fact 1 *For $\mathfrak{C}[\alpha, \beta]$ independent of $\mathcal{K}$, the postulates*

$$\text{If } \mathfrak{C}[\alpha, \beta] \text{ then } I(\mathcal{K} \uplus \{\alpha\}) \geq I(\mathcal{K} \uplus \{\beta\})$$

and

Whenever $\mathcal{K} \uplus \{\alpha\} \neq \mathcal{K} \neq \mathcal{K} \uplus \{\beta\}$, if $\mathfrak{C}[\alpha, \beta]$ then $I(\mathcal{K} \uplus \{\alpha\}) \geq I(\mathcal{K} \uplus \{\beta\})$

are equivalent.

Proof Trivial from the definition of multiset sum. ∎

4.2 An Example of an Enduring Failure of an Inconsistency Measure

Contrariwise, Example 3 [16] involves no duplicates hence the violation of (Dominance) exhibited above extends to the multiset version.

Example 3 *[16] Let $K = \{p, p \wedge r, \neg q\}$. Let $\alpha = p \wedge r \wedge (\neg p \vee q)$ and $\beta = \neg p \vee q$. Since $K \cup \{\alpha\}$ has a single minimal inconsistent subset $\{\neg q, p \wedge r \wedge (\neg p \vee q)\}$ while $K \cup \{\beta\}$ has two minimal inconsistent subsets, which are $\{p, \neg q, \neg p \vee q\}$ and $\{p \wedge r, \neg q, \neg p \vee q\}$, it happens that $I_{MI}(K \cup \{\alpha\}) = 1 < 2 = I_{MI}(K \cup \{\beta\})$ —by definition [6], $I_{MI}(K)$ is the cardinality of $MUS(K)$. (Dominance) thus fails here and so does (taking $I_{MI}(\mathcal{K})$ to be the cardinality of $MIS(\mathcal{K})$) its multiset variant: If $\alpha \vdash \beta$ and $\alpha \nvdash \bot$ then $I(\mathcal{K} \uplus \{\alpha\}) \geq I(\mathcal{K} \uplus \{\beta\})$.*

Since sets are multisets, it is tempting to think that a failure of the set version of a postulate becomes ipso facto a failure of the multiset version of the postulate. In general, this does not hold: for an illustration, look again at Example 2, i.e., $K = \{p, \neg q, p \wedge q\}$ with α being $p \wedge q$ and β being $\neg p \vee q$. $K = K \cup \{\alpha\} = \{p, \neg q, p \wedge q\}$ has one minimal inconsistent subset $\{\neg q, p \wedge q\}$ while $K \cup \{\beta\} = \{p, \neg q, p \wedge q, \neg p \vee q\}$ has two minimal inconsistent subsets, i.e., $\{\neg q, p \wedge q\}$ and $\{p, \neg q, \neg p \vee q\}$, it happens that $\boldsymbol{I}(K \cup \{\alpha\}) \geq \boldsymbol{I}(K \cup \{\beta\})$ is violated by I_{MI} (it returns the cardinality of MUS) as follows

$$I_{MI}\left(\left\{\begin{array}{c} p, \\ \neg q, \\ p \wedge q \end{array}\right\}\right) = 1 < 2 = I_{MI}\left(\left\{\begin{array}{c} p, \\ \neg q, \\ p \wedge q, \\ \neg p \vee q \end{array}\right\}\right)$$

whereas the multiset case is such that $MIS(\mathcal{K} \uplus \{\alpha\})$ consists of two copies of $\{\neg q, p \wedge q\}$ hence $\boldsymbol{I}(\mathcal{K} \uplus \{\alpha\}) \geq \boldsymbol{I}(\mathcal{K} \uplus \{\beta\})$ is satisfied by I_{MI} (it returns the cardinality of MIS) for the instance

$$I_{MI}\left(\left\{\begin{array}{c} p, \\ \neg q, \\ p \wedge q \\ p \wedge q \end{array}\right\}\right) = 2 \geq 2 = I_{MI}\left(\left\{\begin{array}{c} p, \\ \neg q, \\ p \wedge q, \\ \neg p \vee q \end{array}\right\}\right)$$

4.3 Equivalently Using Multiset Sum or Multiset Difference

Theorem 2 also holds in the multiset framework, in the form below. Once more, $\mathfrak{C}[\alpha, \beta]$ stands for any condition independent of $\mathcal{K}$.

Theorem 3 *If $?_=$ is uniformously replaced by $\geq$ or $\leq$ or $=$ in both (∗) and (∗∗) below, then*

$$\text{If } \mathfrak{C}[\alpha, \beta] \text{ then } \boldsymbol{I}(\mathcal{K} \uplus \{\alpha\}) \; ?_= \; \boldsymbol{I}(\mathcal{K} \uplus \{\beta\}) \qquad (*)$$

and

$$\text{If } \mathfrak{C}[\alpha, \beta] \text{ then } \boldsymbol{I}(\mathcal{K} - \{\beta\}) \; ?_= \; \boldsymbol{I}(\mathcal{K} - \{\alpha\}) \qquad (**)$$

are equivalent.

Proof As $\alpha = \beta$ makes (∗) and (∗∗) tautological, assume $\alpha \neq \beta$ throughout. [(∗) → (∗∗)] Applying (∗) wrt $\mathcal{K} - \{\alpha, \beta\}$, it happens that $\mathfrak{C}[\alpha, \beta]$ implies $\boldsymbol{I}((\mathcal{K} - \{\alpha, \beta\}) \uplus \{\alpha\}) \; ?_= \; \boldsymbol{I}((\mathcal{K} - \{\alpha, \beta\}) \uplus \{\beta\})$. In view of $\alpha \neq \beta$, it follows that $\mu_{(\mathcal{K}-\{\alpha,\beta\}) \uplus \{\alpha\}}(\alpha) = \mu_{\mathcal{K}}(\alpha) - 1 + 1 = \mu_{\mathcal{K}}(\alpha)$ while $\mu_{(\mathcal{K}-\{\alpha,\beta\}) \uplus \{\alpha\}}(\beta) = \mu_{\mathcal{K}}(\beta) - 1$ and $\mu_{(\mathcal{K}-\{\alpha,\beta\}) \uplus \{\alpha\}}(\varphi) = \mu_{\mathcal{K}}(\varphi)$ for every $\varphi \neq \beta$. In other words,

$(\mathcal{K} - \{\alpha, \beta\}) \uplus \{\alpha\}$ is $\mathcal{K} - \{\beta\}$. Similarly, $(\mathcal{K} - \{\alpha, \beta\}) \uplus \{\beta\}$ is $\mathcal{K} - \{\alpha\}$. Then, the expected consequence holds: $\mathfrak{C}[\alpha, \beta]$ implies $\boldsymbol{I}(\mathcal{K} - \{\beta\}) ?_= \boldsymbol{I}(\mathcal{K} - \{\alpha\})$.
$[(**) \to (*)]$ Applying $(**)$ wrt $\mathcal{K} \uplus \{\alpha, \beta\}$, it happens that $\mathfrak{C}[\alpha, \beta]$ implies $\boldsymbol{I}((\mathcal{K} \uplus \{\alpha, \beta\}) - \{\beta\}) ?_= \boldsymbol{I}((\mathcal{K} \uplus \{\alpha, \beta\}) - \{\alpha\})$. Due to $\alpha \neq \beta$, it ensues that $\mu_{(\mathcal{K} \uplus \{\alpha, \beta\}) - \{\beta\}}(\beta) = \mu_\mathcal{K}(\beta) + 1 - 1 = \mu_\mathcal{K}(\beta)$ and $\mu_{(\mathcal{K} \uplus \{\alpha, \beta\}) - \{\beta\}}(\alpha) = \mu_\mathcal{K}(\alpha) + 1$ and $\mu_{(\mathcal{K} \uplus \{\alpha, \beta\}) - \{\beta\}}(\varphi) = \mu_\mathcal{K}(\varphi)$ for $\varphi \neq \alpha$. So, $(\mathcal{K} \uplus \{\alpha, \beta\}) - \{\beta\} = \mathcal{K} \uplus \{\alpha\}$. Similarly, $(\mathcal{K} \uplus \{\alpha, \beta\}) - \{\alpha\} = \mathcal{K} \uplus \{\beta\}$. All this leads us to the conclusion that $\mathfrak{C}[\alpha, \beta]$ implies $\boldsymbol{I}(\mathcal{K} \uplus \{\alpha\}) ?_= \boldsymbol{I}(\mathcal{K} \uplus \{\beta\})$. ∎

There is no need to express a counterpart to Theorem 1 because it is an immediate consequence of Theorem 3 due to the absence of proviso in the multiset versions $(*)$–$(**)$.

5 Detailing Inconsistency Postulates for Multisets of Formulas

5.1 Core Postulates: Consistency Null, Unit Monotony, Dominance

As already mentioned at the beginning of Section 4, the first two postulates from Hunter and Konieczny can be replicated in multiset guise quite literally:
- $\boldsymbol{I}(\mathcal{K}) = 0$ iff $\mathcal{K} \not\Vdash \bot$ $\hfill$ (Consistency Null)^M
- $\boldsymbol{I}(\mathcal{K} \uplus \{\alpha\}) \geq \boldsymbol{I}(\mathcal{K})$ $\hfill$ (Unit Monotony)^M

This is also the case for the third postulate, namely (Dominance).
- If $\alpha \Vdash \beta$ and $\alpha \not\Vdash \bot$ then $\boldsymbol{I}(\mathcal{K} \uplus \{\alpha\}) \geq \boldsymbol{I}(\mathcal{K} \uplus \{\beta\})$ $\hfill$ (Dominance)^M

A notable consequence of Fact 1 is that the instance of (Dominance)^M where $\mathcal{K}$ is a set *fails* to reduce to the original postulate.

Lemma 4 *(Dominance)^M is equivalent to the following postulate.*
- *If $\alpha_i \Vdash \beta_i$ for $i = 1, \ldots, n$ and $\{\alpha_1, \ldots, \alpha_n\} \not\Vdash \bot$ then $\boldsymbol{I}(\mathcal{K} \uplus \{\alpha_1, \ldots, \alpha_n\}) \geq \boldsymbol{I}(\mathcal{K} \uplus \{\beta_1, \ldots, \beta_n\})$*

Proof $\boldsymbol{I}(\mathcal{K} \uplus \{\alpha_1, \ldots, \alpha_n\}) \geq \boldsymbol{I}(\mathcal{K} \uplus \{\alpha_1, \ldots, \alpha_{n-1}\} \uplus \{\beta_n\})$ by (Dominance)^M as $\alpha_n \Vdash \beta_n$ and $\alpha_n \not\Vdash \bot$. Iterating, $\boldsymbol{I}(\mathcal{K} \uplus \{\beta_n, \beta_{n-1}, \ldots, \beta_{i+1}\} \uplus \{\alpha_1, \ldots, \alpha_i\}) \geq \boldsymbol{I}(\mathcal{K} \uplus \{\beta_n, \beta_{n-1}, \ldots, \beta_i\} \uplus \{\alpha_1, \ldots, \alpha_{i-1}\})$ in view of $\alpha_i \Vdash \beta_i$ and $\alpha_i \not\Vdash \bot$. Conversely, (Dominance)^M is simply the case $n = 1$. ∎

Lemma 5 *The following postulate is a consequence of (Dominance)^M but not equivalent with it.*
- *If $\alpha \wedge \beta \not\Vdash \bot$ then $\boldsymbol{I}(\mathcal{K} \uplus \{\alpha \wedge \beta\}) \geq \boldsymbol{I}(\mathcal{K} \uplus \{\beta\})$*

Lemma 3 does not extend to the multiset framework: *For $\alpha \in \mathcal{K}$, if $\alpha \not\Vdash \bot$ and $\alpha \Vdash \beta$ then $\boldsymbol{I}(\mathcal{K} \uplus \{\beta\}) = \boldsymbol{I}(\mathcal{K})$* is undesirable, and fails to be entailed by (Dominance)^M together with (Unit Monotony)^M.

5.2 Removal Postulates: Free Formula Dilution and Penalty

In [16], the set version of the following postulate is introduced which is shown to be an immediate consequence of (Monotony).

- If α is free for $\mathcal{K}$ then $\boldsymbol{I}(\mathcal{K}) \geq \boldsymbol{I}(\mathcal{K} - \{\alpha\})$ (Free Formula Dilution)^M

Lemma 6 *(Unit Monotony)^M entails (Free Formula Dilution)^M.*

Proof Since $\mathcal{K} = \{\alpha\} \uplus (\mathcal{K} - \{\alpha\})$ for $\alpha \in \mathcal{K}$, an instance of (Unit Monotony) is $\boldsymbol{I}(\mathcal{K}) \geq \boldsymbol{I}(\mathcal{K} - \{\alpha\})$ for $\alpha \in \mathcal{K}$. For $\alpha \notin \mathcal{K}$, it is trivial that $\mathcal{K} = \mathcal{K} - \{\alpha\}$. Combining both cases, (Unit Monotony) implies that $\boldsymbol{I}(\mathcal{K}) \geq \boldsymbol{I}(\mathcal{K} - \{\alpha\})$ holds for all $\alpha \in F_\mathfrak{L}$. ∎

The set version of the following postulate has been presented in [20].

- If $\alpha \in \mathcal{K}$ is not free for $\mathcal{K}$ then $\boldsymbol{I}(\mathcal{K}) > \boldsymbol{I}(\mathcal{K} - \{\alpha\})$ (Penalty)^M

As a "set"-postulate, (Penalty) ensures that *every* non-free *formula* counts for an increase in the amount of inconsistency. Its multiset version (Penalty)^M insists that *every duplicate* of a non-free formula counts for an increase in the amount of inconsistency.

Whenever α is not free for $\mathcal{K}$, Definition 8 gives $\alpha \in \bigcup MUS(C(\mathcal{K}) \cup \{\alpha\})$. Hence, there exists $S \in MUS(C(\mathcal{K}) \cup \{\alpha\})$ such that $\alpha \in S$. (Penalty)^M therefore ensures that duplicates (whenever involved in an inconsistency) cause an actual increase in amount of inconsistency due to the case that $\mu_\mathcal{K}(\alpha) > 1$ holds —since $\mu_{\mathcal{K}-\{\alpha\}}$ is then the same as $\mu_\mathcal{K}$ except for $\mu_{\mathcal{K}-\{\alpha\}}(\alpha) = \mu_\mathcal{K}(\alpha) - 1$.

Lemma 7 *(Free Formula Dilution)^M and (Penalty)^M entail (Monotony)^M.*

Proof Let $\mathcal{K}' = \{\alpha_1, \alpha_2, \alpha_3, \ldots, \alpha_n\}$. Either α_1 is free for $(\mathcal{K} \uplus \mathcal{K}') - \{\alpha_1\}$ or α_1 is not free for $(\mathcal{K} \uplus \mathcal{K}') - \{\alpha_1\}$. Considering the first alternative, it is easy to check that (Free Formula Dilution)^M gives $\boldsymbol{I}(\mathcal{K} \uplus \mathcal{K}') \geq \boldsymbol{I}((\mathcal{K} \uplus \mathcal{K}') - \{\alpha_1\})$. By the other alternative, (Penalty)^M entails $\boldsymbol{I}(\mathcal{K} \uplus \mathcal{K}') \geq \boldsymbol{I}((\mathcal{K} \uplus \mathcal{K}') - \{\alpha_1\})$. Iterating, $\boldsymbol{I}(\mathcal{K} \uplus \mathcal{K}') \geq \boldsymbol{I}((\mathcal{K} \uplus \mathcal{K}') - \mathcal{K}')$. That is, $\boldsymbol{I}(\mathcal{K} \uplus \mathcal{K}') \geq \boldsymbol{I}(\mathcal{K})$. ∎

As a final word about these postulates, the reader may notice that (Penalty) as stated in [20] omits the requirement $\alpha \in \mathcal{K}$. This discrepancy is *not* due to lifting the postulate to multisets: The requirement already occurs in the "set"-postulate when the notion chosen [5] for a free formula is such that a formula α may fail to be free for a set K while $\alpha \notin K$ (in [20], the notion chosen [6] has a side effect which indeed is that non-free formulas of K are in K).

---

[5] Here, the notion of a free formula is given in the Reminder at the start of Section 2.1.

[6] In counterbalance, [20] must recast (Free Formula Independence) in terms of set difference instead of expressing it in terms of set union.

5.3 Normalization Postulates

Two postulates about normalization were given in [7]. In the multiset version, they are as follows.

- $0 \leq I(\mathcal{K}) \leq 1$ (Normalization)^M
- If $\mathcal{K}' \in MIS(\mathcal{K})$ then $I(\mathcal{K}') = 1$ (MI-Normalization)^M

Taken together, these two postulates mean that MUSes have the highest degree of inconsistency thereby expressing that duplicates do *not* increase the amount of inconsistency. Thus, they form a pair conflicting with (Penalty)^M which indeed means that if a non-free formula α is duplicated in $\mathcal{K}$ then every duplicate of α does increase the amount of inconsistency. $\mathcal{K}'$ in (MI-Normalization)^M is ipso facto a set because members of $MIS(\mathcal{K})$ are sets. Whenever $MIS(\mathcal{K})$ consists of singleton sets only, then (MI-Normalization)^M conflicts with the following postulate that has been introduced in [16].

- $I(\mathcal{K}) = 1$ iff $\mathcal{K}' \Vdash \bot$ for all non-empty $\mathcal{K}' \Subset \mathcal{K}$ (Contradiction)^M

In particular, (Contradiction)^M states that $I(\mathcal{K}) = 1$ iff $MIS(\mathcal{K}) = (2^{F_\mathfrak{L}}, \mu)$ where $\mu(\{\varphi\}) = \mu_\mathcal{K}(\varphi)$ for each (inconsistent) $\varphi \in \mathcal{K}$ and $\mu(X) = 0$ for all other subsets of $F_\mathfrak{L}$. Thus, $I(\mathcal{K}) = 1$ no matter *how many* duplicates of an inconsistent φ are in $\mathcal{K}$, and this contradicts (Penalty)^M.

(Contradiction)^M as well as (MI-Normalization)^M only makes sense in the context of (Normalization)^M which sets 1 as maximum.

5.4 Independence Postulates

Both postulates about tautologies can be straightforwardly reformulated in the multiset version.

- if $\vdash \alpha$ then $I(\mathcal{K} \uplus \{\alpha\}) = I(\mathcal{K})$ (Tautology Independence)^M
- if $\vdash \alpha$ then $I(\mathcal{K} \uplus \{\alpha \wedge \beta\}) = I(\mathcal{K} \uplus \{\beta\})$ (⊤-conjunct Independence)^M

Lemma 2 expresses that (Tautology Independence) essentially follows from (⊤-conjunct Independence) but this fails in the multiset framework; see the next example.

Example 4 *Define*

$$I(\mathcal{K}) = \begin{cases} \text{card}(\mathcal{K}) & \text{if card}(MIS(\mathcal{K})) > 0 \\ 0 & \text{else.} \end{cases}$$

It is easy to show that I fails (Tautology Independence)^M: Simply take $\mathcal{K} = \{p, \neg p\}$ and $\alpha = \top$.[7] Thus, $I(\mathcal{K}) = 2 \neq 3 = I(\mathcal{K} \uplus \{\alpha\})$.

---
[7] Define $\top$ as $\neg \bot$.

It can be shown that $\boldsymbol{I}$ satisfies (Consistency Null)^M, (Unit Monotony)^M, (Conjunction Dominance)^M, and (⊤-conjunct Independence)^M; the details are as follows. If $\mathcal{K}$ is empty then $\boldsymbol{I}(\mathcal{K}) = 0$ else card($\mathcal{K}$) > 0 hence $\boldsymbol{I}(\mathcal{K}) = 0$ iff $MUS(C(\mathcal{K})) = \emptyset$, i.e., $\mathcal{K} \not\vdash \bot$. Thus, $\boldsymbol{I}$ enjoys (Consistency Null)^M. As to (Unit Monotony)^M, it holds because $MIS(\mathcal{K}) \subseteq MIS(\mathcal{K} \uplus \{\alpha\})$. Turning to (Conjunction Dominance)^M, if $\mathcal{K} \uplus \{\alpha\}$ is inconsistent then so is $\mathcal{K} \uplus \{\alpha \wedge \beta\}$. However, card($MIS(\mathcal{K} \uplus \{\varphi\})$) > 0 iff $\mathcal{K} \uplus \{\varphi\}$ is inconsistent. Two consequences of card($MIS(\mathcal{K} \uplus \{\alpha\})$) > 0 are $\boldsymbol{I}(\mathcal{K} \uplus \{\alpha\}) = $ card($\mathcal{K} \uplus \{\alpha\}$) and $\boldsymbol{I}(\mathcal{K} \uplus \{\alpha \wedge \beta\}) = $ card($\mathcal{K} \uplus \{\alpha \wedge \beta\}$). Thus, $\boldsymbol{I}(\mathcal{K} \uplus \{\alpha \wedge \beta\}) = \boldsymbol{I}(\mathcal{K} \uplus \{\alpha\})$. The case that $\mathcal{K} \uplus \{\alpha\}$ is consistent clearly gives $\boldsymbol{I}(\mathcal{K} \uplus \{\alpha \wedge \beta\}) \geq \boldsymbol{I}(\mathcal{K} \uplus \{\alpha\})$. Accordingly, (Conjunction Dominance)^M holds. There only remains to deal with (⊤-conjunct Independence)^M. It is trivial that $\mathcal{K} \uplus \{\beta\}$ is inconsistent iff $\mathcal{K} \uplus \{\alpha \wedge \beta\}$ is inconsistent, that is, card($MIS(\mathcal{K} \uplus \{\alpha \wedge \beta\})$) > 0 iff card($MIS(\mathcal{K} \uplus \{\beta\})$) > 0. As card($\mathcal{K} \uplus \{\alpha \wedge \beta\}$) = card($\mathcal{K} \uplus \{\beta\}$), it follows that $\boldsymbol{I}(\mathcal{K} \uplus \{\alpha \wedge \beta\}) = \boldsymbol{I}(\mathcal{K} \uplus \{\beta\})$.

In fact, $\boldsymbol{I}$ in Example 4 fails the *set* version of (⊤-conjunct Independence). Here is a counter-example. Let $K = \{p, \neg p\}$, take $\alpha = \top$ and $\beta = p$. By the set version of (⊤-conjunct Independence), $\boldsymbol{I}(\{p, \neg p\} \cup \{\top \wedge p\}) = \boldsymbol{I}(\{p, \neg p\} \cup \{p\})$. Since $\{p, \neg p\} \cup \{p\}$ is exactly the *same* set as $\{p, \neg p\}$, the set version actually demands $\boldsymbol{I}(\{p, \neg p\} \cup \{\top \wedge p\}) = \boldsymbol{I}(\{p, \neg p\})$. Now, card($\{p, \neg p\} \cup \{\top \wedge p\}$) = 3 while card($\{p, \neg p\}$) = 2. Hence, $\boldsymbol{I}(\{p, \neg p\} \cup \{\top \wedge p\}) \neq \boldsymbol{I}(\{p, \neg p\} \cup \{p\})$. Nothing similar happens with the multiset version because $\{p, \neg p\} \uplus \{p\}$ is the multiset $\{p, p, \neg p\}$ whose cardinality is 3, not 2.

The proof of Lemma 2 does not extend to the multiset framework, as it relies upon a non-guarded instance of (⊤-conjunct Independence). The main reason seems rather clear: (Tautology Independence)^M imposes an equality for the amount of inconsistency to two multisets of different cardinalities whereas (⊤-conjunct Independence)^M imposes such an equality for the amount of inconsistency to two multisets having the very same cardinality —and multiset sum always increases cardinality (except for the case of the empty set, of course).

Other independence postulates have been proposed, based on special properties of formulas. All these postulates can be written in multiset guise without any difficulty, it suffices here to reproduce the most well-known items:

- If α is free for $\mathcal{K}$ then $\boldsymbol{I}(\mathcal{K} \uplus \{\alpha\}) = \boldsymbol{I}(\mathcal{K})$ (Free Formula Independence)^M
- If α is safe[8] for $\mathcal{K}$ then $\boldsymbol{I}(\mathcal{K} \uplus \{\alpha\}) = \boldsymbol{I}(\mathcal{K})$ (Safe Formula Independence)^M

5.5 Postulates on Conjunction

A "set"-postulate has been introduced in [2] to the effect that inconsistency does not increase when a conjunction is replaced by any one of its conjuncts.

---

[8] A formula φ is safe for $\mathcal{X}$ iff $\varphi \not\vdash \bot$ and no propositional variable occur both in φ and in $\mathcal{X}$.

This conveys the same idea as (Monotony): extra information (whether in the form of an extra formula or in the form of an extra conjunct) cannot make the amount of inconsistency to decrease.

- $I(\mathcal{K} \uplus \{\alpha \wedge \beta\}) \geq I(\mathcal{K} \uplus \{\alpha\})$ (Conjunction Dominance)^M

(Conjunction Dominance) need not be guarded because multisets always increase multiplicity. In any event, multisets seem more suited to the principle that a conjunct can be cut off its conjunction without any problem. This idea supports the following postulate.

- $I(\mathcal{K} \uplus \{\alpha, \beta\}) = I(\mathcal{K} \uplus \{\alpha \wedge \beta\})$ (Adjunction Invariance)^M

No matter how natural in the multiset framework, such conjunction postulates are still incompatible with inconsistency measures based on MUSes: should a conjunction be split, if it has a conjunct involved in an inconsistency then the minimal inconsistent subsets change, and in many cases, so does the number of minimal inconsistent subsets.

In contradistinction to all the other postulates investigated in this chapter, postulates mentioning conjunction do not apply to all inconsistency measures. Obviously, inconsistency measures that assume a clausal form for the formulas in the knowledge base cannot be evaluated against such postulates. The reader may also be concerned with a more general issue, that postulates mentioning conjunction lack generality inasmuch as they seem to apply to specific formulas. This is only apparent, because the underlying logic is classical logic hence simple syntactical moves are available that turn an intuitively appropriate formula into a formula governed by conjunction (this would not be possible for some logics, mainly logics in which conjunction captures a fusion operation).

The choice of underlying logic makes a difference in conjunction postulates; for an illustration see [21] about (Adjunction Invariance).

5.6 Additivity

The simplest postulate comes from [20].

- If $\mathcal{K}$ and $\mathcal{K}'$ are disjoint then $I(\mathcal{K} \uplus \mathcal{K}') \geq I(\mathcal{K}) + I(\mathcal{K}')$ (Super-Additivity)^M

In [7], the idea of a partition of $MUS(K \cup K')$ is used in order to provide a condition under which the sum of the amount of inconsistency of the subbases is a correct way to compute the amount of inconsistency of the whole base. Intuitively, $MIS(\mathcal{K})$ and $MIS(\mathcal{K}')$ must form a partition of $MIS(\mathcal{K} \uplus \mathcal{K}')$ in the sense that $MIS(\mathcal{K} \uplus \mathcal{K}')$ must be obtained by putting together $MIS(\mathcal{K})$ and $MIS(\mathcal{K}')$ while $MIS(\mathcal{K})$ and $MIS(\mathcal{K}')$ have no element in common.

Definition 9 $\{MIS(\mathcal{K}), MIS(\mathcal{K}')\}$ *is a partition of* $MIS(\mathcal{K} \uplus \mathcal{K}')$ *iff both conditions below are satisfied*

- $MIS(\mathcal{K} \uplus \mathcal{K}') = (2^{F_{\mathfrak{L}}}, \mu)$
  where $\mu(X) = \mu_{MIS(\mathcal{K})}(X) + \mu_{MIS(\mathcal{K}')}(X)$ for all $X \subseteq F_{\mathfrak{L}}$,
- there exist no $X \subseteq F_{\mathfrak{L}}$ such that $\mu_{MIS(\mathcal{K})}(X) \geq 1$ and $\mu_{MIS(\mathcal{K}')}(X) \geq 1$.

The multiset version of the postulate introduced in [7] can now be written

- If $\{MIS(\mathcal{K}), MIS(\mathcal{K}')\}$ is a partition of $MIS(\mathcal{K} \uplus \mathcal{K}')$
  then $\boldsymbol{I}(\mathcal{K} \uplus \mathcal{K}') = \boldsymbol{I}(\mathcal{K}) + \boldsymbol{I}(\mathcal{K}')$ \hfill (MI-Separability)^M

5.7 Other Postulates

In principle, it is possible to reformulate any "set"-postulate from the literature into a multiset version. Nevertheless, there would be no point to listing all of these here, hence this short browsing can be terminated with the following postulates requiring that superficial variations must have no impact on the amount of inconsistency. The first postulate guarantees that using a propositional symbol or another for naming makes no difference in the amount of inconsistency. In other words, inconsistency measures should preserve free interpretation of symbols as enjoyed by classical logic.

- If σ and σ' are substitutions such that $\sigma\mathcal{K} = \mathcal{K}'$ and $\sigma'\mathcal{K}' = \mathcal{K}$
  then $\boldsymbol{I}(\mathcal{K}) = \boldsymbol{I}(\mathcal{K}')$ \hfill (Variant Equality)^M

The last postulate to be given explicitly in this chapter takes care of minor modifications of a formula, e.g., swapping two disjuncts.

- If β is a prenormal[9] form of α then $\boldsymbol{I}(\mathcal{K} \uplus \{\alpha\}) = \boldsymbol{I}(\mathcal{K} \uplus \{\beta\})$ (Rewriting)^M

A useful case for applying this postulate in the multiset version is mentioned in the introduction: $\boldsymbol{I}(\{\alpha \wedge \beta, \gamma, \ldots, \upsilon\} \uplus \{\beta \wedge \alpha\}) = \boldsymbol{I}(\{\alpha \wedge \beta, \gamma, \ldots, \upsilon\} \uplus \{\alpha \wedge \beta\})$ and similarly, $\boldsymbol{I}(\{\alpha, \beta, \ldots, \upsilon\} \uplus \{\neg\neg\alpha\}) = \boldsymbol{I}(\{\alpha, \beta, \ldots, \upsilon\} \uplus \{\alpha\})$, etc.

5.8 In General

Sections 5.1-5.7 presented the multiset version of 18 inconsistency postulates. They were formulated exactly as the original "set"-postulate, although now applying to multisets. That is, it seems that nothing happens at the level of *expressing* an inconsistency postulate in the multiset case. As to properties, Section 4 shows that properties with a conclusion of the form $\boldsymbol{I}(\mathcal{K}) = \boldsymbol{I}(\mathcal{K} \uplus \mathcal{K}')$ may hold for "set"-postulates and fail for their multiset versions: cf Lemma 3 for example. Properties with a conclusion of the form $\boldsymbol{I}(\mathcal{K} \uplus \mathcal{K}') \geq \boldsymbol{I}(\mathcal{K} \uplus \mathcal{K}'')$ are more likely to lift to multisets, on the condition that the cardinality of $\mathcal{K}''$

---

[9] Define α' to be a *prenormal form* of α if α' results from α by applying (repeatedly whenever needed) one or more of these principles: commutativity, associativity and distribution for $\wedge$ and $\vee$, De Morgan laws, double negation equivalence.

3.2 Features about "Set"-postulates

3.2.1 Guarded "Set"-postulates

Letting $\mathfrak{C}[\alpha, \beta]$ to stand for a condition on α and β, a number of "set"-postulates are of the form

$$\text{If } \mathfrak{C}[\alpha, \beta] \text{ then } \boldsymbol{I}(K \cup \{\alpha\}) = \boldsymbol{I}(K \cup \{\beta\}). \tag{$\|$}$$

As evidenced in [3], undesirable effects can arise in the event that $\alpha \in K$ but $\beta \notin K$ (or vice versa). Next is an illustration.

Example 1 *Consider* $K = \{p \wedge q, \neg p, \neg q, \neg r\}$. *Take* α *to be* $p \wedge q$ *and take* β *to be* $\neg(\neg p \vee \neg q) \vee r$. *Trivially,* $\alpha \not\vdash \bot$ *and* $\alpha \vdash \beta$. *Then, (Dominance) applies:* $\boldsymbol{I}(K \cup \{\alpha\}) \geq \boldsymbol{I}(K \cup \{\beta\})$. *However,* $\alpha \in K$. *So, (Dominance) here actually imposes* $\boldsymbol{I}(K) \geq \boldsymbol{I}(K \cup \{\beta\})$. *That is,*

$$I\left(\left\{\begin{array}{c} p \wedge q, \\ \neg p, \\ \neg q, \\ \neg r \end{array}\right\}\right) \geq I\left(\left\{\begin{array}{c} p \wedge q, \\ \neg p, \\ \neg q, \\ \neg r, \\ \neg(\neg p \vee \neg q) \vee r \end{array}\right\}\right)$$

Now, all inconsistencies in K *are in* $K \cup \{\beta\}$ *but the latter displays an inconsistency involving* r *whereas* K *does not. In view of this,* $\boldsymbol{I}(K) \geq \boldsymbol{I}(K \cup \{\beta\})$ *is dubious, making (Dominance) questionable for the case that* α *is in* K.

Indeed, for most postulates of the form ($\|$), $\boldsymbol{I}(K) = \boldsymbol{I}(K \cup \{\beta\})$ is not an intended consequence even in the case that $\alpha \in K$ for α such that $\mathfrak{C}[\alpha, \beta]$ holds. In other words, what is intended is actually [3]

Whenever $K \cup \{\alpha\} \neq K \neq K \cup \{\beta\}$, if $\mathfrak{C}[\alpha, \beta]$ then $\boldsymbol{I}(K \cup \{\alpha\}) = \boldsymbol{I}(K \cup \{\beta\})$
(Guarded postulate)

Therefore, "set"-postulates must in fact be rewritten in such guarded versions taking care of ineffective set union for a singleton set contained in the set it is unioned with. It is obvious that the same need also arises for postulates having $\geq$ instead of $=$, except that $K \neq K \cup \{\beta\}$ may be omitted because the case $\alpha \notin K$ while $\beta \in K$ need not be harmful —it amounts to a consequence of (Unit Monotony). An easy illustration is (Dominance) that calls for the guarded version

- Whenever $K \neq K \cup \{\alpha\}$, if $\alpha \vdash \beta$ and $\alpha \not\vdash \bot$ then $\boldsymbol{I}(K \cup \{\alpha\}) \geq \boldsymbol{I}(K \cup \{\beta\})$
([Guarded] Dominance)

---

[3] Of course, such a clumsy proviso "Whenever ..." in (Guarded postulate) can be equivalently rewritten "Whenever $\alpha \notin K$ and $\beta \notin K$" (as done in Section 3.2.2) but the current formulation is retained for ease of comparison in Section 4.1.

not exceed the cardinality of $\mathcal{K}'$: see Lemmas 4-7 for example. As to being more general, it seems difficult to find workable conditions that would ensure a class of properties to lift from the set level to the multiset level.

As a reviewer puts it:

> *All existing postulates seem reasonable somehow, but do we really need them? One can probably come up with dozens of other postulates, but where do we draw the line?*

It is all too easy to go astray when answering these questions, based on the prejudice that a postulate for inconsistency measures is to play a normative role. It should not, and this is witnessed by the existing inconsistency measures as is reported by Thimm in his insightful [22]: There is only *one* postulate that all of the 15 measures examined in [22] agree upon, namely (Consistency Null). Another finding by Thimm which is relevant here is that a number of these 15 measures in [22] fail *most* of the postulates. Can these measures be so wrong? Can these postulates be so spurious? Or is there something else to satisfying a postulate for inconsistency measures? Distinguishing different classes of inconsistency measures is a better role for the postulates and the reader is urged to refrain from ascribing a normative character to these postulates. There are inconsistency measures of highly different nature, enjoying different properties, as they may be tailored to a particular domain of application. It would be so absurd to cast a postulate dismissing such a class of inconsistency measures, since they inherit their properties from a common application domain: it makes no sense to dismiss an application domain.

6 Conclusion

This chapter takes a look at inconsistency measuring for multisets of formulas, from the perspective of postulates for inconsistency measures. It appears that "set"-postulates give rise to multiset postulates in a really straightforward way. In fact, even more straightforward than could have been expected: a number of "set"-postulates admit defective instances, e.g., with a conclusion of the form $\boldsymbol{I}(K \cup \{\alpha\}) \geq \boldsymbol{I}(K \cup \{\beta\})$, that turns out to be problematic when α is in K. Although these "set"-postulates need to be amended with a proviso restricting application to $\alpha \notin K$ for example, no such move is needed in the multiset case.

This chapter additionally checks that various relationships between "set"-postulates extend to their multiset counterparts. A couple of counter-examples in the form of Lemmas 2-3 are provided to give a feel of the kind of properties that fail to carry over. Importantly, examination is conducted from the form of postulates which is argued to be much less diverse in the literature than is theoretically possible.

This deals with postulates that can be stated (although not always giving items with similar effects) in set as well as multiset guise. What about multiset postulates with no corresponding "set"-postulate? Paradoxically, the simplest such example is a postulate —that would amount to the truth $x \leq x$ if rewritten for sets— whose effect is to require that an inconsistency measure ignore duplicates altogether:

- $I(\mathcal{K} \uplus \{\alpha,\alpha\}) = I(\mathcal{K} \uplus \{\alpha\})$ (Duplicate Annihilation)^M

A more significant example is a postulate that would amount to a contradiction $0 < 0$ in the case of sets, a postulate expressing that the increase in the amount of inconsistency lessens with the number of duplicates, i.e., *the more duplicates of a given non-free formula, the lesser the effect of an extra copy of this formula*:

- If $\alpha \in \mathcal{K}$ is not free for $\mathcal{K}$ then $I(\mathcal{K} \uplus \{\alpha,\alpha\}) - I(\mathcal{K} \uplus \{\alpha\}) < I(\mathcal{K} \uplus \{\alpha\}) - I(\mathcal{K})$
                                                                                             (Mitigation)^M

A reviewer has suggested to touch on the question of the behaviour of existing inconsistency measures in the multiset case. The answer depends on how an existing measure is to be extended to the multiset case. In particular, (Duplicate Annihilation)^M provides a generic answer: all existing inconsistency measures can be extended to multisets using the principle that duplicates are not taken into account for the value returned by the inconsistency measure. In symbols, $I^M(\mathcal{K}) \stackrel{\text{def}}{=} I(C(\mathcal{K}))$. However, there are many more possibilities. An example is an inconsistency measure defined in [16] that combines the number of MUSes of a given cardinality with the number of consistent subsets with the same cardinality: although it seems crucial that duplicates are taken into account for the former number, perhaps is it more appropriate not to take them into account for the second number.

Acknowledgements The author is most grateful to both reviewers for their insightful comments and suggestions.

References

[1] M. Ammoura, B. Raddaoui, Y. Salhi, and B. Oukacha. On an MCS-based Inconsistency Measure. *Journal of Approximate Reasoning*, 80:443–459, 2017.

[2] P. Besnard. Revisiting Postulates for Inconsistency Measures. In *14th European Conference on Logics in Artificial Intelligence (JELIA'14)*, volume 8761 of *Lecture Notes in Computer Science*, pages 383–396, 2014.

[3] P. Besnard. Basic Postulates for Inconsistency Measures. In *Transactions on Large-Scale Data- and Knowledge-Centered Systems* XXXIV, volume 10620 of *Lecture Notes in Computer Science*, pages 1–12. 2017.

[4] G. De Bona and A. Hunter. Localising Iceberg Inconsistencies. *Artificial Intelligence*, 246:118–151, 2017.

[5] J. Grant and A. Hunter. Analysing Inconsistent Information using Distance-based Measures. *Journal of Approximate Reasoning*, 89:3–26, 2017.

[6] A. Hunter and S. Konieczny. Measuring Inconsistency through Minimal Inconsistent Sets. In *11th Conference on Principles of Knowledge Representation and Reasoning (KR'08)*, pages 358–366, 2008.

[7] A. Hunter and S. Konieczny. On the Measure of Conflicts: Shapley Inconsistency Values. *Artificial Intelligence*, 174(14):1007–1026, 2010.

[8] A. Hunter, S. Parsons, and M. Wooldridge. Measuring Inconsistency in Multi-Agent Systems. *Künstliche Intelligenz*, 28:169–178, 2014.

[9] S. Jabbour, Y. Ma, B. Raddaoui, and L. Saïs. Quantifying Conflicts in Propositional Logic through Prime Implicates. *Journal of Approximate Reasoning*, 89:27–40, 2017.

[10] S. Jabbour, Y. Ma, B. Raddaoui, L. Saïs, and Y. Salhi. A MIS Partition based Framework for Measuring Inconsistency. In *15th International Conference on Principles of Knowledge Representation and Reasoning (KR'16)*, pages 84–93, 2016.

[11] S. Konieczny and R. Pino Pérez. Logic based Merging. *Journal of Philosophical Logic*, 40(2):239–270, 2011.

[12] W. Liu and K. Mu, editors. *Special Issue on Theories of Inconsistency Measures and their Applications*, volume 89 of *Journal of Approximate Reasoning*. Elsevier, 2017.

[13] M. V. Martinez, A. Pugliese, G. I. Simari, V. S. Subrahmanian, and H. Prade. How Dirty is your Relational Database? An Axiomatic Approach. In *9th European Conference on Symbolic and Quantitative Approaches to Reasoning and Uncertainty (ECSQARU'07)*, volume 4724 of *Lecture Notes in Computer Science*, pages 103–114, 2007.

[14] K. McAreavey, W. Liu, and P. Miller. Computational Approaches to Finding and Measuring Inconsistency in Arbitrary Knowledge Bases. *Journal of Approximate Reasoning*, 55(8):1659–1693, 2014.

[15] K. Mu, W. Liu, and Z. Jin. A General Framework for Measuring Inconsistency through Minimal Inconsistent Sets. *Journal of Knowledge and Information Systems*, 27(1):85–114, 2011.

[16] K. Mu, W. Liu, Z. Jin, and D. Bell. A Syntax-based Approach to Measuring the Degree of Inconsistency for Belief Bases. *Journal of Approximate Reasoning*, 52(7):978–999, 2011.

[17] K. Mu, K. Wang, and L. Wen. Approaches to Measuring Inconsistency for Stratified Knowledge Bases. *Journal of Approximate Reasoning*, 55(2):529–556, 2014.

[18] N. Potyka and M. Thimm. Probabilistic Reasoning with Inconsistent Beliefs using Inconsistency Measures. In *24th International Joint Conference on Artificial Intelligence (IJCAI'15)*, pages 3156–3163, 2015.

[19] A. Syropoulos. Mathematics of Multisets. In *Multiset Processing, Mathematical, Computer Science, and Molecular Computing Points of View: Workshop on Multiset Processing (WMP'00)*, volume 2235 of *Lecture Notes in Computer Science*, pages 347–358, 2000. Published in 2001.

[20] M. Thimm. Inconsistency Measures for Probabilistic Logics. *Artificial Intelligence*, 197:1–24, 2013.

[21] M. Thimm. Measuring Inconsistency with Many-valued Logics. *Journal of Approximate Reasoning*, 86:1–23, 2017.

[22] M. Thimm. On the Compliance of Rationality Postulates for Inconsistency Measures: A more or less Complete Picture. *Künstliche Intelligenz*, 31(1):31–39, 2017.

[23] M. Thimm. Stream-based Inconsistency Measurement. *Journal of Approximate Reasoning*, 68:68–87, 2017.

[24] M. Ulbricht, M. Thimm, and G. Brewka. Measuring Inconsistency in Answer Set Programs. In *15th European Conference on Logics in Artificial Intelligence (JELIA'16)*, volume 10021 of *Lecture Notes in Computer Science*, pages 577–583, 2016.

Inconsistency Measures and Paraconsistent Consequence

Bryson Brown

Department of Philosophy
University of Lethbridge, Canada
brown@uleth.ca

Abstract

The risk that a large database will include inconsistent information is unavoidable.[2, 16, 27] Many different measures of inconsistency, aimed (intuitively) at evaluating *how inconsistent* different databases are, have been proposed.[27] Proposed applications of inconsistency measures include quantifying and comparing the inconsistency of databases, and evaluating ways of reducing or eliminating inconsistencies. Here we focus on a different application of inconsistency measures: a *preservationist* treatment of consequence relations[23] shows that the preservation of inconsistency measures can give rise to paraconsistent logics which constrain how we reason with inconsistent premises (and also how we treat conclusions that cannot be consistently denied). Along the way, we argue for a somewhat liberal understanding of inconsistency measures in general: while definitions of inconsistency measures that have been proposed impose a linear ordering [27], the inconsistency measures that underlie preservationist semantics for the paraconsistent logics *LP*, Priest's *logic of paradox*[5] and first degree entailment [14, 9] don't involve a linear ordering, as the preservationist treatment of these logics shows[9]. We also argue for this more liberal view on general grounds. Our aim is to illustrate the connection between inconsistency measures and a *preservationist* approach to consequence relations [23], which suggest a potentially fruitful relation between preservationist approaches to consequence and paraconsistency, on one hand, and inconsistency measures on the other.

1 Introduction

The risk that a reasonably large database will be inconsistent is inescapable. Kyburg's 'lottery paradox'[18] provides a simple illustration of how difficult it is to avoid this risk by pointing out that any acceptance rule based on a probability threshold provides a clear path to inconsistency: in a fair and sufficiently large one-winner lottery, any probability threshold for acceptance of sentences leads to acceptance of all sentences of the form 'ticket i will not win'. But we can also have evidence at least as strongly supporting the claim that the lottery is fair, in which case the same rule also leads us to accept 'some ticket will win'. Thus we wind up having accepted an inconsistent set of sentences. And Kyburg's point obviously extends beyond simple probabilistic acceptance rules: all cognitive systems are vulnerable to erroneous input, ranging from typographical errors to misinterpretations, misjudgements and exposure to deliberate misinformation.

Kyburg's response was to reject conjunction introduction as a rule under which rationally accepted sentences should be closed. In that spirit, we might propose limits on the application of conjunction introduction in machine reasoning with databases; more generally, we need to avoid closure under the classical ⊨. The specific trouble underlying Kyburg's concern arises from logical *aggregation*, involving inferences that take us from inconsistent collections of individually consistent and highly probable premises to contradictions. Kyburg's critique of closure under conjunction suggests we can avoid trivialization by seeking constrained approaches to *aggregation*, that is, inferences in which we derive consequences that depend on more than one premise. In general, since the risk of error cannot be eliminated, the principle of *explosion* needs to be brought under better control; as we will see, the paraconsistent logics we consider here all weaken aggregation, though in different ways.

Inconsistency measures are intended to provide a measure of how inconsistent a data base or collection of sentences in some formal language is. Many such measures have been proposed; here we focus on measures of inconsistency for sets of sentences belonging to a classical propositional language, and in particular, on some measures that are closely related to paraconsistent logics.

In section 2 we introduce the familiar problem of the trivialization of inconsistent premise sets, together with the parallel (sometimes neglected) problem of conclusions or conclusion sets that cannot be consistently denied. Section 3 presents a very general perspective on consequence relations, drawing on Scott's multiple-conclusion approach in [22]. Section 4 gives a (brief) account of the *preservationist* approach to avoiding the trivialization of inconsistent/unsatisfiable premises (as well as conclusions that cannot be consistently denied or *dis-satisfied*)[23], while Section 5 discusses some motivations for developing and adopting different paraconsistent logics. Finally, Section 6 presents treatments of three paraconsistent logics based on inconsistency measures: Schotch and Jennings' forcing relation, Priest's logic of paradox (LP) and Anderson and Belnap's first degree entailment (FDE), closing with some comments on Henry Kyburg's concerns about the closure of cognitive commitments under conjunction.

Throughout the paper, we will use T and F for the classical truth values, B and N for the non-classical values 'both' and 'neither', upper case Greek letters for sets of sentences, lower case Greek letters for sentences, lower case Latin letters for atoms and, on occasion, other upper case Latin letters to name formal languages.

2 Brief Reflections on Consequence and Trivialization

A sound consequence relation on a language L (and the rules used to specify it) is often said, in semantic terms, to be *truth preserving*. But it's obvious that the standard of correctness for inferences is stronger than mere truth-preservation; what we endorse, logically speaking, are not inferences that just happen to preserve truth, such as:

1. The sky is blue, therefore water is H_2O.

The inferences we really want to endorse are (all and only) those inferences *guaranteed* to preserve truth, inferences such as:

2. The sky is blue and water is H_2O, therefore water is H_2O.

The modal character of the guarantee here can be 'cashed in' by shifting from talk of truth-preserving inferences to talk of satisfiability-preserving and consistency-preserving inferences. A set of premises Γ is standardly said to be *satisfiable* if and only if there is at least one model of Γ, and unsatisfiable if there is no model of Γ.[1] This general concept of satisfiability is central to semantic thinking about logical consequence: the standard definition of the semantic consequences of Γ tells us they are the sentences that are satisfied in every model of Γ, or, in multiple conclusion logics, the set of sets of sentences such that at least one member of each set is satisfied in every model of Γ. When there is no model of Γ, we say Γ is not satisfiable; on the usual account of semantic consequence, it follows that Γ is trivial: every sentence in the language follows from Γ. Obviously enough, this is not particularly helpful. But an explanation of why this is the standard response helps to point us in a better direction.

In general, given inference rules for a language L together with a semantics for L, we say that $\Gamma \vdash \phi$ iff for every consistent extension of Γ, Γ', $\Gamma' \cup \phi$ is also consistent, and we say that $\Gamma \vDash \phi$ iff for every satisfifable extension of Γ, Γ', $\Gamma' \cup \phi$ is a *satisfiable* extension of Γ'. Conversely, in a multiple conclusion system, we require that any consistently deniable (or, speaking semantically, *dis-satisfiable*) extension–that is, an extension all of whose members can be consistently denied (assigned the value 0)–of a consistently deniable conclusion set is also a consistently deniable (dis-satisfiable) extension of the closure of that conclusion set under the inverse of $\vdash$ (respectively, $\vDash$).)

---

[1] The point is general, applying to classical two-valued propositional models as well as to models of other propositional logics such as the (standard presentations of) LP and FDE.

From this perspective, we say that inference 1 above fails the tests because, while "water is not H_2O" is a consistent (and satisfiable) extension of the set {"the sky is blue"}, "water is not H_2O" is *not* a consistent or satisfiable extension of the consistent set of sentences {"The sky is blue", "Water is H_2O"}. So "water is H_2O" is *not* a consistent extension of every consistent extension of {"The sky is blue"}. Conversely, "water is H_2O" can be consistently denied while not denying "the sky is blue." But inference 2 passes the tests, because "water is not H_2O" is neither a consistent nor a satisfiable extension of {"The sky is blue and water is H_2O"}, and (equivalently, but reasoning from right to left) we cannot deny "water is H_2O" without denying "the sky is blue and water is H_2O". So we think of correct inferences as those that guarantee the preservation of consistency, syntactically, and satisfiability, semantically, from left to right, or, equivalently, that guarantee the preservations of consistent deniability and dis-satisfiability from right to left. We say that a given sentence α is a consequence of a set of sentences Γ iff α is a consistent/satisfiable extension of every consistent/satisfiable extension of Γ.

However, this approach to consequence relations has an important limitation: it follows immediately that unsatisfiable/inconsistent premises are trivial. Every sentence in the language follows from them because 'preserving' the satisfiability/consistency of the satisfiable/consistent extensions of any such set is just too easy: every sentence of the language 'preserves' what remains of these properties– that is, nothing whatsoever. So every sentence in the language is a consequence of such sets. The same goes from right to left: the consistent deniability (or, semantically, the *dis-satisfiability*) of a conclusion set that cannot be consistently denied (dis-satisfied) is trivially 'preserved' by every premise set.

The logical trivialization of inconsistent premises has been said, tongue-in-cheek, to be logic's way of telling us to go and find better premises. But all too often this is easier said than done. In real life, as Kyburg argued, our premises may all be well-supported, and, from a practical perspective, we may even regularly and successfully *rely* on each of them, applying them in a wide range of cases with perfectly satisfactory results. But they may still turn out to be inconsistent. Setting aside such everyday, practical concerns, we also know that we can't demonstrate the consistency of a theory of arithmetic within a formalized arithmetic, without running into the unhappy corollary that the theory is inconsistent.

A mirror-image concern arises in the other direction. Given a multiple-conclusion consequence relation we can also reason from the denial of all of some collection of sentences to the denial of at least one member of any set of premises the set follows from. Here again, trivialization threatens: in a multiple conclusion logic conclusion sets that cannot be consistently denied follow trivially from any premises, and denying them all 'commits' us to denying every sentence in the language.

A practical question arises here: when our commitments get us into this kind of trouble, and we don't know what premises or conclusions to surrender, how should we reason in the meanwhile? It would be helpful to have new constraints on the consequence relation that would allow it to continue to guide our reasoning in a non-trivial way. The response adopted by many paraconsistent logics has been to revise our standards of consistency and satisfiability (and, for those like *FDE* that also address the problem of reasoning from denials to further denials, to revise our standards of triviality and dis-satisfiability as well).

But there is an alternative: rather than change these standards, we can look for *other* properties of premises and conclusions that are worth preserving, and develop a new consequence relation that preserves these properties as well. This alternative is highlighted by a broad perspective on consequence relations due to Dana Scott, which opens the door to considering consequence relations grounded in new properties, and to logics that take us beyond consistency, satisfiability and their duals, consistent deniability and dis-satisfiability. But we need not *replace* the familiar properties with these new properties, as dialetheists (among others) have proposed.[23] Instead, they can be treated as *adding* new constraints on reasonable inference that allow us to cope better with flawed information.

3 A Broader View of Consequence

In [22] Dana Scott presents a general account of logical consequence relations. His account, presented in multiple-conclusion fashion, with sets of items appearing on both sides of the turnstile, relied on three formal properties of such relations: reflexivity, monotonicity and transitivity. Scott demonstrated that these properties guarantee a striking result: any reflexive, monotonic and transitive relation on the subsets a given set S is determined by the preservation of some property (we might give it the mnemonic label "T") of the set's members from left to right (or equivalently, of T's complement, F, from right to left). That is to say that, given the allowed T/F valuations on the sentences of L:

If $\Gamma... \vDash \Delta$, then for any valuation on the language V, $V(\gamma_i) = T$ for all $\gamma_i \in \Gamma \to \exists \delta_i \in \Delta$ such that $V(\delta_i) = T$.

Using $\vDash$ to express such a consequence relation on a language, Λ, upper-case Greek letters for subsets of Λ and lower case letters α, β, etc. for members of Λ, we express these general properties of Scott consequence relations as follows:

$R : \alpha \in \Gamma$ and $\alpha \in \Delta \Rightarrow \Gamma \vDash \Delta$
$M : \Gamma \vDash \Delta, \Gamma \subseteq \Gamma' \ \& \ \Delta \subseteq \Delta' \Rightarrow \Gamma' \vDash \Delta'$
$T : \Gamma, \alpha \vDash \Delta, \Gamma \vDash \Delta, \alpha \Rightarrow \Gamma \vDash \Delta$

Thus in this multiple conclusion treatment of consequence, R (for reflexivity), tells us that the singleton set of any sentence in Γ follows from Γ; M

(for monotonicity), tells us that whenever Δ follows from Γ, any superset of Δ follows from any superset of Γ, and T (for transitivity), tells us that whenever Δ follows from $\Gamma \cup \alpha$ and $\{\alpha\}$ follows from Γ, then Δ follows from Γ.

Assuming that, whatever else we might say about it, the property "truth" is preserved from left to right and the dual property "falsehood" is preserved from right to left in a formal semantic consequence relation $\vDash$, the properties R, M and T provide clear formal constraints on these two semantic properties.

Scott's result also helps illuminate the value of multiple conclusion consequence relations: the closure of a set of sentences under a single-conclusion consequence relation is the *intersection* of the sets of sentences satisfied by the models of the premises. Thus a single-conclusion consequence relation only tells us what all the models of the premises *agree on*. A corollary to his result shows that the sets following from a given premise set under a multiple conclusion consequence relation capture, in the form of the least sets intersecting every such conclusion set, all the maximal satisfiable extensions of the premise set. Symmetrically, from right to left while holding the conclusion set fixed, we also capture the maximal dissatisfiable extensions of the conclusion set.[8]

Scott's result shows that any Tarski consequence relation–that is, any reflexive, monotonic and transitive relation on a set S–can be given a 1/0 semantics, that is, any such relation is determined by a set of allowed binary valuations on S.[22] It follows that some property or other of the sentences is *preserved* by the relation (while the complementary property is preserved by its converse).[2]

4 Logical Explosion and Preservationism

The key point in this diagnosis of logical 'explosion' is that when we define a consequence relation R in terms of the *preservation of some property from left to right*, that is, when R is such that any assignment at which the members of a set Γ on the left have 'the property' is one in which at least one item of any set Δ such that $R(\Gamma, \Delta)$ also has the property, our definition of R provides no constraint on what belongs on the right of the relation when there is no assignment at which the property in question applies to all the items on the left. If (*but only if*) preserving that property is all that matters to us, we have no reason to reject any candidate 'consequence' as following from such dreadful premise sets: having lost the property (our measure of the 'goodness' of the

---

[2] As a sidelight here, from this perspective Gödel's incompleteness result should not be interpreted as showing that while the truth of a complete arithmetic remains 'out there,' a consistent and complete theory of it is beyond our finite reach. Instead, it shows something more abstract: that the language of a theory strong enough to express arithmetic cannot be bi-partitioned into a maximal consistent set of sentences, the 'truths' of arithmetic, and its complement, a maximal *consistently deniable* set of sentences, which we could call the 'falsehoods' of arithmetic, because such a language must include a 'diagonal' sentence which, if arithmetic is to be consistent, must be such that neither it nor its negation is provable.

set on the left), no extension of the set has the property. So we have no reason to prefer some extensions over others; no extension makes things *worse* than they already are, because things are already as bad as they can be: the only property we value has been lost, unless and until we drop one or more members from Γ.

In a more hopeful vein, this observation also suggests a preservationist response to logical explosion: if the property whose preservation we rely on to constrain our inferences from various premise sets is lacking in some set of sentences we want to reason with, the right thing to do is to find another property worth preserving. Here we turn to P.K. Schotch and R.E. Jennings' proposals for a modest response to logical explosion:

1. Don't make things worse. (Schotch)
2. Find something you like about your premises, and preserve it. (Jennings)

Schotch's slogan hints at the common-sense recognition that forming the closure of $p, \neg p$ under $\vdash$ or $\vDash$ to arrive at the set of all sentences in our language does indeed (at least intuitively speaking) make things worse. But from a classical perspective things are already as bad as they can be: our premises are inconsistent and (equivalently, given soundness and completeness) unsatisfiable. Since our rules for reasoning preserve only consistency and satisfiability, such premise sets leave us with nothing for our rules to preserve. But speaking intuitively (for now), this seems wrong. In particular, there seems to be something much more bearable, something that should be *much more manageable*, about an inconsistency involving just one atom, or one allowing for a simple bi-partition of the premises into two consistent/satisfiable subsets, than there is about a radically metastatic inconsistency that commits us to accepting every sentence in the language.

Jennings' slogan makes the preservationist response explicit. We may not like the inconsistency/unsatisfiability of $\alpha, \neg \alpha$, but inferring every sentence in our language only makes things worse. Preserving some new property that respects this intuition can provide a systematic way of avoiding disaster, enabling us to reason in a constrained and perhaps even *useful* way, with the inconsistent information at hand.[11] From this perspective, every paraconsistent logic faces the same challenge, whether it invokes a heterodox understanding of 'consistency' and 'satisfiability', as proposed by dialetheic logicians, for example, in the usual interpretations of first-degree entailment (FDE) or Priest's logic of paradox (LP), or makes an explicit departure from grounding logical consequence in the preservation of consistency and satisfiability alone. Here, then, is room for inconsistency measures to do helpful logical work: given a measure of how inconsistent a set of premises is (and conversely, of how far a conclusion set has departed from consistent deniability), we can seek a consequence relation that will preserve that measure from left to right (conversely, from right to left).

5 Motivating Paraconsistent Logics in General

Ensuring the consistency of a large data base is a very difficult thing to do. The focus of Kyburg's concern in his discussion of the lottery paradox was rational decision making and the role of 'acceptance' in that context: relying on probability thresholds (which in turn can be based on a valuation of the 'stakes') to decide what sentences to accept as guides to choice-making makes perfect sense given constraints on the stakes. But as Kyburg argued, rational acceptance based on a probability threshold cannot be closed under logical consequence; in particular, it will not be closed under conjunction, since conjoining independent sentences with probability less than one always produces sentences with still lower probability. Thus adopting a probability threshold as a standard for acceptance (with its value to be determined by an evaluation of what's at stake in a decision) makes good sense from the perspective of decision-theory. But adopting this approach undermines our understanding of the role of *reasoning* with the sentences we accept: we can't assume probabilistically acceptable claims as premises and then draw conclusions that follow from them as further 'input' for decision making without the risk of relying on conclusions that are highly improbable, even contradictory.[18]

Yet the attraction of finding systematic ways to reason with such commitments persists. In particular, when we consider a data base system that aims to go beyond systematically storing and accessing information explicitly fed into it by carrying out some kind of reasoning with that information, any inconsistency in the information threatens to trivialize our reasoning unless we reason in a way that blocks trivialization, that is, unless we are reasoning paraconsistently, whether we choose to adopt a new, paraconsistent logic or simply refuse to draw conclusions that do follow, logically, from our premises. If we insist instead on reasoning according to such classically or intuitionistically valid inference patterns as disjunctive syllogism and reductio ad absurdum, we are committed to drawing arbitrary conclusions from any inconsistency.

Paraconsistent logics offer a way out of this trap; in [2], Nuel Belnap suggested that reasoning computers should rely on first-degree entailment (FDE) rather than "explosive" (non-paraconsistent) logics. As we've already noted, P.K. Schotch and R.E. Jennings have proposed [24] a preservationist response to the logical 'explosion' of inconsistent premise sets, arguing that we should avoid reasoning in ways that *make things worse*, and proposing that we *do* this by identifying some valuable property of our premises and reasoning in a way that preserves it. [6, 9] present preservationist interpretations of Priest's "logic of paradox" LP and first degree entailment, based on the preservation of *ambiguity measures*. These measures are based on the least sets of atoms whose ambiguity is sufficient to enable the 'projection' of *consistent images* of premise sets and consistently *deniable* images of conclusion sets. In Section 6 we will draw on these preservationist logics and the inconsistency measures underlying them, to illustrate how finding inconsistency measures and consequence relations can help us reason in useful ways with inconsistent information. These inconsistency measures (present, though below the surface in the standard multi-valued treatments of FDE and its close relative LP, and explicit in Schotch and Jennings's "forcing" relation) provide us with useful substitutes for the preservation of consistency (and its semantic fellow-traveller, satisfiability) when we need to reason with inconsistent information.

6 Examples and Discussion

In the remainder of the paper, we present treatments of three paraconsistent logics based on inconsistency-measures. We begin with Schotch and Jennings' forcing relation, which explicitly identifies the new properties that are to be preserved by the relation: the *levels*, left and right, of premise and conclusion sets. We then apply the same approach to two logics standardly presented as dialetheic paraconsistent logics, Priest's *Logic of Paradox* (LP) and Belnap and Anderson's *First Degree Entailment* (FDE), and end with a probability-threshold preserving approach to Kyburg's concerns.

Forcing is a system of consequence relations proposed by Schotch and Jennings, using weakened aggregation to prevent the derivation of contradictions from inconsistent premise sets. The aggregation allowed is determined by the level of a given premise set, Γ, defined as follows: $\ell(\Gamma)$ is the least natural n such that Γ can be divided into n consistent subsets. Note, however, that if Γ includes a contradiction, Γ cannot be divided into consistent subsets. In that case, Γ has no proper level in the naturals; to ensure that some level is assigned to every subset of our language, we define $\ell(\Gamma) = \infty$ to be the 'level' of such premise sets. For a premise set Γ of level ℓ, we replace conjunction-introduction ($\wedge - introduction$) with the rule $2/(\ell+1)$. This weaker aggregation rule allows us to infer from any $\ell+1$ sentences in Γ the disjunction of pairwise conjunctions amongst them. For example, applying $2/(\ell+1)$ to a set of level 2, $\{p, \neg p, q\}$ allows us to infer $(p \wedge \neg p) \vee ((p \wedge q) \vee (\neg p \vee q))$. $2/(\ell+1)$ is sound and complete for the Schotch-Jennings forcing relation [1], as indeed is any aggregation rule based on a non n-colourable but n+1-colourable hypergraph–see the appendix of [12].

Our focus here is on explaining the concept of level as a measure of inconsistency: given a premise set Γ such that $\ell(\Gamma) > 1$, classical logic (like any non-paraconsistent logic) trivializes Γ. But if Γ has a well-defined level (i.e. if $\ell(\Gamma)$ is a natural number, using the weaker aggregative rule $2/(\ell+1)$ in place of $\wedge$-intro ensures that we derive no contradictions, and thus avoid trivializing our premises; better still, this rule preserves the *level* of our premises.[1] Level preservation ensures that we don't trivialize premise sets of any level, up to the countably infinite level at which all aggregation fails. This logic, unlike the next two, was explicitly proposed as a preservationist logic, that is, as a logic that preserves a property (the level of premise sets) that is clearly distinct from satisfiability and consistency.[24]

In contrast, Priest's LP is normally presented as a three-valued logic based on the truth-tables of Kleene's three-valued logic. Kleene's tables for the familiar values T and F are classical, but the third value, which Priest labels B (standing for "both") is a *fixed point* for negation, that is $V(\alpha) = B \leftrightarrow V(\neg\alpha) = B$. But Priest treats the third value as a designated value; this 'paradoxical' value B allows for non-trivial models satisfying classically inconsistent premise sets, preventing their trivialization. A *preservationist* account of LP emerges from a measure of inconsistency based on what I call *ambiguity* sets: the ambiguity set for a set of sentences Γ is the set of least sets of atoms whose ambiguity is sufficient to 'project' a consistent image of Γ. Γ' is an *acceptable* extension of Γ if and only if Γ' 'preserves' the ambiguity set of Γ, that is, if and only if at least one member of $Amb(\Gamma)$ is enough to allow the projection of a consistent image of Γ'. Finally, $\Gamma \vDash_L P\alpha$ if and only if the intersection of the maximal acceptable extensions of Γ includes α.

For example, consider the set of wffs, $\Gamma = \{(a \wedge b) \to c, a, b, \neg c\}$. Treating both '$a$' and '$b$' as ambiguous allows us to project a *consistent image* of Γ: by replacing instances of each of the atoms a and b with one or another of two new atoms, a_1, a_2, and b_1, b_2, we can produce a consistent set, for example, Γ': $\{(a_1 \wedge b_1) \to c, a_2, b_2, \neg c\}$ We can also produce a consistent image of Γ by treating 'c' as ambiguous, replacing its instances with one or another of c_1, c_2, for example, $\{(a \wedge b) \to c_1, a, b, \neg c_2\}$. No proper subsets of the sets $\{a, b\}$ and $\{c\}$ can do the job, so each of the sets of atoms $\{a, b\}$ and $\{c\}$ is *minimally sufficient* for projecting a consistent image of Γ.

From the perspective of deciding how best to reason with Γ, and absent further information regarding the *meaning* or pragmatic importance of the atoms in question, we consider the consequences of resolving the assumed ambiguity in both ways. In general, $Amb(\Gamma)$ serves as a measure of which extensions of Γ make the *inconsistency* of Γ worse: for any extension of Γ, Γ^*, if Γ^* is such that no member of $Amb(\Gamma)$ is a *sufficient* basis for projecting a consistent image of Γ^*, then the inconsistency of Γ^* is *greater* than that of Γ.

There is a close parallel here with more familiar consequence relations: one way to arrive at the consequences of a premise set Γ under the classical $\vdash$ is by forming the intersection of the maximal consistency-preserving *extensions* of Γ. Similarly, we obtain the LP consequences of Γ by forming the intersection of all the maximal ambiguity-preserving extensions of Γ, that is all the maximal extensions of Γ that have consistent images when all the atoms belonging to one member of Γ's ambiguity set, the set of sets of atoms each of whose members is sufficient to allow the projection of consistent image of Γ, by the replacement of all instances of each atom in the set with one or another of a pair of new atoms). Thus the LP consequence relation is determined by the *preservation* of this inconsistency measure, viz., the minimal ambiguities sufficient to generate a consistent image of Γ.[5]

Some features of this measure of inconsistency and its role in our paraconsistent consequence relation are worth commenting on. Defined as the set of least sets of atoms each of which is sufficient to project a consistent, ambiguous image of Γ, the ambiguity set for a given set of premises Γ may include both large sets of atoms and small sets of atoms (though our definition requires that none will be subsets of any others). Any extension of Γ that preserves the sufficiency of one or more of these sets as a basis for projecting a consistent, ambiguous image of Γ is regarded as an *acceptable* extension of Γ, i.e. as not *making things worse*. But this measure of inconsistency does not produce a linear ordering of the inconsistency of subsets of L. Thus it differs, in particular, from the approach to LP found in [28], where the inconsistency measure characteristic of LP is defined semantically, as the cardinality of the set of atoms assigned the value B in an LP valuation on the language.

The advantage of this preservationist, LP-based inconsistency measure is that it leads directly to the LP consequence relation, via a preservationist approach to consequence relations and (in particular) Jenning's slogan, *don't make things worse*. If we *value* the fact that the sets of atoms belonging to $Amb(\Gamma)$ are sufficient for projecting of consistent images of Γ, and consider as *acceptable* all and only those extensions of Γ which preserve the sufficiency of at least one such set, then, just as with more familiar consequence relations, we should accept the intersection of the maximal acceptable extensions as the 'consequences' of Γ, the sentences we are committed to accepting given that we have accepted the sentences of Γ. This approach links the LP inconsistency measure to the LP consequence relation, providing a conservative and non-trivial account of the consequences of Γ as the sentences belonging to all the *maximal acceptable extensions* of Γ.

In comparison, while the least number of atoms that must be assigned the value B in an LP valuation satisfying Γ provides a reasonable measure of inconsistency, especially in the context of data bases where data are entered as atomic sentences, it does not capture the property that the LP consequence relation *preserves*. On a more philosophical note, it's also worth pointing out that in general, we don't have grounds for assuming that every atom in the language is equally important. For example, if we can project a consistent image of Γ by treating the atom γ ambiguously, but we can also project a consistent image of Γ by treating both atoms α and β ambiguously, then we might prefer either approach to eliminating the inconsistency, depending on the practical significance of the atoms. So we don't have general *logical* grounds for preferring the first approach to 'correcting' Γ's inconsistency over the second. On the other hand, if some extension of Γ, Γ' were such that we would need to treat all of β, γ and δ ambiguously in order to project a consistent image of Γ', Γ' would and should be regarded as 'making things worse'.

A very similar story applies to FDE. But in this case we treat the logic as a set-set consequence relation, and apply ambiguity in two different, incompatible ways. We begin by replacing the ambiguity sets of atoms used in our treatment of LP with disjoint, ordered pairs of sets of atoms, $\langle P, C \rangle$. The first set must be minimally sufficient to enable the projection of consistent images of the premise sets, just as in the case of LP above. Dually, the second must minimally sufficient to allow the projection of *consistently deniable* images of conclusion sets.[7, 9] (We say that a set of sentences, Δ is consistently deniable if and only if there is a valuation on the atoms that assigns 0 as the value of every member of Δ.) This allows FDE to block *both* the trivialization of inconsistent premise sets and the trivialization (from right to left) of conclusion sets that cannot be consistently *denied*. So long as Γ is consistent and Δ is consistently deniable, $\Gamma \vDash_{FDE} \Delta$ if and only if Γ classically entails Δ, i.e. *iff* $\Gamma \vDash_{CL} \Delta$. But FDE blocks the two trivial cases, in which either the members of Γ cannot all be assigned the value 1, or the members of Δ cannot all be assigned the value 0.

When the premise and conclusion sets are (respectively) consistent and consistently deniable, the only pair of sets appearing in the ambiguity set is $\langle \emptyset, \emptyset \rangle$, and the consequence relation is classical. But things are different, for example, when we consider the pair of sets $\{A, \neg A\}, \{(B \vee \neg B)\}$ as our premise and conclusion sets. In LP, $\{B \vee \neg B\}$ is a consequence of $\{A, \neg A\}$, because every minimally ambiguous, consistent projection of $\{A, \neg A\}$ will have $\{B \vee \neg B\}$ as a consequence: given any consistent set of sentences Γ, at least one of $B, \neg B$ must be a consistent extension of Γ. This is trivialization on the right of the turnstile, rather than on the left. But in FDE, the sets of atoms minimally sufficient for each of these purposes provide measures of both the inconsistency of asserting the premises *and* the inconsistency of denying the conclusions, both of which are to be preserved. So the ambiguity set for $\{(B \vee \neg B)\}$ on the right is $\{B\}$. Treating B as ambiguous, of course, allows for valuations falsifying $(B \vee \neg B)$, so, just as in the standard four-valued semantics, our ambiguity-based semantics for FDE ensures that $\{A, \neg A\} \nvDash \{B \vee \neg B\}$.

Finally, Kyburg's focus on rational decision-making and the role of probability in such decisions suggests another property worth preserving: the probability of the conjunctions of sentences we actually rely on in particular instances of practical reasoning. Decision making, even when conducted in a very formal way, focuses on small ('relevant') subsets of our cognitive commitments. So the (very low) probability that our full set of beliefs is true is not the right measure of the reliability of individual conclusions that follow from proper subsets of those beliefs. Kyburg's response was to think in terms of probability thresholds, determined in light of the 'stakes' involved in particular choices. But, as suggested in [6, p. 281], we can construct a 'consequence relation'–though not a formal one–based on the largest conjunctions whose probabilities surpass the threshold; the classical consequences of these conjunctions could then be *individually* rationally accepted as grounds for action.

7 Conclusion

The preservation of measures of inconsistency from left to right, and of their duals (which we might call *measures of triviality*) from right to left of the ⊢ relation (naturally paralleled semantically by the preservation of measures of unsatisfiability from left to right and measures of what we might call *non-disatisfiability* from right to left of the ⊨ relation) leads to interesting approaches to reasoning with inconsistent (unsatisfiable) premises and trivial (unfalsifiable) conclusions. More generally, the identification of new consequence relations determined by the preservation of properties other than consistency and satisfiability on the left, and consistent deniability and 'dis-satisfiability' on the right, is a promising approach to generalizing the concept of a consequence relation.

These preservation relations emerged from applying inconsistency measures on the left, together with duals of those measures on the right of the turnstile. The tools required to generate such consequence relations are conceptually economical, and the resulting, close connection between the preservation of certain inconsistency measures and paraconsistent consequence relations is elegant and logically interesting. Other measures of inconsistency in the literature(possibly in combination with dual measures of the triviality of conclusion sets) may be helpful in the development of new paraconsistent consequence relations, broadening the scope of Belnap's early proposal for non-explosive logics and their potential applications to computer reasoning in [2].

References

[1] P. Apostoli and B. Brown "A Solution to the Completeness Problem for Weakly Aggregative Modal Logic," *The Journal of Symbolic Logic* **60**, 3: 832-842, 1995.

[2] N. Belnap "How a Computer Should Think" in [21]: 30-56, 1977.

[3] R. Benham, C. Mortensen and G. Priest "Chunk and Permeate III: The Dirac Delta Function" *Synthese* **191**, 13: 3057-3062, 2014.

[4] Proceedings of the Third Workshop. Bimbo, Katalin, ed., London, College Publications, 2017.

[5] B. Brown "Yes, Virginia, There Really are Paraconsistent Logics" *Journal of Philosophical Logic* 28: 489-500 (October, 1999)

[6] B. Brown "Adjunction and Aggregation" *Nous* **33**, 2: 273-283, 1999.

[7] B. Brown "Ambiguity Games and Preserving Ambiguity Measures," in [23].

[8] B. Brown "Refining Preservation", in *Paraconsistency: Logic and Applications* Dordrecht, Springer, 123-140 2012.

[9] B. Brown "A Preservationist Perspective on Relevance Logic," in [4].

[10] B. Brown and G. Priest "Chunk and Permeate" *The Journal of Philosophical Logic* **33**, 4: 379-388, 2004.

[11] B. Brown and G. Priest "Chunk and Permeate II: the Bohr Atom" *European Journal for the Philosophy of Science* DOI 10.1007/s13194-014-0104-7, January, 2015.

[12] B. Brown and P. K Schotch "Logic and Aggregation," *Journal of Philosophical Logic*, **28** 265-287 1999.

[13] R. Dordrecht *Induction, Acceptance and Rational Belief*, 1970.

[14] J. M. Dunn "Intuitive Semantics for First-degree Entailments and 'Coupled Trees' " *Philosophical Studies* **29**, 149-168, 1976.

[15] M. Forbes and D. Hull and K. Okruhlik (eds.) *PSA 1992: Proceedings of the 1992 Biennial Meeting of the Philosophy of Science Association, Vol. 2* East Lansing, MI: Philosophy of Science Association 1993.

[16] J. Grant and A. Hunter "Analyzing Inconsistent Information using Distance-based Measures" *International Journal of Approximate Reasoning* 89 3-26 2017.

[17] J. Grant and M.V. Martinez Editors. Measuring Inconsistency in Information. College Publications, 2018.

[18] H.E. Kyburg Jr. "Conjunctivitis" 55-82 in [13] 1970.

[19] G. Payette "Getting the most out of inconsistency" *Journal of Philosophical Logic* 44 573-592 2015.

[20] N. Rescher and R. Manor "On Inference from Inconsistent Premises" *Theory and Decision* 1, 179-217 1970.

[21] G. Ryle (ed.), *Contemporary Aspects of Philosophy*, Stockfield, Oriel Press, 1977.

[22] D. Scott "Completeness and Axiomatizability in Many-valued Logic" *Proceedings of the Tarski Symposium* (Proceedings of Symposia in Pure Mathematics, vol. 25, University of California, Berkeley California, 1971) Providence, R.I. 411-435 1974.

[23] P. K. Schotch and B. Brown and R.E Jennings *On Preserving: Essays on Preservationism and Paraconsistent Logic* University of Toronto Press, Toronto, 2009.

[24] P. K. Schotch and R.E Jennings "On Detonating" 306-327 in [25].

[25] R. Sylvan and G. Priest and J. Norman *Paraconsistent Logic: Essays on the Inconsistent* Philosophia Verlag, München, 1989.

[26] M. Thimm "Inconsistency Measures for Probabilistic Logics" *Artificial Intelligence* 197 1-24 2013.

[27] M. Thimm "On the Expressivity of Inconsistency Measures" *Artificial Intelligence* 234 120-151 2016.

[28] M. Thimm "On the Evaluation of Inconsistency Measures", in [17].

Inconsistency Measurement in Probabilistic Logic

Glauber De Bona[1], Marcelo Finger[2], Nico Potyka[3],
Matthias Thimm[4]

[1]Department of Computer Science
University College London, UK
glauberbona@gmail.com,
[2]Department of Computer Science
University of Sao Paulo, Brazil
mfinger@ime.usp.br
[3]University of Osnabruck, Germany
nico.potyka@uni-osnabrueck.de
[4]Institute for Web Science and Technologies
University of Koblenz-Landau, Germany
thimm@uni-koblenz.de

Abstract

We survey the state of the art of inconsistency measurement in probabilistic logics. Compared to the setting of inconsistency measurement in classical logic, the incorporation of probabilistic assessments brings new challenges that have to be addressed by computational accounts to inconsistency measures. For that, we revisit rationality postulates for this setting and discuss the intricacies of probabilistic logics. We give an overview on existing measures and discuss their compliance with the rationality postulates. Finally, we discuss the relationships of inconsistency measures for probabilistic logic with Dutch books from economics.

1 Introduction

In this chapter, we will focus on measuring inconsistency in probabilistic logics. As opposed to classical knowledge bases, we enrich formulas with probabilities. A formula with probability 1 is supposed to be true (like a classical true formula), a formula with probability 0 is supposed to be false (like a classical negated true formula). Probabilities between 0 and 1 express our uncertainty about the truth state. In particular, we are often interested in conditional probabilities.

Example 1. Suppose we want to design an expert system for medical decision support. A group of medical experts is asked about their beliefs about the relationships between particular diseases and corresponding symptoms. Let us assume that the experts state the following beliefs about disease d and symptoms s_1, s_2:

- the probability of a patient with disease d exhibiting both symptom s_1 and symptom s_2 is at least 60%;

- the probability of a patient with disease d exhibiting symptom s_1 but not symptom s_2 is at least 50%;

- the probability of a patient with disease d exhibiting symptom s_1 is at most 80%.

Taken together, the experts' beliefs are inconsistent: according to the first two items, the probability of symptom s_1, given disease d, should be at least 110%. We have to adapt the beliefs in order to restore consistency. How should we proceed? Which pieces of information should be changed to restore consistency? Moreover, which pieces are to blame for the inconsistency, and to which degree? Once chosen which statement to change, should it be deleted or adapted by raising or lowering the probability in order to approximate consistency? These are the kind of questions an inconsistency measure for probabilistic logic can help to answer. □

In the following, we will mainly focus on propositional probabilistic knowledge bases to keep things simple. Our logical language is similar to the ones considered in [28, 29, 24, 22]. Many ideas can be transferred to relational languages. We will give some examples and further references as we proceed.

We will start our discussion with a quick introduction to propositional probabilistic logics. Subsequently, we will introduce a collection of rationality postulates in Section 3. While many of these properties are direct translations from the classical setting, we will consider additional postulates that take the role of probabilities into account. One important postulate in this context is *Continuity*, which basically states that minor changes in probabilities should not yield major changes in the degree of inconsistency. Interestingly, Continuity is in conflict with some classical postulates. In Section 4, we will then discuss six approaches to measure inconsistency in probabilistic logics. We will start with measures that are inspired by classical approaches and then look at several measures that make stronger use of the probabilities in the knowledge bases. We compare these measures with respect to the postulates that they satisfy or violate. In Section 5, we will sketch some applications of inconsistency measures. We will briefly discuss how they can be used to repair inconsistent knowledge bases and to reason with knowledge bases that contain conflicts of different kinds.

2 Preliminaries

We consider a propositional language built up over a finite set of *atomic propositions (atoms)* $\mathcal{A} = \{a_1, \ldots, a_n\}$ in the usual way. That is, formulas are constructed inductively by connecting atomic propositions with logical connectives like $\neg, \wedge, \vee, \rightarrow$. $\mathcal{L}_\mathcal{A}$ denotes the set of all well-formed propositonal formulas over $\mathcal{A}$. Additionally, $\top$ denotes a tautology $a \vee \neg a$ for some $a \in \mathcal{A}$, and $\bot$ denotes a contradiction $\neg\top$.

A *possible world* w over $\mathcal{A}$ is a conjunction of $|\mathcal{A}| = n$ literals containing either a or $\neg a$ for each $a \in \mathcal{A}$. For instance, if $\mathcal{A} = \{a, b, c\}$, then $a \wedge b \wedge c$ and $a \wedge b \wedge \neg c$ are two of the $2^3 = 8$ possible worlds over $\{a, b, c\}$. We denote by $\mathcal{W}_\mathcal{A} = \{w_1, \ldots, w_{2^n}\}$ the set of all possible worlds over $\mathcal{A}$. $w \in \mathcal{W}_\mathcal{A}$ *satisfies* $a \in \mathcal{A}$ ($w \models a$) iff a is a positive literal in w (a is not negated in w). For instance, $a \wedge b \wedge \neg c$ satisfies a and b but falsifies c. $\models$ can be extended to all $\varphi \in \mathcal{L}_\mathcal{A}$ recursively as usual. For instance, $w \models \neg\varphi$ iff $w \models \varphi$ does not hold, $w \models \varphi_1 \wedge \varphi_2$ iff $w \models \varphi_1$ and $w \models \varphi_2$ and so on.

A *probabilistic conditional* (or simply *conditional*) is an expression of the form $(\varphi|\psi)[\underline{q}, \bar{q}]$, where $\varphi, \psi \in \mathcal{L}_\mathcal{A}$ are propositional formulas and $\underline{q}, \bar{q} \in [0, 1] \cap \mathbb{Q}$ are rational numbers with $\underline{q} \leq \bar{q}$. Intuitively, $(\varphi|\psi)[\underline{q}, \bar{q}]$ says that "the probability that φ is true given that ψ is true lies within the interval $[\underline{q}, \bar{q}]$". $(\mathcal{L}_\mathcal{A} \mid \mathcal{L}_\mathcal{A})$ denotes the set of all conditonals over $\mathcal{A}$. A conditional $(\varphi|\psi)[q, q]$ with equal lower and upper bound is called *precise* and denoted by $(\varphi|\psi)[q]$ to improve readability. Similarly, if the condition is tautological, we write $(\varphi)[\underline{q}, \bar{q}]$ rather than $(\varphi|\top)[\underline{q}, \bar{q}]$ and call $(\varphi)[\underline{q}, \bar{q}]$ an *unconditional probabilistic assessment*.

A *probabilistic interpretation* $\pi : \mathcal{W}_\mathcal{A} \to [0, 1]$, with $\sum_j \pi(w_j) = 1$, is a probability mass over the set of possible worlds. Each probabilistic interpretation π induces a probability measure $P_\pi : \mathcal{L}_\mathcal{A} \to [0, 1]$ by means of $P_\pi(\varphi) = \sum_{w_j \models \varphi} \pi(w_j)$. Let $\mathcal{P}(\mathcal{A})$ be the set of all probabilistic interpretations $\pi : \mathcal{W}_\mathcal{A} \to [0, 1]$. A conditional $(\varphi|\psi)[\underline{q}, \bar{q}]$ is *satisfied* by π, also denoted by $\pi \models (\varphi|\psi)[\underline{q}, \bar{q}]$ iff $P_\pi(\varphi \wedge \psi) \geq \underline{q}P_\pi(\psi)$ and $P_\pi(\varphi \wedge \psi) \leq \bar{q}P_\pi(\psi)$. Note that when $P_\pi(\psi) > 0$, a probabilistic conditional $(\varphi|\psi)[\underline{q}, \bar{q}]$ is constraining the conditional probability of φ given ψ; but any π with $P_\pi(\psi) = 0$ trivially[1] satisfies the conditional $(\varphi|\psi)[\underline{q}, \bar{q}]$ (this semantics is adopted by Halpern [16], Frisch and Haddawy [13] and Lukasiewicz [24], for instance). For a conditional c we denote by $\mathsf{Mod}(c)$ the set of models of c, i.e., $\mathsf{Mod}(c) = \{\pi \mid \pi \models c\}$.

In order to take account of information that was presented multiple times, we regard a *knowledge base* as a finite multiset κ of probabilistic conditionals. Formally, a knowledge base κ is defined by a *multiplicity function* $M_\kappa : (\mathcal{L}_\mathcal{A} \mid \mathcal{L}_\mathcal{A}) \to \mathbb{N}$ such that $M_\kappa(C) > 0$ for only finitely many $C \in (\mathcal{L}_\mathcal{A} \mid \mathcal{L}_\mathcal{A})$. $M_\kappa(C)$ is the number of occurrences of C in κ. Let $\mathbb{K}$ denote the set of all knowledge

---

[1] An approach that does not trivialize when $P_\pi(\psi) = 0$ can be found in [6].

bases. We write $C \in \kappa$ iff $M_\kappa(C) > 0$. For two knowledge bases κ_1, κ_2 we define multiset union $\cup$, multiset intersection $\cap$, and multiset difference $\setminus$ via

$$M_{\kappa_1 \cup \kappa_2}(C) = M_{\kappa_1}(C) + M_{\kappa_2}(C)$$
$$M_{\kappa_1 \cap \kappa_2}(C) = \min\{M_{\kappa_1}(C), M_{\kappa_2}(C)\}$$
$$M_{\kappa_1 \setminus \kappa_2}(C) = \max\{0, M_{\kappa_1}(C) - M_{\kappa_2}(C)\}$$

for all $C \in (\mathcal{L}_\mathcal{A} \mid \mathcal{L}_\mathcal{A})$. The cardinality of κ is $|\kappa| = \sum_{C \in (\mathcal{L}_\mathcal{A} \mid \mathcal{L}_\mathcal{A})} M_\kappa(C)$. If κ is such that M_κ is non-zero exactly for $C_1, \ldots, C_m$, we denote κ by $\{C_1 : M_\kappa(C_1), \ldots, C_m : M_\kappa(C_m)\}$.

Example 2. $\kappa = \{(a)[0.2] : 2, a[0.8] : 1\}$ denotes the knowledge base that contains two instances of the conditional $(a)[0.2]$, one instance of the conditional $a[0.8]$ and no other conditionals. We have

$$\{(a)[0.2] : 2, (a)[0.8] : 1\} \cup \{(a)[0.8] : 1, (b)[0.5] : 1\}$$
$$= \{(a)[0.2] : 2, (a)[0.8] : 2, (b)[0.5] : 1\}$$

and

$$\{(a)[0.2] : 2, (a)[0.8] : 1\} \setminus \{(a)[0.2] : 1\} = \{(a)[0.2] : 1, (a)[0.8] : 1\}.$$

If κ is an ordinary set in the sense that $M_\kappa(C) = 1$ for all $C \in \kappa$, we omit the postfix $: 1$ to improve readability. For each knowledge base κ we assume some arbitrary but fixed *canonical enumeration* $\langle \kappa \rangle = (c_1, \ldots, c_m)$ that represents κ as a sequence of its elements (including duplicates).

A *probabilistic interpretation* $\pi : \mathcal{W}_\mathcal{A} \to [0, 1]$ satisfies a knowledge base κ, denoted by $\pi \models \kappa$, if $\pi \models c$ for all $c \in \kappa$. We let $\mathsf{Mod}(\kappa) = \{\pi \mid \pi \models \kappa\}$ and call the elements of $\mathsf{Mod}(\kappa)$ the models of κ. A knowledge base κ is *consistent* (or *satisfiable*) if $\mathsf{Mod}(\kappa) \neq \varnothing$. κ is precise if all conditionals in κ are precise.

Knowledge bases κ_1, κ_2 are *extensionally equivalent*, denoted by $\kappa_1 \equiv^e \kappa_2$, if and only if $\mathsf{Mod}(\kappa_1) = \mathsf{Mod}(\kappa_2)$. Knowledge bases κ_1, κ_2 are *semi-extensionally equivalent* [45], denoted by $\kappa_1 \equiv^s \kappa_2$, if and only if there is a bijection $\rho_{\kappa_1, \kappa_2} : \kappa_1 \to \kappa_2$ such that $c \equiv^e \rho_{\kappa_1, \kappa_2}(c)$ for every $c \in \kappa_1$. This means that two knowledge bases κ_1 and κ_2 are semi-extensionally equivalent if we find a mapping between the conditionals of both knowledge bases such that each conditional of κ_1 is extensionally equivalent to its image in κ_2. Note that $\kappa_1 \equiv^s \kappa_2$ implies $\kappa_1 \equiv^e \kappa_2$ [45].

3 Rationality Postulates

There are many ways of measuring the inconsistency of a set of formulas in some formal language. For example, we may have an idiosyncratic measurement that maps every consistent set to -3.2 and every inconsistent set to 7. Besides the arbitrariness of those values, such a measurement does not allow one to express that one theory is "more inconsistent" than another. And yet, this is something one may want to express. For instance, the knowledge base $\kappa_1 = \{(p)[0.5], (\neg p)[0.5001]\}$ seems "less inconsistent" than $\kappa_2 = \{(p)[1], (\neg p)[1]\}$ because the probabilities in κ_2 need to be adjusted more drastically than those in κ_1. If we are mainly interested in measuring the inconsistency of a knowledge base, it is reasonable to postulate that every consistent knowledge base is associated to the same measurement, say 0; similarly, we would expect that the degree of inconsistency of κ_1 is lower than the one of κ_2. In the following, we will discuss a collection of *rationality postulates* that have been proposed for inconsistency measures. As we will see, not all of these postulates can be satisfied simultaneously.

To begin with, we introduce inconsistency measures for probabilistic logics and some additional terminology. Inconsistency measures for probabilistic knowledge bases are defined analogously to those for classical knowledge bases.

Definition 1. An inconsistency measure $\mathcal{I}$ is a function $\mathcal{I} : \mathbb{K} \to [0, \infty)$.

The value $\mathcal{I}(\kappa)$ for a knowledge base κ is called the *inconsistency value* of κ with respect to $\mathcal{I}$. Minimal inconsistent sets of probabilistic knowledge bases are defined analogously to their classical counterparts.

Definition 2. A set $\mathcal{M}$ of probabilistic conditionals is *minimal inconsistent* if $\mathcal{M}$ is inconsistent and every $\mathcal{M}' \subsetneq \mathcal{M}$ is consistent.

We let $\mathsf{MI}(\kappa)$ denote the set of the minimal inconsistent subsets of $\kappa \in \mathbb{K}$. Intuitively, the conditionals in a minimal inconsistent subset of a knowledge base are those that are responsible for an atomic conflict.

Example 3. Consider the knowledge base

$$\kappa = \{(a)[0.2] : 1, (a)[0.8] : 1, (a \wedge b)[0.6] : 1, (b \vee d)[1] : 1, (c)[0.5] : 1\}$$

Here, we have two minimal inconsistent subsets $\{(a)[0.2] : 1, (a)[0.8] : 1\}$ and $\{(a)[0.2] : 1, (a \wedge b)[0.6] : 1\}$.

Conditionals that do not take part in such a conflict are called *free*.

Definition 3. A probabilistic conditional $c \in \kappa$ is *free* in κ if and only if $c \notin \mathcal{M}$ for all $\mathcal{M} \in \mathsf{MI}(\kappa)$.

For a conditional or a knowledge base C let $\mathcal{A}(C) \subseteq \mathcal{A}$ denote the set of atoms appearing in C. A conditional is safe with respect to a knowledge base κ if it does not share any atoms with κ [45].

Definition 4. A probabilistic conditional $c \in \kappa$ is *safe* in κ if and only if $\mathcal{A}(c) \cap \mathcal{A}(\kappa \setminus \{c\}) = \varnothing$.

The notion of a free conditional is more general than the notion of a safe conditional [45].

Proposition 1. If c is safe in κ then c is free in κ.

Example 4. We continue Example 3. Here, $(b \vee d)[1] : 1$ is a free formula and $(c)[0.5] : 1$ is both free and safe.

We first consider a set of qualitative postulates from [45] that have direct counterparts for classical knowledge bases. κ, κ' denote knowledge bases and c a probabilistic conditional.

Consistency κ is consistent if and only if $\mathcal{I}(\kappa) = 0$

Monotonicity $\mathcal{I}(\kappa) \leq \mathcal{I}(\kappa \cup \{c\})$

Super-additivity If $\kappa \cap \kappa' = \varnothing$ then $\mathcal{I}(\kappa \cup \kappa') \geq \mathcal{I}(\kappa) + \mathcal{I}(\kappa')$

Weak independence If $c \in \kappa$ is safe in κ then $\mathcal{I}(\kappa) = \mathcal{I}(\kappa \setminus \{c\})$

Independence If $c \in \kappa$ is free in κ then $\mathcal{I}(\kappa) = \mathcal{I}(\kappa \setminus \{c\})$

Penalty If $c \in \kappa$ is not free in κ then $\mathcal{I}(\kappa) > \mathcal{I}(\kappa \setminus \{c\})$

Irrelevance of syntax If $\kappa_1 \equiv^s \kappa_2$ then $\mathcal{I}(\kappa_1) = \mathcal{I}(\kappa_2)$

MI-separability If $\mathsf{MI}(\kappa_1 \cup \kappa_2) = \mathsf{MI}(\kappa_1) \cup \mathsf{MI}(\kappa_2)$ and $\mathsf{MI}(\kappa_1) \cap \mathsf{MI}(\kappa_2) = \varnothing$ then $\mathcal{I}(\kappa_1 \cup \kappa_2) = \mathcal{I}(\kappa_1) + \mathcal{I}(\kappa_2)$

Normalisation $\mathcal{I}(\kappa) \in [0,1]$

The property *consistency* demands that $\mathcal{I}(\kappa)$ takes the minimal value 0 if and only if κ is consistent. *Monotonicity* demands that $\mathcal{I}$ is non-decreasing under the addition of new information. *Super-additivity* strengthens this condition for disjoint knowledge bases. The properties *weak independence* and *independence* say that the inconsistency value should remain unchanged when adding "innocent" information. *Penalty* is the counterpart of *independence* and demands that adding inconsistent information increases the inconsistency value. *Irrelevance of syntax* states that the inconsistency value should not depend on the syntactic representation of conditionals. We use the equivalence relation $\equiv^s$ here since all inconsistent knowledge bases are equivalent with respect to $\equiv^e$. For an inconsistency measure $\mathcal{I}$, imposing irrelevance of syntax to hold in terms of $\equiv^e$ would yield $\mathcal{I}(\kappa) = \mathcal{I}(\kappa')$ for every two inconsistent knowledge bases κ, κ'. The property *MI-separability* states that determining the value of $\mathcal{I}(\kappa_1 \cup \kappa_2)$ can be split into determining the values of $\mathcal{I}(\kappa_1)$ and $\mathcal{I}(\kappa_2)$ if the minimal inconsistent subsets of $\kappa_1 \cup \kappa_2$ correspond to the disjoint union of those of κ_1 and κ_2. *Normalisation* states that inconsistency values should be bounded from above by one.

The following proposition states some relationships between these properties. The proof can be found in [45].

Proposition 2. Let $\mathcal{I}$ be an inconsistency measure and let κ, κ' be some knowledge bases.

1. If $\mathcal{I}$ satisfies *super-additivity* then $\mathcal{I}$ satisfies *monotonicity*.
2. If $\mathcal{I}$ satisfies *independence* then $\mathcal{I}$ satisfies *weak independence*.
3. If $\mathcal{I}$ satisfies **MI**-*separability* then $\mathcal{I}$ satisfies *independence*.
4. $\kappa \subseteq \kappa'$ implies $\mathsf{MI}(\kappa) \subseteq \mathsf{MI}(\kappa')$.
5. If $\mathcal{I}$ satisfies *independence* then $\mathsf{MI}(\kappa) = \mathsf{MI}(\kappa')$ implies $\mathcal{I}(\kappa) = \mathcal{I}(\kappa')$.
6. If $\mathcal{I}$ satisfies *independence* and *penalty* then $\mathsf{MI}(\kappa) \subsetneq \mathsf{MI}(\kappa')$ implies $\mathcal{I}(\kappa) < \mathcal{I}(\kappa')$.

Our qualitative properties do not take the crucial role of probabilities into account. In order to account for these we need some further notation. Let κ be a knowledge base. For $\vec{x} \in [0,1]^{2|\kappa|}$ we denote by $\kappa[\vec{x}]$ the knowledge base that is obtained from κ by replacing the probabilities of the conditionals in κ by the values in $\vec{x}$. More precisely, if $\langle \kappa \rangle = ((\varphi_1|\psi_1)[q_1, \bar{q}_1], \ldots, (\varphi_n|\psi_n)[q_n, \bar{q}_n])$ then $\langle \kappa[\vec{x}] \rangle = ((\varphi_1|\psi_1)[x_1, x_2], \ldots, (\varphi_n|\psi_n)[x_{2n-1}, x_{2n}])$ for $\vec{x} = \langle x_1, \ldots, x_{2n} \rangle \in [0,1]^{2n}$. Similarly, for a single probabilistic conditional $c = (\varphi|\psi)[q, \bar{q}]$ and $x_1, x_2 \in [0,1]$ we abbreviate $c[x_1, x_2] = (\varphi|\psi)[x_1, x_2]$. The characteristic function of a knowledge base κ takes a probability vector $\vec{x} \in [0,1]^{2|\kappa|}$ and replaces the probabilities in κ accordingly. The formal definition makes use of the order on the probabilistic conditionals of a knowledge base that we discussed in Section 2.

Definition 5. Let $\kappa \in \mathbb{K}$ be a knowledge base. The function $\Lambda_\kappa : [0,1]^{2|\kappa|} \to \mathbb{K}$ with $\Lambda_\kappa(\vec{x}) = \kappa[\vec{x}]$ is called the *characteristic function* of κ.

The characteristic inconsistency function is composed of the characteristic function and an inconsistency measure and shows how different probability vectors $\vec{x} \in [0,1]^{|\kappa|}$ affect the inconsistency value.

Definition 6. Let $\mathcal{I}$ be an inconsistency measure and let $\kappa \in \mathbb{K}$ be a knowledge base. The function

$$\theta_{\mathcal{I},\kappa} : [0,1]^{2|\kappa|} \to [0,\infty)$$

with $\theta_{\mathcal{I},\kappa} = \mathcal{I} \circ \Lambda_\kappa$ is called the *characteristic inconsistency function* of $\mathcal{I}$ and κ.

The following property *continuity* [45] describes our main demand for continuous inconsistency measurement, i.e., a "slight" change in the knowledge base should not result in a "vast" change of the inconsistency value.

Continuity For any $\vec{y} \in [0,1]^{2|\kappa|}$, $\lim_{\vec{x} \to \vec{y}} \theta_{\mathcal{I},\kappa}(\vec{x}) = \theta_{\mathcal{I},\kappa}(\vec{y})$

The above property demands a certain *smoothness* of the behavior of $\mathcal{I}$. Given a fixed set of probabilistic conditionals this property demands that changes in the *quantitative* part of the conditionals trigger a continuous change in the inconsistency value. Note that we require the *qualitative* part of the conditionals, i.e. premises and conclusions of the conditionals, to be fixed. This makes this property not applicable for the classical setting. In the probabilistic setting satisfaction of this property is helpful for the knowledge engineer in restoring consistency. Observe that for every knowledge base $\kappa \in \mathbb{K}$ there is always a $\vec{x} \in [0,1]^{2|\kappa|}$ such that $\kappa[\vec{x}]$ is consistent, cf. [43].

Even though the property *continuity* is a natural requirement in the probabilistic setting, it is incompatible with two of our qualitative postulates.

Proposition 3. Let $\mathcal{I}$ be an inconsistency measure and let κ, κ' be some knowledge bases.

1. There is no $\mathcal{I}$ that satisfies *consistency*, *independence*, and *continuity*.

2. There is no $\mathcal{I}$ that satisfies *consistency*, *MI-separability*, and *continuity*.

The proof can be found in [9]. These incompatibility results suggest that, in order to drive the rational choice of an inconsistency measure for probabilistic knowledge bases, we must abandon at least one postulate among *consistency*, *independence* and *continuity* (recall that *MI-separability* entails *independence*). The property *consistency* seems to be indisputable since the least one can expect from an inconsistency measure is that it separates inconsistent from consistent cases, or some inconsistency from none. Therefore, we should give up either *independence* or *continuity*. A simple solution is to give up *independence* for its weaker version *weak independence* that is compatible with *consistency* and *continuity* [9].

In fact, there are more compelling reasons for giving up *independence* rather than *continuity*. The property *independence* was introduced based on the notion that minimal inconsistent subsets are the purest form of inconsistency [20], capturing all its causes in a knowledge base [19]. This notion can be traced back to the work of Reiter on the diagnosis problem [37] and to the standard AGM framework of belief revision [1], where minimal inconsistent subsets have a central role. Nevertheless, in the probabilistic case, minimal inconsistent subsets may fail to detect all causes of inconsistency, as the next example illustrates.

Example 5. Recall the situation in Example 1, formalized into the knowledge base $\kappa = \{(s_1 \wedge s_2)[0.6, 1], (s_1 \wedge \neg s_2)[0.5, 1], (s_1)[0, 0.8]\}$. Suppose we want to schedule a meeting among the 3 different experts responsible for these assignments in order to reconcile them. To save resources, we plan to invite only the physicians whose probabilistic assessments are somehow causing the inconsistency. If minimal inconsistent subsets are supposed to capture such causes, the third physician would not be invited, for $\mathsf{MI}(\kappa) = \{\{(s_1 \land s_2)[0.6, 1], (s_1 \land \neg s_2)[0.5, 1]\}\}$, and $(s_1)[0, 0.8]$ is free in κ. Suppose the expert who elicited the first conditional, $(s_1 \land s_2)[0.6, 1]$, admits that the lower bound is rather high and relaxes it to $(s_1 \land s_2)[0.5, 1]$, being compatible with $(s_1 \land \neg s_2)[0.5, 1]$. Nonetheless, the updated knowledge base $\kappa' = \{(s_1 \land s_2)[0.5, 1], (s_1 \land \neg s_2)[0.5, 1], (s_1)[0, 0.8]\}$ would still be inconsistent, for the first two conditionals imply $(s_1)[1]$, contradicting the third one. □

The arguments above indicate that *continuity* is a more natural requirement than *independence* while measuring inconsistency in probabilistic knowledge bases, and one can still consistently demand *weak independence*.

4 Approaches

In the following we survey approaches to inconsistency measures for probabilistic logics from the literature.

4.1 Classical Approaches

We start with existing approaches to inconsistency measurement for classical logic and adapt those to the probabilistic case, see also [45] where those measures were adapted to probabilistic conditional logic with precise probabilities. In particular, we have a look at the drastic inconsistency measure, the MI inconsistency measure, the MI^C inconsistency measure, and the η-inconsistency measure, see e.g. [20, 23] for the classical definitions. What these approaches have in common, due to their origin, is that they concentrate on the qualitative part of inconsistency rather than the quantitative part, i.e. the probabilities.

The simplest approach to define an inconsistency measure is by just differentiating whether a knowledge base is consistent or inconsistent.

Definition 7. Let $\mathcal{I}_{\mathrm{drastic}} : \mathbb{K} \to [0, \infty)$ be the function defined as

$$\mathcal{I}_{\mathrm{drastic}}(\kappa) = \begin{cases} 0 & \text{if } \kappa \text{ is consistent} \\ 1 & \text{if } \kappa \text{ is inconsistent} \end{cases}$$

for $\kappa \in \mathbb{K}$. The function $\mathcal{I}_{\mathrm{drastic}}$ is called the *drastic inconsistency measure*.

The drastic inconsistency measure allows only for a binary decision on inconsistencies and does not quantify the severity of inconsistencies. One thing to note is that $\mathcal{I}_{\mathrm{drastic}}$ is the upper bound for any inconsistency measure that satisfies *consistency* and *normalization*, i.e., if $\mathcal{I}$ satisfies *consistency* and *normalization* then $\mathcal{I}(\kappa) \leq \mathcal{I}_{\mathrm{(}}\kappa)$ for every $\kappa \in \mathbb{K}$ [44].

The next inconsistency measure quantifies inconsistency by the number of minimal inconsistent subsets of a knowledge base.

Definition 8. Let $\mathcal{I}_{\mathrm{MI}} : \mathbb{K} \to [0, \infty)$ be the function defined as

$$\mathcal{I}_{\mathrm{MI}}(\kappa) = |\mathsf{MI}(\kappa)|$$

for $\kappa \in \mathbb{K}$. The function $\mathcal{I}_{\mathrm{MI}}$ is called the MI *inconsistency measure*.

The definition of the MI inconsistency measure is motivated by the intuition that the more minimal inconsistent subsets the greater the inconsistency.

Only considering the number of minimal inconsistent subsets may be too simple for assessing inconsistencies in general. Another indicator for the severity of inconsistencies is the size of minimal inconsistent subsets. A large minimal inconsistent subset means that the inconsistency is distributed over a large number of conditionals. The more conditionals involved in an inconsistency the less severe the inconsistency is. Furthermore, a small minimal inconsistent subset means that the participating conditionals strongly represent contradictory information. The following inconsistency measure is from [20] and aims at differentiating between minimal inconsistent sets of different size.

Definition 9. Let $\mathcal{I}_{\mathrm{MI}}^{C} : \mathbb{K} \to [0, \infty)$ be the function defined as

$$\mathcal{I}_{\mathrm{MI}}^{C}(\kappa) = \sum_{\mathcal{M} \in \mathsf{MI}(\kappa)} \frac{1}{|\mathcal{M}|}$$

for $\kappa \in \mathbb{K}$. The function $\mathcal{I}_{\mathrm{MI}}^{C}$ is called the MI *inconsistency measure*.

Note that $\mathcal{I}_{\mathrm{MI}}^{C}(\kappa) = 0$ if $\mathsf{MI}(\kappa) = \varnothing$. The MI^C inconsistency measure sums over the reciprocal of the sizes of all minimal inconsistent subsets. In that way, a large minimal inconsistent subset contributes less to the inconsistency value than a small minimal inconsistent subset.

The work [23] employs probability theory itself to measure inconsistency in classical theories by considering probability measures on classical interpretations. Those ideas can be extended for measuring inconsistency in probabilistic logics by considering probabilistic interpretations on probabilistic interpretations. Let $\hat{\pi} : \mathcal{P}(\mathcal{A}) \to [0, 1]$ be a probabilistic interpretation on $\mathcal{P}(\mathcal{A})$ such that $\hat{\pi}(\pi) > 0$ only for finitely many $\pi \in \mathcal{P}(\mathcal{A})$. Let $\mathcal{P}^2(\mathcal{A})$ be the set of those probabilistic interpretations. Then define the probability measure $P_{\hat{\pi}}$ analogously via

$$P_{\hat{\pi}}(c) = \sum_{\pi \in \mathcal{P}(\mathcal{A}), \pi \models c} \hat{\pi}(\pi) \qquad (1)$$

for a conditional c. This means that the probability (in terms of $\hat{\pi}$) of a conditional is the sum of the probabilities of probabilistic interpretations that satisfy c. Note also that by restricting $\hat{\pi}$ to assign a non-zero value only to finitely many $\pi \in \mathcal{P}(\mathcal{A})$, the sum in (1) is well-defined.

Now consider the following definition of the η-inconsistency measure.

Definition 10. Let $\mathcal{I}_\eta : \mathbb{K} \to [0, \infty)$ be the function defined as
$$\mathcal{I}_\eta(\kappa) = 1 - \max\{\eta \mid \exists \hat{\pi} \in \mathcal{P}^2(\mathcal{A}) : \forall c \in \kappa : \hat{\pi}(c) \geq \eta\}$$
for $\kappa \in \mathbb{K}$. The function $\mathcal{I}_\eta$ is called the *η-inconsistency measure*.

The idea of the η-inconsistency measure is that it looks for the largest probability that can be consistently assigned to the conditionals of a knowledge base and defines the inconsistency value inversely proportional to this probability.

Example 6. We continue Example 5 and consider
$$\kappa = \{(s_1 \wedge s_2)[0.6, 1], (s_1 \wedge \neg s_2)[0.5, 1], (s_1)[0, 0.8]\}$$
Recall that $\mathsf{MI}(\kappa) = \{\{(s_1 \wedge s_2)[0.6, 1], (s_1 \wedge \neg s_2)[0.5, 1]\}\}$ and therefore
$$\mathcal{I}_{\text{drastic}}(\kappa) = 1$$
$$\mathcal{I}_{\text{MI}}(\kappa) = 1$$
$$\mathcal{I}_{\text{MI}}^C(\kappa)1/2$$
Finally, consider $\hat{\pi} \in \mathcal{P}^2(\mathcal{A})$ such that $\hat{\pi}(\pi_1) = \hat{\pi}(\pi_2) = 1/2$ for

$\pi_1(s_1 \wedge s_2) = 0.8 \quad \pi_1(\neg s_1 \wedge s_2) = 0.2 \quad \pi_1(s_1 \wedge \neg s_2) = 0 \quad \pi_1(\neg s_1 \wedge \neg s_2) = 0$
$\pi_2(s_1 \wedge s_2) = 0 \quad \pi_2(\neg s_1 \wedge s_2) = 0 \quad \pi_2(s_1 \wedge \neg s_2) = 0.8 \quad \pi_2(\neg s_1 \wedge \neg s_2) = 0$

Observe that

$\pi_1 \models (s_1 \wedge s_2)[0.6, 1] \qquad \pi_1 \models (s_1)[0, 0.8]$
$\pi_2 \models (s_1 \wedge \neg s_2)[0.5, 1] \qquad \pi_2 \models (s_1)[0, 0.8]$

and therefore
$$\hat{\pi}((s_1 \wedge s_2)[0.6, 1]) = 0.5$$
$$\hat{\pi}((s_1 \wedge \neg s_2)[0.5, 1]) = 0.5$$
$$\hat{\pi}((s_1)[0, 0.8]) = 1$$

and thus $\hat{\pi}(c) \geq 0.5$ for all $c \in \kappa$. It can be easily seen that there cannot be some $\hat{\pi}'$ with $\hat{\pi}(c) > 0.5$ for all $c \in \kappa$ and therefore $\mathcal{I}_\eta(\kappa) = 0.5$.

The inconsistency measures discussed above were initially developed for inconsistency measurement in classical theories and therefore allow only for a "discrete" measurement. Hence, all of the above discussed inconsistency measures do not satisfy *continuity*. In particular, we have the following results [45].

Proposition 4. $\mathcal{I}_{\text{drastic}}$ satisfies *consistency, monotonicity, irrelevance of syntax, weak independence, independence*, and *normalisation*.

Proposition 5. $\mathcal{I}_{\text{MI}}$ satisfies *consistency, monotonicity, super-additivity, irrelevance of syntax, weak independence, independence*, **MI-separability**, and *penalty*.

Proposition 6. $\mathcal{I}_{\text{MI}}^C$ satisfies *consistency, monotonicity, super-additivity, irrelevance of syntax, weak independence, independence*, **MI-separability**, and *penalty*.

Proposition 7. $\mathcal{I}_\eta$ satisfies *consistency, monotonicity, irrelevance of syntax, weak independence, independence*, and *normalisation*.

However, satisfaction of *continuity* is crucial for an inconsistency measure in probabilistic logics in order to assess inconsistencies in a meaningful manner [45]. In the following, we continue with a survey of inconsistency measures that take the probabilities of conditionals into account and therefore address the postulate *continuity*.

4.2 Distance-based Approaches

The measures presented in [45] rely on distance measures defined as follows.

Definition 11 (Distance measure). A function $d : \mathbb{R}^n \times \mathbb{R}^n \to [0, \infty)$ (with $n \in \mathbb{N}^+$) is called a *distance measure* if it satisfies the following properties (for all $\vec{x}, \vec{y}, \vec{z} \in \mathbb{R}^n$):

1. $d(\vec{x}, \vec{y}) = 0$ if and only if $\vec{x} = \vec{y}$ (*reflexivity*)
2. $d(\vec{x}, \vec{y}) = d(\vec{y}, \vec{x})$ (*symmetry*)
3. $d(\vec{x}, \vec{y}) \leq d(\vec{x}, \vec{z}) + d(\vec{z}, \vec{y})$ (*triangle inequality*)

The simplest form of a distance measure is the *drastic distance measure* d_0 defined as $d_0(\vec{x}, \vec{y}) = 0$ for $\vec{x} = \vec{y}$ and $d_0(\vec{x}, \vec{y}) = 1$ for $\vec{x} \neq \vec{y}$ (for $\vec{x}, \vec{y} \in \mathbb{R}^n$ and $n \in \mathbb{N}^+$). A more interesting distance measure is the *p-norm distance*.

Definition 12. Let $n, p \in \mathbb{N}^+$. The function $d_p : \mathbb{R}^n \times \mathbb{R}^n \to [0, \infty)$ defined via

$$d_p(\vec{x}, \vec{y}) = \sqrt[p]{|x_1 - y_1|^p + \ldots + |x_n - y_n|^p}$$

for $\vec{x} = (x_1, \ldots, x_n), \vec{y} = (y_1, \ldots, y_n) \in \mathbb{R}^n$ is called the *p-norm distance*.

Special cases of the *p*-norm distance include the *Manhattan distance* (for $p = 1$) and the *Euclidean distance* (for $p = 2$).

Now we can define the "severity of inconsistency" in a knowledge base by the minimal distance of the knowledge base to a consistent one. As we are able to identify knowledge bases of the same qualitative structure in a vector space, we can employ distance measures for measuring inconsistency.

Definition 13. Let d be a distance measure. Then the function $\mathcal{I}_d : \mathbb{K} \to [0, \infty)$ defined via

$$\mathcal{I}_d(\kappa) = \inf\{d(\vec{x}, \vec{y}) \mid \kappa = \kappa[\vec{x}] \text{ and } \kappa[\vec{y}] \text{ is consistent}\} \quad (2)$$

for $\kappa \in \mathbb{K}$ is called the *d-inconsistency measure*.

The idea behind the d-inconsistency measure is that we look for a consistent knowledge base that both 1.) has the same qualitative structure as the input knowledge base and 2.) is as close as possible to the the input knowledge base. That is, if the input knowledge base is $\kappa[\vec{x}]$ we look at all $\vec{y} \in [0,1]^{2|\kappa|}$ such that $\kappa[\vec{y}]$ is consistent and $\vec{x}$ and $\vec{y}$ are as close as possible with respect to the distance measure d. While using the drastic distance measure gives us drastic inconsistency measure—that is $\mathcal{I}_{d_0} = \mathcal{I}_{\text{drastic}}$ [45]—using p-norms gives us genuinely novel inconsistency measures.

Example 7. We continue Example 5 and consider

$$\kappa = \{(s_1 \wedge s_2)[0.6, 1], (s_1 \wedge \neg s_2)[0.5, 1], (s_1)[0, 0.8]\}$$

Then we have

$$\mathcal{I}_{d_1}(\kappa) = 0.3$$
$$\mathcal{I}_{d_2}(\kappa) \approx 0.55$$

In order to see $\mathcal{I}_{d_1}(\kappa) = 0.3$ observe that changing the interval of $(s_1 \wedge s_2)[0.6, 1]$ to $[0.5, 1]$ and of $(s_1)[0, 0.8]$ to $[0, 1]$ restores consistency. More precisely, the probabilistic interpretation π defined via

$$\pi(s_1 \wedge s_2) = 0.5$$
$$\pi(s_1 \wedge \neg s_2) = 0.5$$
$$\pi(\neg s_1 \wedge s_2) = \pi(\neg s_1 \wedge \neg s_2) = 0$$

satisfies $(s_1 \wedge s_2)[0.5, 1]$, $(s_1 \wedge \neg s_2)[0.5, 1]$, and $(s_1)[0, 1]$. Note that we changed the probability interval of the first conditional by a value of 0.1 and that of the last interval by 0.2, yielding a sum of 0.3. It can be seen that no smaller modification yields a consistent knowledge base, so we have $\mathcal{I}_{d_1}(\kappa) = 0.3$.

In order to see $\mathcal{I}_{d_2}(\kappa) \approx 0.55$ consider the knowledge base κ' given via

$$\kappa' = \{(s_1 \wedge s_2)[0.5, 1], (s_1 \wedge \neg s_2)[0.4, 1], (s_1)[0, 0.9]\}$$

Observe κ' can be obtained from κ' by changing each probability interval by a value of 0.1, thus the distance between the probability intervals wrt. the Euclidean distance d_2 is $\sqrt{0.1^2 + 0.1^2 + 0.1^2} \approx 0.55$. It can also easily be seen that κ' is consistent and no other knowledge base with smaller Euclidean distance is consistent.

Taking results from [45, 9] into account we obtain the following picture regarding compliance with rationality postulates[2]. For reasons of simplicity,

---
[2] Note that [9] corrects some false claims from [45].

we only consider the instantiation of $\mathcal{I}_d$ with the p-norm distance d_p, see [45] for a more general treatment.

Proposition 8. Let $p \geq 1$. $\mathcal{I}_{d_p}$ satisfies *consistency, monotonicity, super-additivity* (only for $p = 1$), *irrelevance of syntax, weak independence,* and *continuity.*

For a variant of the above measure that also works with infinitesimal probabilities we refer to [26].

4.3 Violation-based Approaches

The next family of inconsistency measures is based on the idea of measuring the violation of the numerical constraints that define the satisfaction relation for conditionals. For this purpose, it is convenient to represent the conditional satisfaction constraints in matrix notation [28, 21]. We assume an arbitrary but fixed order on $\mathcal{W}_\mathcal{A}$ (e.g., lexicographically) so that we can identify each probabilistic interpretation π with the vector $(\pi(w_1), \ldots, \pi(w_n))'$, $n = |\mathcal{W}_\mathcal{A}|$, where the worlds are enumerated according to our order.

Recall that $\pi \models (\varphi|\psi)[\underline{q}, \bar{q}]$ iff $P_\pi(\varphi \wedge \psi) \geq \underline{q} P_\pi(\psi)$ and $P_\pi(\varphi \wedge \psi) \leq \bar{q} P_\pi(\psi)$. Subtracting the right-hand-side of the inequalities and putting in the definition of P_π, we notice that these are linear inequalities over π. For instance, $P_\pi(\varphi \wedge \psi) \geq \underline{q} P_\pi(\psi)$ is true if and only if

$$0 \leq P_\pi(\varphi \wedge \psi) - \underline{q} P_\pi(\psi) = P_\pi(\varphi \wedge \psi) - \underline{q}\big(P_\pi(\varphi \wedge \psi) + P_\pi(\neg\varphi \wedge \psi)\big)$$
$$= (1 - \underline{q}) P_\pi(\varphi \wedge \psi) - \underline{q} P_\pi(\neg\varphi \wedge \psi)$$
$$= (1 - \underline{q}) \sum_{w_j \models \varphi \wedge \psi} \pi(w_j) - \underline{q} \sum_{w_j \models \neg\varphi \wedge \psi} \pi(w_j)$$
$$= \sum_{j=1}^{n} \big((1 - \underline{q}) \cdot 1_{\{\varphi \wedge \psi\}}(w_j) - \underline{q} \cdot 1_{\{\neg\varphi \wedge \psi\}}(w_j)\big) \cdot \pi(w_j)$$
$$= a\,\pi,$$

where the indicator function $1_{\{F\}} : \mathcal{W}_\mathcal{A} \to \{0, 1\}$ yields 1 if the argument satisfies the formula $F \in \mathcal{L}_\mathcal{A}$ and 0 otherwise, and a is a n-dimensional row vector whose i-th component is $a_j = (1 - \underline{q}) \cdot 1_{\{\varphi \wedge \psi\}}(w_j) - \underline{q} \cdot 1_{\{\neg\varphi \wedge \psi\}}(w_j)$. Note that $a\pi \geq 0$ is equivalent to $b\pi \leq 0$, where $b = -a$. We can write all constraints for our knowledge base in this form. Thus, given some knowledge base κ, we can arrange the vectors corresponding to the constraints in a matrix A_κ such that an interpretation π satisfies κ if and only if $A_\kappa \pi \leq 0$. We call A_κ the *constraint matrix of* κ.

If κ is inconsistent, the system of inequalities $Ax \leq 0$ cannot be satisfied by a probability vector. However, we can relax the constraint to $A\pi \leq \epsilon$ for some non-negative vector ϵ. The minimum size of a vector that makes the system satisfiable is the *minimum violation value* of the knowledge base and can be understood as an inconsistency measure [30, 9].

Formally, the minimal violation value of a knowledge base is defined by an optimization problem. The problem is parameterized by a vector norm that measures the size of ϵ.

Definition 14 (Minimal Violation Value with respect to $\|.\|$). Let κ be a knowledge base with corresponding constraint matrix A_κ (of size $m \times n$). Let $\|.\|$ be some continuous vector norm. The *minimal violation value of κ with respect to $\|.\|$* is defined by

$$\min_{(x,\epsilon) \in \mathbb{R}^{n+m}} \|\epsilon\| \quad (3)$$

$$\text{subject to} \quad A_\kappa x \leq \epsilon,$$

$$\sum_{i=1}^{n} x_i = 1,$$

$$x \geq 0,$$

$$\epsilon \geq 0.$$

We denote the minimal violation value of κ by $\mathcal{I}_{\|.\|}(\kappa)$.

Proposition 9. $\mathcal{I}_{\|.\|}$ is an inconsistency measure and satisfies Consistency, Monotonicity, Weak Independence, Irrelevance of syntax and Continuity.

$\mathcal{I}_{\|.\|}$ satisfies neither Independence nor MI-Separability.

Proofs for these results can be found in [30, 9]. Measuring inconsistency by measuring the error in the system of inequalities is conceptually less intuitive than measuring the distance to a consistent knowledge base. However, it has some computational advantages as we discuss soon and still measures the degree of inconsistency continuously.

Example 8. Let $\kappa_\rho = \{(a)[\rho], (a)[1-\rho]\}$ for $0 \leq \rho \leq 0.5$. For instance, $\kappa_{0.5} = \{(A)[0.5], (A)[0.5]\}$ is consistent and $\kappa_0 = \{(A)[0], (A)[1]\}$ is inconsistent. Intuitively, the degree of inconsistency of κ_ρ should increase as ρ decreases from 0.5 to 0. We let $\mathcal{I}_1$, $\mathcal{I}_2$ and $\mathcal{I}_\infty$ denote the minimal violation measure with respect to the Manhattan norm $\|x\|_1 = \sum_{i=1}^{n} |x_i|$, Euclidean norm $\|.\|_2 = \sqrt{\sum_{i=1}^{n} x_i^2}$ and Maximum norm $\|x\|_\infty = \max\{|x_1|, \ldots, |x_n|\}$, respectively. Table 1 shows the corresponding inconsistency values.

Inspecting (3) shows that computing minimal violation values is a convex optimization problem (the objective function is convex and all constraints are linear). This has the principal advantage that we do not have to deal with non-global local optima and have polynomial runtime guarantees with respect to

	$\kappa_{0.5}$	$\kappa_{0.49}$	$\kappa_{0.4}$	$\kappa_{0.2}$	κ_0
$\mathcal{I}_1$	0	0.02	0.2	0.6	1
$\mathcal{I}_2$	0	0.014	0.141	0.424	0.707
$\mathcal{I}_\infty$	0	0.01	0.1	0.3	0.5

Table 1: Some minimal violation inconsistency values (Example 8).

the number of optimization variables and constraints. However, the number of optimization variables is still exponential in the number of atomic propositions (recall that each optimization variable corresponds to the probability of an interpretation).

We can do slightly better when using the Euclidean norm, which gives us a quadratic optimization problem. When using the Manhattan or the Maximum norm, the minimal violation value can actually be computed by a linear optimization problem. For the Manhattan norm $\|x\|_1 = \sum_{i=1}^{n} |x_i|$, we can immediately rewrite (3) as a linear program because the vector ϵ is constrained to be non-negative. Therefore, the absolute value function can be ignored.

Proposition 10. The minimal violation value $\mathcal{I}_{\|\cdot\|_1}(\kappa)$ with respect to the Manhattan norm can be computed by solving the following linear program:

$$\min_{(x,\epsilon)\in\mathbb{R}^{n+m}} \sum_{i=1}^{m} \epsilon_i \qquad (4)$$

$$\text{subject to} \quad A_\kappa x \leq \epsilon,$$

$$\sum_{i=1}^{n} x_i = 1,$$

$$x \geq 0,$$

$$\epsilon \geq 0.$$

For the Maximum norm $\|x\|_\infty = \max\{|x_1|, \ldots, |x_n|\}$, we can replace the relaxing vector ϵ with a scalar because we are only interested in the maximum component. Again, we can ignore the absolute value function due to the non-negativity of ϵ.

Proposition 11. The minimal violation value $\mathcal{I}_{\|\cdot\|_\infty}(\kappa)$ with respect to the

Maximum norm can be computed by solving the following linear program:

$$\min_{(x,\epsilon)\in\mathbb{R}^{n+1}} \epsilon \qquad (5)$$

$$\text{subject to} \quad A_\kappa x \leq \epsilon \cdot \vec{1},$$

$$\sum_{i=1}^{n} x_i = 1,$$

$$x \geq 0,$$

$$\epsilon \geq 0,$$

where $\vec{1} \in \mathbb{R}^m$ denotes the column vector that contains only ones.

Let us relate these linear programs to the probabilistic satisfiability problem [14, 21], that is, to the problem of deciding whether a given knowledge base is consistent. This problem comes down to finding a solution $x \in \mathbb{R}^n$ of the following system of linear inequalities:

$$A_\kappa x \leq 0,$$

$$\sum_{i=1}^{n} x_i = 1,$$

$$x \geq 0.$$

A standard way to solve this system is to apply Phase 1 of the Simplex algorithm. This comes down to solving a linear program like

$$\min_{(x,s)\in\mathbb{R}^{n+1}} s \qquad (6)$$

$$\text{subject to} \quad A_\kappa x \leq 0,$$

$$\sum_{i=1}^{n} x_i + s = 1,$$

$$x \geq 0,$$

$$\epsilon \geq 0,$$

Note that the vector $(0,1) \in \mathbb{R}^{n+1}$ is always a feasible vertex from which we can start the Simplex algorithm for this problem. κ is consistent iff the optimal solution $s^* = 0$. Notice that the structure of (6) is very similar to the structure of (5). We have the same number of optimization variables and constraints and there is only a minor difference in the first and second constraint. (4) does also have a very similar structure even though it adds a number of optimization variables that is linear in the size of the knowledge base. However, since n is exponential in the number of atoms of our language, this difference is usually negligible. In this sense, computing minimal violation measures for Manhattan and Maximum norm is barely harder than performing a probabilistic satisfiability test. Since inconsistency measures generalize probabilistic satisfiability tests, this is some evidence that these measures belong to the most efficient inconsistency measures. Note also that even though the Simplex algorithm has exponential runtime in the worst-case, it usually takes time linear in the number of optimization variables and quadratic in the number of constraints in practice [25]. In fact, due to the similarity between (6), (5) and (4), we can speed up computing minimal violation values for the Manhattan and Maximum norm by using similar techniques like for the probabilistic satisfiability problem. In particular, column generation techniques proved useful in this context [14, 18, 11, 7].

4.4 Dutch-book Measures

In formal epistemology, there is an interest in measuring the incoherence of an agent whose beliefs are given as probabilities over propositions or previsions (expected values) for random variables — a *Bayesian* agent. If we have propositions from classical logic, the formalized problem at hand is exactly the one we are investigating. When the agent's degrees of belief are represented by a knowledge base, to measure the agent's incoherence is to measure the inconsistency of such a knowledge base. Schervish, Kadane and Seidenfeld [40, 41, 39] have proposed ways to measure the incoherence of an agent based on Dutch books.

Dutch books have been proposed by De Finetti as a foundation for probability theory [10]. Dutch book arguments are based on the agent's betting behavior induced by her degrees of belief, typically used to show her irrationality. These arguments rely on an operational interpretation of (imprecise) degrees of belief, in which their lower/upper bounds determines when the agent considers as fair to take part in a gamble, defined as follows:

Definition 15. A *gamble* on $\varphi|\psi$, with $\varphi, \psi \in \mathcal{L}_\mathcal{A}$, is an agreement between an agent and a bettor with two parameters, the *stake* $\lambda \in \mathbb{R}$ and the *relative price* $q \in [0, 1]$, stating that:

- the agent pays $\lambda \times q$ to the bettor if ψ is true and φ is false;

- the bettor pays $\lambda \times (1 - q)$ to the agent if ψ is true and φ is true;

- the gamble is called off, causing neither profit nor loss to the involved parts, if ψ is false.

A *Dutch book* is a set of gambles that will cause the agent a sure loss, no matter which possible world is the case. For instance, suppose an agent is willing to take part in two gambles, on φ and on $\neg \varphi$, both with stake 10 and relative price 0.6. No matter whether φ or $\neg\varphi$ is the case, the agent has to pay $10 \times 0.6 = 6$ to the bettor, while receiving only $10 \times 0.4 = 4$ back, which causes her a net loss of 2.

A central result of De Finetti's theory of probabilities lies in the fact that if an agent respects the laws of probabilities, no Dutch books are possible. That is, a Dutch book (sure loss) is possible only when the agent gambles are inconsistent with the laws or probability.

To relate the epistemic state of an agent to her vulnerability to Dutch books, we need a *willingness-to-gamble assumption*: if an agent believes that the probability of a proposition φ being true given that another proposition ψ is true lies within $[\underline{q}, \bar{q}]$, she finds acceptable (fair) gambles on $\varphi|\psi$ with stake $\lambda \geq 0$ and relative price $\underline{q}$ and gambles with stake $\lambda \leq 0$ and relative price $\bar{q}$. An agent *is vulnerable to (or exposed to)* a given Dutch book if she sees as fair, under the willingness-to-gamble assumption, each of the gambles in the Dutch book. We assume any set of gambles the agent sees as fair contains exactly two gambles on $(\varphi_i|\psi_i)$ per each conditional $(\varphi_i|\psi_i)[\underline{q}_i, \bar{q}_i] \in \kappa$, the base formalizing the agent's beliefs: one with stake $\lambda_i \geq 0$ and relative price $\underline{q}_i$; and the other with stake $-\bar{\lambda}_i \leq 0$ and relative price $\bar{q}_i$. This is not restrictive, since gambles on the same $(\varphi_i|\psi_i)$ with the same relative price can be merged by summing the stakes, and the absence of a gamble is equivalent to a stake equal to zero. We can thus denote any set of gambles the agent finds acceptable simply by the absolute value of its stakes $\lambda_1, \bar{\lambda}_1, \ldots, \lambda_m, \bar{\lambda}_m \geq 0$, where $m = |\kappa|$.

If the set of probabilistic conditionals that represents an agent's epistemic state turns out to be inconsistent, then she is exposed to a Dutch book, and vice-versa [27]. In other words, an agent sees as fair a set of gambles that causes her a guaranteed loss if, and only if, the knowledge base codifying her (conditional) degrees of belief is inconsistent. In this way, Dutch book arguments were introduced to show that a set of degrees of belief must obey the axioms of probability and are a standard proof of incoherence (introductions to Dutch books and their relation to incoherence can be found in [42] and [10]). Hence, a natural approach to measuring an agent's degree of incoherence is through the magnitude of the sure loss she is vulnerable to. The intuition says that, the more incoherent an agent is, the greater the guaranteed loss that can be imposed on her through a Dutch book. Nevertheless, with no bounds on the stakes, such loss would also be unlimited for incoherent agents. To better understand the loss a Dutch book causes to an agent, we formalize it in the following.

Consider the knowledge base $\kappa = \{(\varphi_i|\psi_i)[\underline{q}_i, \bar{q}_i] | 1 \leq i \leq m\}$ representing an agent's epistemic state. Let $\lambda_i, \bar{\lambda}_i \geq 0$ denote gambles on $(\varphi_i|\psi_i)$ the agent sees as acceptable, the first with relative price $\underline{q}_i$ and stake $\lambda_i \geq 0$, the second with relative price $\bar{q}_i$ and stake $-\bar{\lambda}_i \leq 0$, for $1 \leq i \leq m$. A set of gambles the agent sees as fair can then be represented by the vector $\mathcal{G} = \langle \lambda_1, \bar{\lambda}_1, \ldots, \lambda_m, \bar{\lambda}_m \rangle$. If a possible world w_j is the case, the total net loss for the agent is

$$\ell_\kappa(\mathcal{G}, w_j) = \sum_{i=1}^{m} \bar{\lambda}_i(I_{\varphi_i \wedge \psi_i}(w_j) - \bar{q}_i I_{\psi_i}(w_j)) - \lambda_i(I_{\varphi_i \wedge \psi_i}(w_j) - q_i I_{\psi_i}(w_j)).$$

Given the knowledge base κ representing an agent's epistemic state, the set of gambles $\mathcal{G}$ is a Dutch book if $\ell_\kappa(\mathcal{G}, w_j) < 0$ for all $w_j \in \mathcal{W}_A$. When $\mathcal{G}$ is a Dutch book, the *sure loss* is defined as the amount the agent is guaranteed to lose, which is $\ell_\kappa^{sure}(\mathcal{G}) = \min_{w_j \in \mathcal{W}_A} \ell_\kappa(\mathcal{G}, w_j)$. If $\mathcal{G}$ is a not a Dutch book, there is a possible world where the agent does not lose (maybe wins), then $\ell_\kappa^{sure}(\mathcal{G}) = 0$. For an arbitrary set of gambles $\mathcal{G}$ that the agent sees as fair, we thus define $\ell_\kappa^{sure}(\mathcal{G}) = \max\{\min_{w_j \in \mathcal{W}_A} \ell_\kappa(\mathcal{G}, w_j), 0\}$. The maximum Dutch-book sure loss an agent is vulnerable to is given by $\max_\mathcal{G} \ell_\kappa^{sure}(\mathcal{G})$. To see that this maximization in unbounded, note that if $\ell_\kappa^{sure}(\mathcal{G}) = c > 0$ for some $\mathcal{G} = \langle \lambda_1, \bar{\lambda}_1, \ldots, \lambda_m, \bar{\lambda}_m \rangle$, then $\mathcal{G}' = \lambda \langle \lambda_1, \bar{\lambda}_1, \ldots, \lambda_m, \bar{\lambda}_m \rangle$ implies $\ell_{sure}(\mathcal{G}) = \lambda c > 0$ for any positive scalar $\lambda \in \mathbb{R}$. Consequently, any incoherent agent is vulnerable to an unbounded sure loss, and this quantity is not suitable to measure her incoherence.

Different strategies to measure incoherence as a finite Dutch-book loss are found in the Bayesian Statistics and formal epistemology literature. Schervish *et al.* propose a flexible formal approach to normalize the maximum sure loss generating a family of incoherence measures for upper and lower previsions on bounded random variables [41], which we adapt to our case. The simplest measures of this family arise when the sure loss is normalized by either the sum of the absolute values of the stakes, $\|\mathcal{G}\|_1 = \sum_i \lambda_i + \bar{\lambda}_i \leq 1$, or their maximum, $\|\mathcal{G}\|_\infty = \max\{\lambda_i, \bar{\lambda}_i \mid 1 \leq i \leq m\}$. We define the inconsistency measures $\mathcal{I}_{DB}^{sum} : \mathbb{K} \to [0, \infty)$ and $\mathcal{I}_{DB}^{max} : \mathbb{K} \to [0, \infty)$ on knowledge bases as these two incoherence measures on the corresponding agents represented by these knowledge bases [3]:

$$\mathcal{I}_{DB}^{sum}(\kappa) = \max_\mathcal{G} \frac{\ell_\kappa^{sure}(\mathcal{G})}{\|\mathcal{G}\|_1} \quad \text{and} \quad \mathcal{I}_{DB}^{max}(\kappa) = \max_\mathcal{G} \frac{\ell_\kappa^{sure}(\mathcal{G})}{\|\mathcal{G}\|_\infty}$$

Even though incoherence measures based, on one hand, on Dutch books proposed by the formal epistemology community and inconsistency measures based, on the other hand, on violations minimization proposed by Artificial Intelligence researchers may seem unrelated at first, they are actually two sides of the same coin. The linear programs that compute the maximum guaranteed loss an agent is exposed to are technically dual to those that minimize violations to measure inconsistency [9] [4]:

Theorem 1. *For any $\kappa \in \mathbb{K}$, $\mathcal{I}_{DB}^{sum}(\kappa) = \mathcal{I}_{\|\cdot\|_\infty}(\kappa)$ and $\mathcal{I}_{DB}^{max}(\kappa) = \mathcal{I}_{\|\cdot\|_1}(\kappa)$.*

---
[3]If $\|\mathcal{G}\|_1 = \|\mathcal{G}\|_\infty = 0$, then $\ell_\kappa^{sure}(\mathcal{G}) = 0$ and we define $\ell_\kappa^{sure}(\mathcal{G})/\|\mathcal{G}\|_1 = \ell_\kappa^{sure}(\mathcal{G})/\|\mathcal{G}\|_\infty = 0$.

[4]Nau [27] has already investigated this matter, mentioning some similar results.

Theorem 1 gives an operational interpretation, based on betting behavior, for the inconsistency measures $\mathcal{I}_{\|.\|_\infty}$ and $\mathcal{I}_{\|.\|_1}$. Naturally, the result also implies that $\mathcal{I}_{DB}^{sum}$ and $\mathcal{I}_{DB}^{max}$ hold the same properties as $\mathcal{I}_{\|.\|_\infty}$ and $\mathcal{I}_{\|.\|_1}$, respectively.

More elaborate measures proposed by Schervish *et al.* [41] normalize the sure loss by the amount the agent (or bettor) can possibly lose, either per gamble or in total. These quantities are called *escrows*. Let $q = \langle q_1, \ldots, q_m \rangle$, $\bar{q} = \langle \bar{q}_1, \ldots, \bar{q}_m \rangle$ and $1_m = \langle 1 \ldots 1 \rangle$ and be tuples with m elements. In a single gamble with stake $\lambda_i > 0$ and relative price q_i, the agent's (or the bettor's) escrow is $e_{\underline{i}}^a = \lambda_i \times q_i$ (or $e_{\underline{i}}^b = \lambda_i \times (1 - q_i)$). Conversely, in a gamble with stake $-\bar{\lambda}_i < 0$ and relative price $\bar{q}_i$, the agent's (or the bettor's) escrow is $e_{\overline{i}}^a = \bar{\lambda}_i \times (1 - \bar{q}_i)$ (or $e_{\overline{i}}^b = \bar{\lambda}_i \times \bar{q}_i$). Hence, if $\mathcal{G} = \langle \lambda_1, \bar{\lambda}_1, \ldots, \lambda_m, \bar{\lambda}_m \rangle$ is a set of gambles, $e^a(\mathcal{G}) = \langle e_{\underline{1}}^a, e_{\overline{1}}^a, \ldots, e_{\underline{m}}^a, e_{\overline{m}}^a \rangle$ (or $e^b(\mathcal{G}) = \langle e_{\underline{1}}^b, e_{\overline{1}}^b, \ldots, e_{\underline{m}}^b, e_{\overline{m}}^b \rangle$) is the vector containing how much the agent (or the bettor) can lose per each gamble. Normalizing the sure loss by the maximum or the sum of this vector's elements yields inconsistency measures defined as:

$$\mathcal{I}_{DB}^{a,sum}(\kappa) = \max_{\mathcal{G}} \frac{\ell_\kappa^{sure}(\mathcal{G})}{\|e^a(\mathcal{G})\|_1} \; ; \; \mathcal{I}_{DB}^{a,max}(\kappa) = \max_{\mathcal{G}} \frac{\ell_\kappa^{sure}(\mathcal{G})}{\|e^a(\mathcal{G})\|_\infty}$$

$$\mathcal{I}_{DB}^{b,sum}(\kappa) = \max_{\mathcal{G}} \frac{\ell_\kappa^{sure}(\mathcal{G})}{\|e^b(\mathcal{G})\|_1} \; ; \; \mathcal{I}_{DB}^{b,max}(\kappa) = \max_{\mathcal{G}} \frac{\ell_\kappa^{sure}(\mathcal{G})}{\|e^b(\mathcal{G})\|_\infty}$$

These four inconsistency measures ($\mathcal{I}_{DB}^{a,sum}$, $\mathcal{I}_{DB}^{b,sum}$, $\mathcal{I}_{DB}^{a,max}$ and $\mathcal{I}_{DB}^{b,max}$) satisfy most of the rationality postulates [9]:

Proposition 12. $\mathcal{I}_{DB}^{a,sum}$, $\mathcal{I}_{DB}^{b,sum}$, $\mathcal{I}_{DB}^{a,max}$ and $\mathcal{I}_{DB}^{b,max}$ are well-defined, satisfy Consistency, Monotonicity and Weak Indepedence and are continuous for probabilities within $(0,1)$. Furthermore, $\mathcal{I}_{DB}^{a,max}$ and $\mathcal{I}_{DB}^{b,max}$ satisfy Super-additivity, and $\mathcal{I}_{DB}^{a,sum}$ satisfies Normalization.

From their definition, one can see that $\mathcal{I}_{DB}^{a,sum}$, $\mathcal{I}_{DB}^{b,sum}$, $\mathcal{I}_{DB}^{a,max}$ and $\mathcal{I}_{DB}^{b,max}$ also satisfy Irrelevance of syntax, since the net loss of a gamble on $\varphi|\psi$ does not depend on how these formulas are written.

4.5 Fuzzy-logic Measure

In [8], another inconsistency measure on probabilistic knowledge bases is proposed that makes use of Fuzzy concepts. The central notion of [8] is the *candidacy function*. A candidacy function is similar to a fuzzy set [15] as it assigns a degree of membership of a probabilistic interpretation belonging to the models of a knowledge base. More formally, a candidacy function $\mathfrak{C}$ is a function $\mathfrak{C} : \mathcal{P}(\mathcal{A}) \to [0,1]$. A uniquely determined candidacy function $\mathfrak{C}_\kappa$ can be assigned to a (consistent or inconsistent) knowledge base κ as follows. For a probabilistic interpretation $\pi \in \mathcal{P}(\mathcal{A})$ and a set of probabilistic interpretations $S \subseteq \mathcal{P}(\mathcal{A})$ let $d(\pi, S)$ denote the distance of π to S with respect to the Euclidean norm, i.e., $d(\pi, S)$ is defined via

$$d(\pi, S) = \inf \left\{ \sqrt{\sum_{\omega \in \mathcal{W}_\mathcal{A}} (\pi(\omega) - \pi'(\omega))^2} \mid \pi' \in S \right\}.$$

Let $h : \mathbb{R}^+ \to (0, 1]$ be a strictly decreasing, positive, and continuous log-concave function with $h(0) = 1$. Then the candidacy function $\mathfrak{C}_\kappa^h$ for a knowledge base κ is defined as

$$\mathfrak{C}_\kappa^h(\pi) = \prod_{c \in \kappa} h\left(\sqrt{2^{|\mathcal{A}|}} d(\pi, \mathsf{Mod}(\{c\}))\right)$$

for every $\pi \in \mathcal{P}(\mathcal{A})$. Note that the definition of the candidacy function $\mathfrak{C}_\kappa^h$ depends on the size of the signature $\mathcal{A}$. The intuition behind this definition is that a probabilistic interpretation π that is near to the models of each probabilistic conditional in κ gets a high candidacy degree wrt. $\mathfrak{C}_\kappa^h$. It is easy to see that it holds that $\mathfrak{C}_\kappa^h(\pi) = 1$ if and only if $\pi \in \mathsf{Mod}(()\kappa)$. Using the candidacy function $\mathfrak{C}_\kappa^h$ the inconsistency measure $\mathcal{I}_{\mathrm{cand}}^h$ can be defined via

$$\mathcal{I}_{\mathrm{cand}}^h(\kappa) = 1 - \max_{\pi \in \mathcal{P}(\mathcal{A})} \mathfrak{C}_\kappa^h(\pi) \qquad (7)$$

for a knowledge base κ. The following results has been shown in [8].

Proposition 13. $\mathcal{I}_{\mathrm{cand}}^h$ satisfies *consistency, monotonicity, continuity,* and *normalization.*

The function $\mathcal{I}_{\mathrm{cand}}^h$ does not satisfy *super-additivity* as shown in [44].

Example 9. Let $\mathcal{A} = \{a_1, a_2\}$ be a propositional signature and let $\kappa_1 = \{(a_1)[1], (a_1)[0]\}$ and $\kappa_2 = \{(a_2)[1], (a_2)[0]\}$ be knowledge bases and let $\kappa = \kappa_1 \cup \kappa_2$. Note that both κ_1 and κ_2 are inconsistent and $\kappa_1 \cap \kappa_2 = \emptyset$. As $\mathcal{I}_{\mathrm{cand}}^h$ is defined on the semantic level and does not take the names of propositions into account it follows that $\mathcal{I}_{\mathrm{cand}}^h(\kappa_1) = \mathcal{I}_{\mathrm{cand}}^h(\kappa_2)$. As the situations in κ_1 and κ_2 are symmetric and κ_i is symmetric with respect to $(a_i)[1]$ and $(a_i)[0]$ there are probabilistic interpretations π_i with $\mathcal{I}_{\mathrm{cand}}^h(\kappa_i) = 1 - \mathfrak{C}_{\kappa_i}^h(\pi_i)$ for $i = 1, 2$ and

$$d(\pi_1, \mathsf{Mod}(\{(a_1)[1]\})) = d(\pi_1, \mathsf{Mod}(\{(a_1)[0]\}))$$
$$= d(\pi_2, \mathsf{Mod}(\{(a_2)[1]\}))$$
$$= d(\pi_2, \mathsf{Mod}(\{(a_2)[0]\}))$$

Let $x = d(\pi_1, \mathsf{Mod}(\{(a_1)[1]\}))$ and let $h^* : \mathbb{R}^+ \to (0, 1]$ be a strictly decreasing, positive, and continuous log-concave function with $h^*(0) = 1$ and

$h^*\left(\sqrt{2^{|\mathcal{A}|}}x\right) = 0.5$. Then it follows $\mathfrak{C}^{h^*}_{\kappa_1}(\pi_1) = 0.25$ and $\mathcal{I}^h_{\text{cand}}(\kappa_1) = 0.75$. In order to satisfy *super-additivity* $\mathcal{I}^h_{\text{cand}}$ must satisfy

$$\mathcal{I}^h_{\text{cand}}(\kappa) \geq \mathcal{I}^h_{\text{cand}}(\kappa_1) + \mathcal{I}^h_{\text{cand}}(\kappa_2) = 1.5$$

which is a contradiction since $\mathcal{I}^h_{\text{cand}}$ satisfies *normalization*.

On the other hand, $\mathcal{I}^h_{\text{cand}}$ complies with our notion of *irrelevance of syntax*[5]

Proposition 14. $\mathcal{I}^h_{\text{cand}}$ *satisfies irrelevance of syntax.*

Proof. Let κ_1 and κ_2 be such that $\kappa_1 \equiv^s \kappa_2$. Without loss of generality, assume $\kappa_1 = \{c_1, \ldots, c_n\}$ and $\kappa_2 = \{d_1, \ldots, d_n\}$ with $c_i \equiv d_i$ for $i = 1, \ldots, n$. It follows $\mathsf{Mod}(\{c_i\}) = \mathsf{Mod}(\{d_i\})$ for $i = 1, \ldots, n$ and therefore

$$d(\pi, \mathsf{Mod}(\{c_i\})) = d(\pi, \mathsf{Mod}(\{d_i\}))$$

for every π and $i = 1, \ldots, n$. It follows $\mathfrak{C}^h_{\kappa_1} = \mathfrak{C}^h_{\kappa_2}$ and therefore the claim. □

4.6 Entropy Measures

In [38] an inconsistency measure is presented that is based on the notion of *generalized divergence* which generalizes cross-entropy. Given vectors $\vec{y}, \vec{z} \in (0,1]^n$ with $\vec{y} = (y_1, \ldots, y_n)$ and $\vec{z} = (z_1, \ldots, z_n)$, the generalized divergence $D(\vec{y}, \vec{z})$ from $\vec{y}$ to $\vec{z}$ is defined

$$D(\vec{y}, \vec{z}) = \sum_{i=1}^n y_i \log_2 \frac{y_i}{z_i} - y_i + z_i$$

We abbreviate further

$$D^2(\vec{y}, \vec{z}) = D(\vec{y}, \vec{z}) + D(\vec{z}, \vec{y}) = \sum_{i=1}^n y_i \log_2 \frac{y_i}{z_i} + z_i \log_2 \frac{z_i}{y_i}$$

In [38], the measure $\mathcal{I}_{\text{gd}}$ is only defined for conditionals with point probabilities. So let $\kappa = \{c_1, \ldots, c_n\}$ and $c_i = (\psi_i|\varphi_i)[d_i]$ for $i = 1, \ldots, n$. Then the inconsistency measure $\mathcal{I}_{\text{gd}}$ is defined via

$$\mathcal{I}_{\text{gd}}(\kappa) = \min\{D^2(\vec{y}, \vec{z}) \mid \pi \in \mathcal{P}(\mathcal{A}) \text{ and}$$
$$y_i = (1-d_i)P_\pi(\psi_i\varphi_i) \text{ and } z_i = d_i P_\pi(\neg\psi_i \wedge \varphi_i) \text{ for } i = 1, \ldots, n\}$$

---

[5] In [44] it has been shown, however, that $\mathcal{I}^h_{\text{cand}}$ violates a slightly different notion of irrelevance of syntax.

Let $\vec{y}^*, \vec{z}^*, P_{\pi^*}$ be some parameters such that $D^2(\vec{y}^*, \vec{z}^*)$ is minimal and $y_i^* = (1-d_i)P_{\pi^*}(\psi_i\varphi_i)$ and $z_i^* = d_i P_{\pi^*}(\neg\psi_i \wedge \varphi_i)$ are satisfied for $i = 1, \ldots, n$. Then it follows that

$$y_i^* - z_i^* = P_{\pi^*}(\psi_i \wedge \varphi_i) - d_i P_{\pi^*}(\varphi_i)$$

for $i = 1, \ldots, n$. Minimizing $D^2(\vec{y}, \vec{z})$ amounts to finding a probabilistic interpretation π^* such that $\vec{y}^*$ and $\vec{z}^*$ are as close as possible to each other with respect to D^2. In particular, if there is a π^* such that $y^* = z^*$ it follows that $P_{\pi^*}(\psi_i \wedge \varphi_i) - d_i P_{\pi^*}(\varphi_i) = 0$ and therefore $\pi \in \mathsf{Mod}((\psi_i|\varphi_i)[d_i])$ (for $i = 1, \ldots, n$), i.e, κ is consistent. Furthermore, the more y_i^* differs from z_i^* the more $P_{\pi^*}(\psi_i|\varphi_i)$ differs from d_i (for $i = 1, \ldots, n$). The measure $\mathcal{I}_{\text{gd}}$ is similar in spirit to $\mathcal{I}_d$ as they both minimize the distance of a knowledge base to a consistent one. However, the implementation of those measures is different as they use different distance measures. The following result has been shown in [44].

Proposition 15. The function $\mathcal{I}_{\text{gd}}$ satisfies *consistency, monotonicity, super-additivity, weak independence*, and *continuity*.

Another approach to inconsistency measuring based on entropy can be derived from works on consistency repairing in de Finetti's coherence setting [3, 4, 5]. In order to correct incoherent conditional probabilities, a discrepancy based on Kullback-Leibler divergence is minimized, which can be understood as an inconsistency measure. In the following, we adapt the approach of [3, 4, 5] to our semantics, keeping their focus on sets of conditionals with point probabilities.

Consider the knowledgebase $\kappa = \{c_1, \ldots, c_n\}$, with $c_i = (\psi_i|\varphi_i)[q_i]$ and $d_i \in (0,1)$ [6] for $i = 1, \ldots, n$. The approach of [5] is based on the following scoring rule, which can be seen as evaluating the accuracy of a set of probabilistic conditionals in a possible world w:

$$S_{CRV}(\kappa, w) = \sum_{i=1}^{n} 1_{\{\varphi_i \wedge \psi_i\}}(w) \ln q_i + \sum_{i=1}^{m} 1_{\{\neg\varphi_i \wedge \psi_i\}}(w) \ln(1 - q_i). \quad (8)$$

Any probabilistic interpretation $\pi \in \mathcal{P}(\mathcal{A})$, which is a probability mass over the worlds $w \in W_\mathcal{A}$, defines an expected value $E_\pi(S_{CRV}(\kappa, w))$. For a fixed π, the point probabilities $\vec{q} = \langle q_1, \ldots, q_n\rangle$ that maximizes $E_\pi(S_{CRV}(\kappa[\vec{q}], w))$ are given by the vector $\vec{q}^\pi = \langle \frac{P_\pi(\varphi_1 \wedge \psi_1)}{P_\pi(\psi_1)}, \ldots, \frac{P_\pi(\varphi_n \wedge \psi_n)}{P_\pi(\psi_n)}\rangle \in [0,1]^n$. The following discrepancy between a knowledge base κ and a probabilistic interpretation π gives the expected gap in accuracy, when measured by S_{CRV}, between the suboptimal $\kappa = \kappa[\vec{q}]$ and the maximally accurate $\kappa[\vec{q}^\pi]$:

---
[6] In [5], extreme probabilities (0 or 1) are avoided for technical reasons.

$$d_{CRV}(\kappa,\pi) = E_\pi\Big(S_{CRV}(\kappa[\vec{q}^\pi],w) - S_{CRV}(\kappa,w)\Big)$$
$$= \sum_{1\le i\le n, P_\pi(\psi_i)>0} P_\pi(\psi_i)\Big(q_i^\pi \ln\frac{q_i^\pi}{q_i} + (1-q_i^\pi)\ln\frac{1-q_i^\pi}{1-q_i}\Big).$$

This discrepancy directly yields an inconsistency measure for precise probabilistic knowledge bases. In [5], π must be such that $\pi(\bigvee_i \psi_i) = 1$, but we drop that restriction due to the semantics we are adopting.

$$\mathcal{I}_{CRV}(\kappa) = \min\{d_{CRV}(\kappa,\pi) | \pi \in \mathcal{P}(\mathcal{A})\}.$$

Proposition 16. $\mathcal{I}_{CRV}$ satisfies consistency, monotonicity and continuity.

5 Applications

The inconsistency measures introduced in the previous section give us a tool to analyze inconsistent knowledge bases. Our final goal is to reason over these knowledge bases in a sensible way. There are at least two ideas that we can consider for this purpose.

1. Repair the inconsistent knowledge base and apply classical probabilistic reasoning algorithms.

2. Apply paraconsistent reasoning algorithms that can deal with inconsistent knowledge bases.

The distance-based approaches in Section 4.2 are particularly well suited for repairing knowledge bases. In fact, when computing the inconsistency value of the knowledge base, we usually do so by finding a consistent knowledge base that minimizes the selected distance to the original knowledge base. However, there are some obstacles. First of all, if we consider only point probabilities, whether or not a unique closest consistent knowledge base exists depends on the selected norm. For instance, uniqueness is guaranteed for the Euclidean norm, but not for the Manhattan and Maximum norm. As a simple example, consider the knowledge base $\{(a)[0.2], a[0.6]\}$. With respect to the Manhattan norm, each repair $\{(a)[p] : 2\}$ with $p \in [0.2, 0.6]$ is minimal and the choice would be arbitrary without further assumptions.

Second, even if a unique solution exists, repairing the knowledge base means loss of information. To make this clear, consider the knowledge bases $\{(a)[0.4], a[0.6]\}$ and $\{(a)[0.1], a[0.9]\}$. Both knowledge bases have the unique minimal repair $\{(a)[0.5] : 2\}$ with respect to the Euclidean norm. The fact that the second knowledge base has a significantly higher variance is lost. If we think of the knowledge bases as representing the opinions of two different experts, 0.5 is close to both experts' opinion in the first knowledge base, but not in the second.

In cases like this, where we have to shift a huge amount of probability mass to repair the knowledge base, applying paraconsistent reasoning mechanisms can be a better choice. We can derive such reasoning mechanisms from the fuzzy- and violation-based approaches. The idea is to replace the models of a knowledge base with those probabilistic interpretations that are close enough to being a model. That is, if the knowledge base is consistent, we use the usual models to perform reasoning. If it is inconsistent, we use the probabilistic interpretations that are closest to being a model. For the fuzzy-based measures from Section 4.5, this means that we use those interpretations that maximize the candidacy value with respect to $\mathfrak{C}_\kappa^h$ [8]. For the violation-based measures from Section 4.3, we use those interpretations that minimize the violation of the knowledge base [30, 35]. We explain this approach in somewhat more detail for minimum violation measures.

As a first step, we define the *generalized models* of a knowledge base κ as the set of probabilistic interpretations that minimize the violation value (3) in Definition 14. More strictly speaking, we let

$$\mathsf{GMod}(\kappa) = \{\pi \in \mathcal{P}(|)(\pi, \epsilon) \ minimizes \ (3) \ for \ some \ \epsilon \in \mathbb{R}\}.$$

Intuitively, $\mathsf{GMod}(\kappa)$ contains those probability distributions that violate the knowledge base minimally. In particular, if κ is consistent, we have $\mathsf{Mod}(\kappa) = \mathsf{GMod}(\kappa)$. However, often we have some special conditionals that should not be violated at all. We call these conditionals *integrity constraints*. We assume that the integrity constraints are consistent. Now given a knowledge base κ and a set of integrity constraints IC, we define the corresponding generalized models as the set of probabilistic interpretations that satisfy IC and minimally violate κ.

Definition 16. Let κ, IC be knowledge bases such that IC is consistent. Let $\|.\|$ be some continuous vector norm. The set of *generalized models of κ with respect to IC and $\|.\|$* is defined by

$$\mathsf{GMod}_{IC}^{\|.\|}(\kappa) = \{\pi \in \mathsf{Mod}(IC) \mid (\pi, \epsilon) \ minimizes \ (3) \ for \ some \ \epsilon \in \mathbb{R}\}.$$

If IC and $\|.\|$ are clear from the context or not important for the discussion, we will just write $\mathsf{GMod}(\kappa)$ to keep our notation simple. $\mathsf{GMod}(\kappa)$ is guaranteed to be non-empty and has some nice technical properties that allow us to reason as efficiently with generalized models as with classical models in many cases.

There are two major approaches to perform reasoning over consistent probabilistic knowledge bases. In both cases, our final goal is to answer conditional probabilistic queries, that is, to compute the conditional probability of a formula φ given another formula ψ. We denote such queries by $(\varphi \mid \psi)$. The first

Query	$\|\cdot\|_1$	$\|\cdot\|_2$	$\|\cdot\|_\infty$
$(P \mid N)$	$[0.1, 0.9]$	$[0.376, 0.624]$	$[0.366, 0.633]$
$(P \mid Q)$	$[0.1, 0.9]$	$[0.533, 0.679]$	$[0.536, 0.689]$
$(P \mid R)$	$[0.1, 0.9]$	$[0.321, 0.467]$	$[0.314, 0.463]$
(N)	$[1, 1]$	$[0.801, 0.801]$	$[0.789, 0.789]$

Table 2: Generalized entailment results (rounded to 3 digits) for Nixon diamond with $IC = \varnothing$ (Example 10).

approach is to compute upper and lower bounds on the conditional probability of φ given ψ with respect to all models of κ [28, 17]. This approach is often referred to as the *probabilistic entailment problem*. The second approach is a two-stage process. We first select a best model that satisfies the knowledge base and then use this model to compute the conditional probability of φ given ψ [29, 22]. The 'best' model is determined by an evaluation function. For instance, we may be interested in maximizing entropy or minimizing some notion of distance to a prior distribution. We refer to this approach as the *model selection problem*. Both approaches can be easily generalized to inconsistent knowledge bases by just replacing the probabilistic interpretations that satisfy the knowledge base (the classical models) with those that minimally violate the knowledge base (the generalized models) [34, 35]. A detailed description and discussion of both approaches can be found in [36]. The following example illustrates how our generalization of the probabilistic entailment problem to inconsistent knowledge bases can be applied.

Example 10. Let us consider the *Nixon diamond*. We believe that quakers (Q) are usually pacifists (P) while republicans (R) are usually not. However, we know that Nixon (N) was both a quaker and a republican. Let us model our beliefs with the following knowledge base:

$$\kappa = \{(P \mid Q)[0.9], (P \mid R)[0.1], (N)[1], (Q \wedge R \mid N)[1]\}.$$

κ is inconsistent. For instance, its minimal violation value with respect to the Euclidean norm is $\mathcal{I}_{\|\cdot\|_2}(\kappa) \approx 0.42$. Let us set $IC = \varnothing$ and ask for the probability that Nixon was a pacifist. Table 2 shows the result and some additional queries that show in which way the knowledge in κ has been relaxed.

We can in particular see that the Manhattan norm yields the most conservative results in the sense that it provides very large answer intervals. However, the answer is still bounded away from the trivial bounds 0 and 1. For the Euclidean and the Maximum norm, we maintain the knowledge that quakers are probably pacifists and that republicans are probably not (the probabilities are bounded away from 0.5). We also notice in Table 2 that the probability

Query	$\|\cdot\|_1$	$\|\cdot\|_2$	$\|\cdot\|_\infty$
$(P \mid N)$	$[0.1, 0.9]$	$[0.384, 0.615]$	$[0.376, 0.624]$
$(P \mid Q)$	$[0.1, 0.9]$	$[0.517, 0.615]$	$[0.520, 0.624]$
$(P \mid R)$	$[0.1, 0.9]$	$[0.384, 0.482]$	$[0.376, 0.481]$
(N)	$[1, 1]$	$[1, 1]$	$[1, 1]$

Table 3: Generalized entailment results (rounded to 3 digits) for Nixon diamond with $IC = \{(N)[1]\}$ (Example 10).

that the person under consideration is Nixon (N) has also been subject to change. However, since we have no doubts about Nixon's existence, we let $(N)[1]$ become an integrity constraint. That is, we now let $IC = \{(N)[1]\}$ and

$$\kappa = \{(P \mid Q)[0.9], (P \mid R)[0.1], (QR \mid N)[1]\}$$

Table 3 shows the new generalized entailment results.

The generalizations of both the probabilistic entailment problem and the model selection problem satisfy some interesting properties. Intuitively, these properties can be described as follows:

Consistency If $\kappa \cup IC$ is consistent, the generalized reasoning results coincide with the classical reasoning results.

Independence If some subset of $\kappa \cup IC$ is consistent, then generalized reasoning results that depend only on this subset coincide with the classical reasoning results.

Continuity If $\kappa \cup IC$ is topologically close to a consistent knowledge base, then the generalized reasoning results will be close to the classical reasoning results.

A thorough discussion of these properties and their exact preconditions can be found in [32, 36].

In several applications, we want to override general rules by more specific rules. This can be modeled by a knowledge base that is partitioned into subsets with different priorities. If we assume that each subset of the partition is consistent with given integrity constraints, we can consider another form of generalized probabilistic reasoning. Similar as before, we start with the models of the integrity constraints $\mathcal{M}_0$. We then select from $\mathcal{M}_0$ those models that minimally violate the conditionals with highest priority yielding a subset $\mathcal{M}_1$ of $\mathcal{M}_0$. We continue in this way, constructing $\mathcal{M}_{i+1}$ by selecting from $\mathcal{M}_i$ those models that minimally violate the conditionals with the next highest priority. A detailed description of this approach and its properties can be found in [31].

The following example illustrates how this approach can be used to generalize the probabilistic entailment problem to knowledge bases with priorities.

Example 11. We consider a probabilistic version of an access control policy scenario from [2]. Suppose we have different files and different users and want to automatically deduce the probability that a user has access to a file. If the probability is 1, we might grant access immediately, otherwise we might send a confirmation request to the system administrator. If the probability is very low, say smaller than 0.1, we might want to send a warning in addition.

We model this problem using a relational probabilistic language similar to [24, 12]. We build up formulas over a finite set of typed predicate symbols, a finite set of typed individuals and an infinite set of (typed) variables. We allow the usual logical connectives, but do not allow quantifiers.

We use the types *User* and *File* and the predicates *grantAccess(User, File)*, *employee(User)*, *exec(User)*, *blacklisted(User)*, *confidential(File)*, where *exec* abbreviates *executive manager*. Let *alice* and *bob* be individuals of type *User* and let *file1*, *file2* be individuals of type *File*.

Our priority knowledge base has the form $\kappa = (\kappa_1, \kappa_2, \kappa_3, \kappa_4, \kappa_5, IC)$, where a higher index means higher priority. That is, κ_1 has the lowest and κ_5 has the highest priority (disregarding the integrity constraints IC that cannot be violated at all). The subsets of the knowledge base are defined as follows:

$\kappa_1 = \{(grantAccess(U, F))[0], (blacklisted(U))[0.05]\}$

$\kappa_2 = \{(grantAccess(U, F) \mid employee(U))[0.5],$
$\qquad (blacklisted(U) \mid employee(U))[0.01]\}$

$\kappa_3 = \{(grantAccess(U, F) \mid confidential(F))[0]\}$

$\kappa_4 = \{(grantAccess(U, F) \mid exec(U))[0.7],$
$\qquad (blacklisted(U) \mid exec(U))[0.001]\}$

$\kappa_5 = \{(exec(alice))[1], (employee(bob))[1], (confidential(file1))[1]\}$

$IC = \{(employee(U) \mid exec(U))[1], (grantAccess(U, F) \mid blacklisted(U)(F))[0]\}$

On the first level, we define generic knowledge. If no knowledge is available, we do not want to grant access to anybody. Also, we make the assumption that it is rather unlikely that a user is blacklisted. On the second level, we increase the access probability and decrease the blacklist probability for employees. On level 3, we make an exception for confidential files. Afterwards, we further increase access probability and decrease blacklist probability for executive managers on level 4. The last level contains domain knowledge. We know that *alice* is an executive manager, *bob* is an employee and *file1* is confidential. Our integrity constraints state that executive managers are employees and that we do not grant access to blacklisted users.

We have the following rounded reasoning results when using the Euclidean norm to determine our strict priority models:

$grantAccess(alice, file1)[0.7]$ $grantAccess(bob, file1)[0]$
$grantAccess(alice, file2)[0.7]$ $grantAccess(bob, file2)[0.5]$
$blacklisted(alice)[0.0001]$ $blacklisted(bob)[0.01]$.

The results make intuitively sense. For instance, the first query shows that for the executive manager *alice*, the access rule $(grantAccess(U, F) \mid exec(U))[0.7]$ with priority 4 has been applied, while $(grantAccess(U, F) \mid confidential(F))[0]$ with priority 3 and $(grantAccess(U, F))[0]$ with priority 1 have been ignored. Similarly, we can see that for the employee *bob* the rule $(grantAccess(U, F) \mid confidential(F))[0]$ with priority 3 applies because *file1* is confidential and *bob* is not an executive manager.

Generalized reasoning approaches can also be applied in multi-agent systems. For instance, in [33], multi-agent decision problems have been investigated where each agent has individual beliefs and utilities. Generalized Probabilistic Entailment can be used to derive group beliefs from the individual beliefs. Then expected utilities for the group can be computed from these group beliefs. Since this approach yields utility intervals rather than point utilities, one can define different preference relations. These approaches satisfy independence and continuity properties similar to the ones that we discussed after Example 10 and also satisfy some desirable social choice properties [33].

6 Summary

In this chapter, we gave an overview of approaches to measuring inconsistency in probabilistic logics. The most important property that distinguishes measures for probabilistic logics from measures for classical logics is *Continuity*. Continuity guarantees that minor changes of probabilities cannot result in major changes in the inconsistency value. This property seems highly desirable for analyzing inconsistencies in probabilistic logics because conflicts can be resolved by carefully adjusting probabilities in the knowledge base. However, as explained in Section 3, Continuity is actually in conflict with *Independence* and *MI-Separability* that have been considered for classical measures. As argued in Section 3, our position is that these properties should be given up for probabilistic knowledge bases in favor of *Continuity*.

In Section 4, we discussed different approaches for measuring inconsistency in probabilistic logics. The first class of measures was directly *adapted from inconsistency measures for classical logics*. While these measures are able to measure inconsistencies qualitatively, they do not take probabilities into account. *Distance-based measures* attempt to minimize the distance in probabilities from the original knowledge base to a consistent repair. They measure inconsistency continuously and usually yield a repair as a byproduct. However, they can be difficult to compute due to their non-convex nature. *Violation-based measures* attempt to find a better tradeoff between computational and analytic properties. To do so, they do not minimize the distance in probabilities directly, but try to minimize the error in the numerical constraints that correspond to the knowledge base. While this approach is less intuitive than minimizing the distance directly, it still measures inconsistency continuously and can be solved by convex programming techniques in general, and even via linear programming for two specific measures. These two measures are equivalent to some *measures based on Dutch books*, from the Bayesian philosophy/statistics community, which were then presented. Afterwards, we discussed a *measure based on fuzzy logic* that relies on assigning degrees of membership of probabilistic interpretation belonging to models of a knowledge base. Finally, we discussed *measures that rely on the notion of entropy*.

In Section 5, we sketched some applications of inconsistency measures for probabilistic logics in repairing and reasoning with inconsistent knowledge bases. Distance-based measures are well suited for repairing knowledge bases. Adapting the probabilities in the knowledge base in a minimal way seems to be the most intuitive way to repair inconsistent probabilistic knowledge bases. However, by replacing the inconsistent knowledge base with a repair, we may lose information about the variance in the information. So instead, we may want to infer probabilities directly from the inconsistent knowledge base. Violation-based measures are well suited for this purpose. By replacing the models of a knowledge base with those probability distributions that minimally violate the knowledge base, we can transfer reasoning approaches for consistent knowledge bases to inconsistent ones. As we discussed, these generalizations guarantee that classical reasoning results on the consistent part of the knowledge base remain unaffected (Independence) and that reasoning results over knowledge bases that are close to consistent knowledge bases are not too far from the classical reasoning results (Continuity).

References

[1] C. Alchourrón, P. Gärdenfors, and D. Makinson. On the Logic of Theory Change: Partial Meet Contraction and Revision Functions. *The Journal of Symbolic Logic*, 50(02):510–530, 1985.

[2] P. Bonatti, M. Faella, and L. Sauro. Adding Default Attributes to EL++. In *AAAI*, 2011.

[3] A. Capotorti and G. Regoli. Coherent Correction of Inconsistent Conditional Probability Assessments. In *Proceedings of IPMU'08*, pages 891–898, 2008.

[4] A. Capotorti, G. Regoli, and F. Vattari. On the Use of a New Discrepancy Measure to Correct Incoherent Assessments and to Aggregate Conflicting Opinions based on Imprecise Conditional Probabilities. In *Proceedings of the Sixth International Symposium on Imprecise Probability: Theories and Applications, ISIPTA*, 2009.

[5] A. Capotorti, G. Regoli, and F. Vattari. Correction of Incoherent Conditional Probability Assessments. *International Journal of Approximate Reasoning*, 51(6):718–727, 2010.

[6] G. Coletti and R. Scozzafava. *Probabilistic Logic in a Coherent Setting*. Trends in Logic, Vol. 15: Studia Logica Library. Kluwer Academic Publishers, 2002.

[7] F. Cozman and L. Ianni. Probabilistic Satisfiability and Coherence Checking through Integer Programming. In *Proc. Symbolic and Quantitative Approaches to Reasoning with Uncertainty, ECSQARU 2013*, volume 7958 of Lecture Notes in Computer Science, pages 145–156. Springer Berlin Heidelberg, 2013.

[8] L. Daniel. *Paraconsistent Probabilistic Reasoning*. PhD thesis, L'École Nationale Supérieure des Mines de Paris, 2009.

[9] G. De Bona and M. Finger. Measuring Inconsistency in Probabilistic Logic: Rationality Postulates and Dutch Book Interpretation. *Artificial Intelligence*, 227:140–164, 2015.

[10] B. de Finetti. Theory of Probability, 1974.

[11] M. Finger and G. De Bona. Probabilistic Satisfiability: Logic-based Algorithms and Phase Transition. In *IJCAI*, pages 528–533, 2011.

[12] J. Fisseler. First-order Probabilistic Conditional Logic and Maximum Entropy. *Logic Journal of IGPL*, 20(5):796–830, 2012.

[13] A. Frisch and P. Haddawy. Anytime Deduction for Probabilistic Logic. *Artificial Intelligence*, 69(1–2):93–122, 1994.

[14] G. Georgakopoulos, D. Kavvadias, and C. Papadimitriou. Probabilistic Satisfiability. *Journal of Complexity*, 4(1):1–11, 1988.

[15] G. Gerla. *Fuzzy Logic: Mathematical Tools for Approximate Reasoning*. Trends in Logic. Springer-Verlag, 2001.

[16] J. Halpern. An Analysis of First-Order Logics of Probability. *Artificial Intelligence*, 46:311–350, 1990.

[17] P. Hansen and B. Jaumard. Probabilistic Satisfiability. In *Handbook of Defeasible Reasoning and Uncertainty Management Systems*, volume 5, pages 321–367. Springer Netherlands, 2000.

[18] P. Hansen and S. Perron. Merging the local and global approaches to probabilistic satisfiability. *International Journal of Approximate Reasoning*, 47(2):125 – 140, 2008.

[19] A. Hunter and S. Konieczny. Shapley Inconsistency Values. In *10th International Conference on Principles of Knowledge Representation and Reasoning (KR)*, pages 249–259, 2006.

[20] A. Hunter and S. Konieczny. Measuring Inconsistency through Minimal Inconsistent Sets. In *Proceedings of the Eleventh International Conference on Principles of Knowledge Representation and Reasoning (KR'2008)*, pages 358–366, 2008.

[21] B. Jaumard, P. Hansen, and M. Poggi. Column Generation Methods For Probabilistic Logic. *ORSA - Journal on Computing*, 3(2):135–148, 1991.

[22] G. Kern-Isberner. *Conditionals in Nonmonotonic Reasoning and Belief Revision*. Springer, 2001.

[23] K. M. Knight. Measuring Inconsistency. *Journal of Philosophical Logic*, 31:77–98, 2001.

[24] T. Lukasiewicz. Probabilistic Deduction with Conditional Constraints over Basic Events. *J. Artif. Int. Res.*, 10(1):199–241, 1999.

[25] J. Matousek and B. Gärtner. *Understanding and Using Linear Programming*. Universitext (1979). Springer, 2007.

[26] D. P. Muiño. Measuring and Repairing Inconsistency in Probabilistic Knowledge Bases. *International Journal of Approximate Reasoning*, 52(6):828–840, 2011.

[27] R. Nau. Coherent Assessment of Subjective Probability. Technical report, DTIC Document, 1981.

[28] N. Nilsson. Probabilistic Logic. *Artificial Intelligence*, 28:71–88, 1986.

[29] J. B. Paris. *The Uncertain Reasoner's Companion – A Mathematical Perspective*. Cambridge University Press, 1994.

[30] N. Potyka. Linear Programs for Measuring Inconsistency in Probabilistic Logics. In *Proceedings of the 14th International Conference on Principles of Knowledge Representation and Reasoning (KR'14)*, 2014.

[31] N. Potyka. Reasoning over Linear Probabilistic Knowledge Bases with Priorities. In *The 9th International Conference on Scalable Uncertainty Management (SUM'15)*, 2015.

[32] N. Potyka. *Solving Reasoning Problems for Probabilistic Conditional Logics with Consistent and Inconsistent Information*. PhD thesis, FernUniversität in Hagen, Hagen, 2015.

[33] N. Potyka, E. Acar, M. Thimm, and H. Stuckenschmidt. Group Decision Making via Probabilistic Belief Merging. In *Proceedings of the 25th International Joint Conference on Artificial Intelligence (IJCAI'16)*, 2016.

[34] N. Potyka and M. Thimm. Consolidation of Probabilistic Knowledge Bases by Inconsistency Minimization. In *Proceedings of the 21st European Conference on Artificial Intelligence (ECAI'14)*, pages 729–734, August 2014.

[35] N. Potyka and M. Thimm. Probabilistic Reasoning with Inconsistent Beliefs using Inconsistency Measures. In *Proceedings of the 24th International Joint Conference on Artificial Intelligence (IJCAI'15)*, 2015.

[36] N. Potyka and M. Thimm. Inconsistency-tolerant Reasoning over Linear Probabilistic Knowledge bases. *International Journal of Approximate Reasoning*, 2017 (to appear).

[37] R. Reiter. A theory of diagnosis from first principles. *Artificial intelligence*, 32(1):57–95, 1987.

[38] W. Rödder and L. Xu. Elimination of Inconsistent Knowledge in the Probabilistic Expert-System-Shell SPIRIT (in German). In *Operations Research Proceedings: Selected Papers of the Symposium on Operations Research 2000*, pages 260–265, 2001.

[39] M. Schervish, J. Kadane, and T. Seidenfeld. Measures of Incoherence: How not to Gamble if You Must. In *Bayesian Statistics 7: Proceedings of the 7th Valencia Conference on Bayesian Statistics*, pages 385–402, 2003.

[40] M. Schervish, T. Seidenfeld, and J. Kadane. Two Measures of Incoherence: How not to Gamble if You Must. Technical report, Department of Statistics, Carnegie Mellon University, 1998.

[41] M. Schervish, T. Seidenfeld, and J. Kadane. Measuring incoherence. *Sankhyā: The Indian Journal of Statistics, Series A*, pages 561–587, 2002.

[42] A. Shimony. Coherence and the Axioms of Confirmation. *The Journal of Symbolic Logic*, 20(01):1–28, 1955.

[43] M. Thimm. Measuring Inconsistency in Probabilistic Knowledge Bases. In *Proceedings of the Twenty-Fifth Conference on Uncertainty in Artificial Intelligence (UAI'09)*, pages 530–537. AUAI Press, June 2009.

[44] M. Thimm. *Probabilistic Reasoning with Incomplete and Inconsistent Beliefs*. Number 331 in Dissertations in Artificial Intelligence. IOS Press, 2011.

[45] M. Thimm. Inconsistency Measures for Probabilistic Logics. *Artificial Intelligence*, 197:1–24, April 2013.

Measuring Database Inconsistency

Hendrik Decker

PROS, DSIC, Universidad Politécnica de Valencia, Spain
hdecker@pms.ifi.lmu.de

Abstract

First, we revisit background, definition and examples of database inconsistency measures. Then, we recapitulate three applications of database inconsistency measures: integrity checking, relaxing repairs, and repair checking. Database inconsistency measures are used to make these applications inconsistency-tolerant. The degree of database inconsistency need not be computed for any of them. Rather, only an increase, decrease or invariance of inconsistency between consecutive database states across updates has to be evidenced. Last, we take a look at the similarities and differences of mathematical measures, 'classical' inconsistency measures and database inconsistency measures.

1 Introduction

An inconsistency measure is a mathematical or procedural description of a function that outputs the amount of inconsistency of a given input set of formal sentences. Such sets may involve conflicts, and hence infringe the quality and consistency of the information conveyed by the formulas. If the sentences are represented in the language of some formal logic, conflicts correspond to logical inconsistency between formulas, or to the unsatisfiability of formulas.

The history of inconsistency measuring has its roots in John Grant's seminal article [51]. After years of ignorance, the field started to blossom about a quarter century later, sparked by papers such as [63] [57] [58] [64]. An early survey is provided in [59], which also traces precursor topics and related subjects in the fields of information theory [89] [70], knowledge evolution [81] [2] [49], paraconsistency, [18] [82] [23], probabilistic logic [20] and possibilistic logic [47].

Most measures in the literature that serve for assessing the amount of inconsistency in a logic setting are tailored to be applied to what is frequently called "knowledge bases". In most cases, these are either propositional sets of formulas [63] [58] [64] [83] [61] [8] [76] [9] [90] [62] or interpretations of first-order predicate logic theories, i.e., essentially, sets of ground literals that model the given theory [51] [52] [53] [84] [66] [71] [72] [93]. For convenience, let us call these measures *classical inconsistency measures*, or simply *classical measures*.

The measures for assessing database inconsistency as featured in this paper are significantly different from the ones cited above. The main differences in a nutshell: Database logic is not classical logic, and the main use of database inconsistency measures is not to quantify the amount of inconsistency in absolute numbers (although they can do that), but to indicate the relative difference of the amount of inconsistency in consecutive database states, for the purpose of inconsistency-tolerant integrity maintenance across updates.

After this introduction, we revisit some background issues on which this paper is based, in Section 2. In Section 3, we define what is a database inconsistency measure and give some examples that are of interest in the remainder. In Section 4, we recapitulate two characteristic applications of database inconsistency measures: integrity checking and repair checking. They are about monitoring, maintaining and improving the integrity of stored data across updates, and in particular comparing the difference of inconsistency in consecutive database states. In Section 5, we compare and differentiate the basic features of classical inconsistency measures and database inconsistency measures. Apart from this section, related work by other authors on inconsistency measures is also addressed in Section 1 and Subsection 3.2. In Section 6, we recapitulate the paper and give an outlook to future work. The appendix contains an indexed list of all numbered definitions and their respective subjects in the paper.

2 Preliminary Concepts

In this section, we recapitulate some basic issues that underlie the remainder of the paper. In Subsection 2.1, we turn to databases, updates and integrity constraints. In Subsection 2.2, we define, for convenience, a generic class of database methods called 'update checkers'. In Subsection 2.3, we address semantic and syntactic restrictions that may apply to the input of database inconsistency measures and their applications.

As a notational convention, we use the symbols $\Rightarrow$ and $\Leftrightarrow$ for meta-level entailment and, resp., meta-level equivalence, in definitions and result statements.

2.1 Databases, Updates and Integrity Constraints

We assume a basic familiarity with relational and deductive databases, as well as with datalog and first-order predicate logic. For details, the reader may consult, e.g., [1] [85] [5]. We adopt the usual logic programming style notation for database clauses [68]. In particular, we use the symbol $\sim$ for denoting database negation, with the usual interpretation: for a formula F and a database D, $\sim F$

is *true*, resp., *false* if and only if F evaluates to *false* or, resp., *true* according to the semantics associated to D.

Throughout the paper, let symbols D, IC, U and adornments thereof by superscripts or subscripts always stand for a database state, an integrity theory and, resp., an update.

Updates and database integrity are revisited in Subsections 2.1.1 and 2.1.2.

2.1.1 Updates

The dynamics of databases are effected by updates. Formally, an update is a finite bipartite set of database clauses (i.e., facts or deductive rules) to be inserted to or, resp., deleted from a given database state in a single atomic transaction.

Updates may cause and also repair integrity violations. Three characteristic applications of database inconsistency measures are integrity checking (i.e., to check if an update causes any integrity violation), repairing (i.e., to reduce the amount of integrity violations by updates) and repair checking (i.e., to check if an update is a repair). These applications are addressed in detail in Section 4. For repair checking, the following definition will be useful.

Definition 1 (*undo*)
Let $\overline{U}$ denote the *undo* of U: for each element of the form *insert X* or *delete Y* in U, $\overline{U}$ contains *delete X* or, resp., *insert Y*, and nothing else.

For convenience, let D^U denote the updated database state obtained by applying U to D. We denote consecutive updates U, U' of D as D^U and then $D^{UU'}$. Hence, $D^{U\overline{U}} = D$.

2.1.2 Database Consistency and Integrity

There are several different notions of database consistency [86] [44]. In this paper, we take database consistency as synonymous with database integrity [46].

An *integrity theory* is a finite set of first-order predicate logic sentences, known as *integrity constraints* (or simply *constraints*). They capture database properties that are supposed to be satisfied and remain so across state changes. A complementary viewpoint is that constraints embody semantic conditions that are meant to rule out states which would be faulty or meaningless.

Example 1 *We present some typical specimens of integrity constraints.*
a) $\forall x \forall y (p(x,y) \wedge p(x,y') \rightarrow y = y')$ *requires that the values of the first attribute column of the p-relation, on which the second column's values functionally depend, are unique for each row, as is typical for primary key constraints.*

b) $\forall x \forall y(p(x,y) \rightarrow \exists z(q(y,z)))$ *demands that the second attribute value of facts in p also occurs in q, as is typical for foreign key constraints.*

c) $\forall x \forall y(married(x,y) \rightarrow x \neq y)$, *and also the logically equivalent denial clause* $\leftarrow married(x,x)$, *where the empty head of the clause expresses the falsity of its body, stipulate that the* married *relation is anti-reflexive, i.e., that nobody is married to him- or herself.*

For simplicity, we only consider integrity theories that are logically satisfiable. Unwarranted unsatisfiability of integrity theories is dealt with in [11], Subsection 2.1 and Section 3 of [45] and Subsection 5.5 of [42].

Moreover, we assume w.l.o.g. that, from now on, each constraint I is represented in *prenex form* (which includes common disjunctive and conjunctive normal forms as well as denial form), obtained by moving all quantifiers leftmost while maintaining their left-to-right ordering in the original formula, by well-known equivalence-preserving rewrites. Also, we assume a language with sufficiently many variable symbols, such that the variables in I are *standardized apart*, i.e., no two quantifiers in I bind the same variable.

Hence, the constraint $\forall x \forall y(q(x,y) \rightarrow \exists z(p(y,z)))$ in Example 1*b* has to be re-written to some equivalent prenex form, e.g., $\forall x \forall y \exists z(q(x,y) \rightarrow p(y,z))$. Alternatively, this constraint can be represented in extended causal syntax [68] by the denial $\leftarrow q(x,y), \forall z \sim p(y,z)$. Such denial constraints in extended syntax can be conveniently processed by further rewriting them into a set of normal clauses. For instance, the above constraint can be rewritten into the set of clauses $\{\leftarrow q(x,y), \sim aux(y),\ aux(y) \leftarrow p(y,z)\}$, where aux is a fresh, hitherto unused predicate symbol, $\leftarrow q(x,y), \sim aux(y)$ is the top clause of the rewritten constraint, and the clause $aux(y) \leftarrow p(y,z)$ that defines aux is added to the database. Such rewritings are described in more detail in [68] [22].

Integrity constraints usually are interpreted and processed as queries about the consistency of stored data. Their truth or falsity is determined by the semantics of the database, which can be proof-theoretic (e.g., the completion of the current database state [16]) or model-theoretic (e.g., the stable or well-founded model semantics [50] [94]). In particular, the classical logic negation symbol $\neg$ in constraints is interpreted as database negation (cf. first paragraph of Subsection 2.1).

If a constraint I evaluates to *true* in a given state D, then we say that I is *satisfied* in D; if it evaluates to *false*, then we say that I is *violated* in D. If all constraints in an integrity theory IC are satisfied, then we also say that IC is *satisfied* in D, or, synonymously, that (D, IC) is *consistent*. If any constraint in IC is violated in D, then we also say that IC is *violated* in D, or, synonymously, that (D, IC) is *inconsistent*.

We next revisit Definition 3.1 in [42] of 'cases', i.e., instances of constraints in IC, based on substitutions of 'global' variables. Cases are of interest for defining database inconsistency measures (Section 3), as well as for integrity checking (Subsection 4.1).

Definition 2 (*global variable, case, basic case*)
Let I be a constraint of the form QI', where Q is the (possibly empty) quantification prefix of all variables in the quantifier-free formula I'.
a) A variable x in I' is called a *global* variable in I if x is $\forall$-quantified in Q, and no quantifier of the form $\exists y$ occurs left of $\forall x$ in Q.
b) Let $Q = GQ'$, where G is the (possibly empty) quantifier of all global variables in I, and σ be a mapping from the set of variables in G to a set of terms such that no variable in Q' occurs in the image of σ. Then, $I\sigma$ denotes the universal closure of the formula obtained from I' by replacing, for each x in G, each occurrence of x in I' by $\sigma(x)$. Each such $I\sigma$ is called a *case* of I. If there are no global variables in $I\sigma$, then the latter is also called a *basic case* of I.
c) Let $basic(IC)$ denote the set of all basic cases of constraints in IC.

Example 2 *Each of the formulas* $\exists z(\neg q(a,a) \vee p(a,z))$, $\exists z(\neg q(a,b) \vee p(b,z))$, $\forall x \exists z(\neg q(x, f(x)) \vee p(f(x), z))$ *and* $\forall x \exists z(\neg q(x,x) \vee p(x,z))$ *is a case of the constraint* $I = \forall x \forall y \exists z(\neg q(x,y) \vee p(y,z))$. *The first two are also basic cases of* I, *while neither* $\exists z(\neg q(a,b) \vee p(a,z))$ *nor* $\neg q(a,b) \vee p(b,c)$ *is a case of* I. *Each constraint also is a case of itself.*

2.2 Update Checkers

In Section 4, we are going to feature two applications of database inconsistency measures that are related to updates. For that, it is convenient to have the following generic definition of methods for checking if a given update has some property, e.g., if it preserves consistency or reduces inconsistency.

Definition 3 (*update checker*)
A mapping uc of triples (D, IC, U) to $\{yes, no\}$ is called an *update checker*.

Typical update checkers are integrity checkers and repair checkers, as discussed in Section 4.

In the literature, various approaches to inconsistency measuring, integrity checking, repairing and repair checking are proposed, each of them defined for some particular class of databases, constraints or updates. The approach developed in this paper is generic, hence independent of any particular inconsistency measure or update checker. Thus, whenever, for some mapping mp from tuples (D, IC) or triples (D, IC, U) to some range of values, we say, "for each tuple (D, IC) ..." or "for each triple (D, IC, U) ...", we actually mean to tacitly add, "...such that mp is defined for the input (D, IC)" or, resp., "... (D, IC, U)".

2.3 Syntactic and Semantic Restrictions

Certain syntactic or semantic restrictions on tuples (D, IC) or triples (D, IC, U) may apply, which we will state only if necessary in the respective context. The restriction to integrity theories that are satisfiable has already been mentioned in Subsection 2.1.2. Other common restrictions are properties such as the range-restrictedness [78] or the safety [92] of database clauses and constraints. Also the representation of elements in D or IC often is required to obey some syntactic form, e.g., clausal or conjunctive normal form.

As already mentioned, database inconsistency measures are closely linked to certain update checkers, each of which is defined for certain classes of triples (D, IC, U). Thus, the applicability of an inconsistency measure μ is confined by the projection (D, IC) of triples (D, IC, U) for which the update checker that uses μ is defined or applicable.

For simplicity, let us agree on the following general restriction. It avoids having to bother with differences between the satisfiability and the theorem-hood of integrity constraints, or with undefined or third truth values, or with subtle epistemic distinctions. Such issues are dealt with in [87].

In this paper, we assume that the semantics of each triple (D, IC, U) is *binary*, i.e., each $I \in IC$, when queried against D, has a unique *yes/no* answer, and also the answer of I when queried against D^U is always either *yes* or *no*. The semantics of significantly large classes of databases and constraints comply with the binary assumption, for instance all pairs (D, IC) such that D is relational and IC is range-restricted, and also all deductive databases and integrity theories that are acyclic [12] and range-restricted.

3 Database Inconsistency Measures

For checking the preservation, increase or reduction of inconsistency, the latter needs to be measurable, so that the inconsistency before and after updates can be compared. That is the purpose of database inconsistency measures.

In Subsection 3.1, we define database inconsistency measures. In 3.2, we give examples of such measures that are of interest in the remainder.

3.1 Definition of Database Inconsistency Measures

For the purpose of inconsistency-tolerant integrity checking, database inconsistency measures have been introduced in [40] and further developed in [41] [33] [36]. The main application of database inconsistency measures is not, in the first place, to quantify the amount of inconsistency in given databases, but to make the extents of inconsistency in consecutive database states comparable. In fact, the values of inconsistency between database states before and after updates do not necessarily have to be computed in order to become comparable.

Rather, only the increase, decrease, invariance or incomparability of inconsistency between such states is of interest for integrity checking and repairing.

For pairs (D, IC), database inconsistency measures consider the amount of violated integrity constraints or cases thereof, or causes of the violation of constraints, i.e., those parts of D that are responsible for constraint violations in IC [27] [36] [37]. Abstracting away from what exactly is sized for comparing the inconsistency of states, database inconsistency measures are defined as follows.

Definition 4 (*database inconsistency measure*)
A *database inconsistency measure* μ is a mapping from pairs (D, IC) to a partially ordered range of degrees of inconsistency. We denote the ordering by the infix predicate $\leq$; $X{<}Y$ means that $X{\leq}Y$ and $X{\neq}Y$. The negation of $\leq$ is denoted by $\not\leq$, and the negation of $<$ by $\not<$.

The literature about inconsistency measures is replete with discussions of properties and conditions that arguably should be required for inconsistency measures (cf., e.g., [54] [36] [90] [10]). However, except for the property exposed in Definition 5, below, we do not impose any further condition on database inconsistency measures, such that they would have 'nice' properties. Most of these properties are elusive and their desirability tends to be application-specific. Moreover, some of the most popular of such properties do not hold up against the non-monotonicity of database negation, as observed in [33] [34] [36]. Yet, it can be argued that the following property is desirable in general. It is widely adopted in the literature. We also adopt it for each database inconsistency measure considered in this paper. In particular, each measure addressed in Subsection 3.2 is positive-definite.

Definition 5 (*positive-definite measure*)
Let μ be a database inconsistency measure with a least element in its range, denoted by o. We say that μ is *positive-definite* if and only if, for each pair of databases D, D' and each pair of integrity theories IC, IC' such that IC is satisfied in D and IC' is violated in D', $\mu(D, IC) = o$ and $o < \mu(D',IC')$.

3.2 Examples of Database Inconsistency Measures

Let ι and ζ be mappings of tuples (D, IC) which output the set of integrity constraints that are violated in (D, IC) and, resp., the set of constraints in $basic(IC)$ that are violated in (D, IC) [41] [36]. It is easy to see that ι and ζ are database inconsistency measures, and so are the mappings denoted by $|\iota|$ and $|\zeta|$, which output the cardinality of $\iota(D, IC)$ and, resp., $\zeta(D, IC)$. Clearly, the range of $|\iota|$ and $|\zeta|$ is numerical, with least element 0, while the ranges of $\iota(D, IC)$ and, resp., $\zeta(D, IC)$, viz. the powersets of IC and, resp., $basic(IC)$, are not; they are partially ordered by $\subseteq$, and their least element is $\emptyset$.

Example 3 Let $D = \{p(0,1), p(0,2), p(1,0), p(3,3), p(4,2), p(3,4), q(4,4)\}$ and $IC = \{\leftarrow p(x,y), q(y,y), \leftarrow p(x,y), p(y,z)\}$. Clearly, we have
$\iota(D, IC) = \{\leftarrow p(x,y), p(y,z)\}$,
$\zeta(D, IC) = \{\leftarrow p(0,1), p(1,0), \leftarrow p(1,0), p(0,1), \leftarrow p(1,0), p(0,2),$
$\leftarrow p(3,3), p(3,3), \leftarrow p(3,3), p(3,4)\}$.
$|\iota|(D, IC) = 1$, $|\zeta|(D, IC) = 5$.

A border case example of an inconsistency measure is one with a binary range, named β in [36], viz. the mapping that outputs its o-element $\beta(D, IC) = true$ if (D, IC) is totally consistent, else $\beta(D, IC) = false$. The range $\{true, false\}$ of β is ordered by the relationship $true < false$.

With regard to the measured inconsistency values displayed in Example 3, we recall what has been said at the beginning of Subsection 3.1. The absolute values of database inconsistency measures as featured in this paper are only of secondary interest; what matters most is the increase, decrease, invariance or incomparability of those values upon updates. Applications where that matters are going to be addressed in Section 4.

Other examples of database inconsistency measures, based on collections or counts of the *causes* of integrity violations, have been defined in [28] [34] [36]. For relational databases and range-restricted constraints in clausal form, causes are sets of ground literals that correspond to the database facts whose presence or absence violate some constraint [27].

For example, in relational databases, the unique cause of the violation of some basic case $\leftarrow B$ of a conjunctive denial constraint precisely consists of the literals in B [27]. Or, for a basic case $L_1 \vee \ldots \vee L_n$ ($n > 0$) of a constraint in clausal form, where each L_i is a literal with atom A_i, the unique cause of its violation in a relational database is $\{\overline{L_1}, \ldots, \overline{L_n}\}$, where $\overline{L_i} = \neg A_i$ if $L_i = A_i$ and $\overline{L_i} = A_i$ if $L_i = \neg A_i$. Due to the non-monotonicity of database negation, the definition of causes is more involved for deductive databases and more general forms of constraints; it recurs on the only-if halves of the completion of database predicates, as described in [28] [34] [36]. For example, the only cause of the violation of the referential constraint $\forall x \exists y (q(x,x) \rightarrow p(x,y))$ in the database D of Example 3 is $\{q(4,4)\} \cup \{\neg p(4,n) \mid n \in \mathcal{L}\}$ where $\mathcal{L}$ is the set of ground terms in the underlying language.

Example 4, below, features the cause-based measures κ and υ. κ maps (D, IC) to the set of causes of the violations of constraints. Thus, for each cause $C \in \kappa(D, IC)$, there is a constraint $I \in IC$ such that $C \vDash \neg I$, and for each $I' \in basic(IC)$ that is violated in D, there is at least one cause $C \in \kappa(D, IC)$ such that $C \vDash \neg I'$. υ maps (D, IC) to the union of all causes in $\kappa(D, IC)$.

Example 4 For D and IC as in Example 3, we have
$\kappa(D, IC) = \{\{p(0,1), p(1,0)\}, \{p(1,0), p(0,2)\}, \{p(3,3)\}, \{p(3,3), p(3,4)\}\}$
$\upsilon(D, IC) = \{p(0,1), p(1,0), p(0,2), p(3,3), p(3,4)\}$.

Analogously to $|\iota|$ and $|\zeta|$, $|\kappa|$ can be defined. We deal with cause-based measures not only with regard to their possible use in applications as described in Section 4, but also because they are especially apt for applications about query answering in inconsistent databases, as touched upon in [30] [28] and the last paragraph of Subsection 5.5. They are also mentioned in Section 6, with regard to a pending study of how to draw repairs from causes.

Besides the database inconsistency measures addressed in this paper, also some others have been studied in [41] [36] [34], including an adaptation of the inconsistency measure in [52] to pairs (D, IC).

In the literature, there are other database inconsistency measures. Rather than considering constraints of arbitrary generality as in this paper, they serve the special purpose of assessing the amount of violations of either functional dependencies [75] [74] or referential integrity constraints [80] in relational databases, or topological constraints [77] in spatial databases.

These measures have been conceived with applications such as data quality management, data cleansing and data integration in mind, but not specifically for monitoring the dynamics of database updates. Yet, they can be used, and are indeed meant to be used, for comparing the inconsistency of different database instances, in order to choose between them with regard to avoiding unnecessary inconsistency and preferring versions with lower degrees of inconsistency. However, the values of these measures have to be computed for enabling a comparison of the degrees of inconsistency of different input states. This effectively means that all constraint violations have to be uncovered, which may be a daunting task for databases with large extensions.

In contrast to that, the values of ι, ζ, κ, $|\iota|$, $|\zeta|$, $|\kappa|$ do not necessarily have to be computed, for the applications outlined in Section 4. Rather, it suffices to find out if their values would decrease or remain invariant upon updates, for perceiving that a given update does not increase the degree of database inconsistency, or that an update is a valid repair. Another difference is that the non-monotonicity of database negation is not taken into consideration in the cited papers about special purpose measures, in contrast to this paper (cf. Subsection 5.2 and Example 10).

4 Applications of Database Inconsistency Measures

In this section, we recapitulate three characteristic applications of database inconsistency measures. The first and the third are update checking applications, viz. integrity checking (Subsection 4.1) and repair checking (Subsection 4.3). Both are defined with regard to some database inconsistency measure μ. For each input triple (D, IC, U), the former checks if U preserves consistency, i.e., if $\mu(D, IC, U) \leq \mu(D, IC)$, while the latter checks if U decreases inconsistency, i.e., if $\mu(D, IC, U) < \mu(D, IC)$. As will become clearer in Subsection 4.2, measure-based repair checking involves a measure-based generalization of the conventional notion of 'repair' as defined in [3].

4.1 Inconsistency-tolerant Integrity Checking

An update checker, the purpose of which is to filter out updates that would violate some integrity constraint, is called an *integrity checker*. If $ic(D, IC, U)$ = *yes*, we say that U is *accepted* by *ic*. If $ic(D, IC, U) = no$, we say that U is *rejected* by *ic*. By the way, neither acceptance nor rejection of U necessarily would determine or preclude any further action by the database system, its administrator, its user or its application, of what to do with the output of *ic*. By default, U is rejected if $ic(D, IC, U) = no$. But also a modification of D or IC or U such that the modified triple becomes acceptable is an option, as in active databases, belief revision or truth maintenance systems. Or, updates that cause tolerable violations of "soft" constraints may be waived through. However, any decisions or actions taken after or triggered by the output of *ic* are out of the scope of integrity checking.

For simplicity, we assume that, for each integrity checker *ic*, there is a well-defined domain of triples (D, IC, U) for which *ic* is defined as a total mapping. Thus, we do not have to be concerned with subtle differences between satisfiability and theoremhood of constraints nor with undefined or non-binary truth values, nor with the non-termination of processing constraints as queries.

Conventional integrity checkers postulate that the database to be updated is totally consistent with its constraints. They accept updates only if they preserve total consistency. The idea behind inconsistency-tolerant integrity checking has been to realize a form of integrity checking that allows that both the database state to be updated and the updated state can be inconsistent [39] [40] [41] [42]. This is captured by the following definition.

Definition 6 (*inconsistency-tolerant measure-based integrity checker*)
Let *ic* be an update checker and μ an inconsistency measure. *ic* is called a *sound*, resp., *complete* μ-*based integrity checker* if (*), resp., (**) holds, for each triple (D, IC, U):

(*) $ic(D, IC, U)$ = yes $\Rightarrow$ $\mu(D^U, IC) \leq \mu(D, IC)$,

(**) $\mu(D^U, IC) \leq \mu(D, IC)$ $\Rightarrow$ $ic(D, IC, U)$ = yes.

In words, (*) means that U is accepted by *ic* only if U either decreases or does not change the amount of inconsistency as measured by μ. Thus, the contrapositive $\mu(D^U, IC) \nleq \mu(D, IC) \Rightarrow ic(D, IC, U) = no$ of (*) means that *ic* rejects U if U neither decreases inconsistency nor does it leave it invariant. In particular, if $\mu(D, IC) < \mu(D^U, IC)$, i.e., if U increases inconsistency, then each sound μ-based integrity checker *ic* outputs $ic(D, IC, U) = no$.

Conversely, completeness of ic as formalized in (∗∗) means that, whenever U decreases the amount of consistency or leaves it invariant, then ic accepts U. If ic is complete and $ic(D, IC, U) = no$, then, by the contrapositive of (∗∗), $\mu(D^U, IC) \not\leq \mu(D, IC)$ follows. Thus, U is rejected because U would neither decrease the measured amount of inconsistency nor leave it unchanged. If, additionally, $\leq$ is a total order, then $\mu(D^U, IC) \not\leq \mu(D, IC)$ is equivalent to $\mu(D, IC) < \mu(D^U, IC)$. If ic is not complete, then ic might over-cautiously reject an update U even if U decreases or does not change the amount of integrity violation in (D, IC).

Obviously, conventional integrity checking does not depend on any particular μ. However, by Definitions 5 and 6, it is easy to see that, for each μ whatsoever, each conventional integrity checker vacuously is a μ-based integrity checker, since it only accepts input triples (D, IC, U) such that (D, IC) is consistent. Thus, Definition 6 properly generalizes conventional integrity checking.

By results in [39] [40] [42], it turns out that, for each $\mu \in \{\iota, |\iota|, \zeta, |\zeta|, \kappa, |\kappa|\}$, many (though not all) known integrity checkers, whenever applied to triples (D, IC, U) such that (D, IC) is inconsistent, are sound μ-based integrity checkers, e.g., the ones in [79] [21] [69] [88].

The completeness of integrity checkers is less frequently preserved when they are applied to triples (D, IC, U) such that (D, IC) is inconsistent. In fact, some integrity checkers are incomplete even when applied only to triples (D, IC, U) such that (D, IC) is consistent, e.g., the one in [55]. However, for relational databases and range-restricted constraints in clausal form, the method in [79] has been shown in [40] [42] to be a complete ζ-based integrity checker.

Inconsistency in (D, IC) may be complex or opaque or unknown. Hence, the computation of $\mu(D, IC)$ may be unfeasible, in particular if D is big. Thus, we are interested in integrity checkers that do not have to compute the measure on which they are based. Fortunately, many such methods are known, such as those in [79], [21], [69], [88], the domains of which can be soundly extended to triples (D, IC, U) such that (D, IC) is not necessarily consistent. They accept U only if U does not introduce any new constraint violation, but are ignorant of the actual amount of integrity violation in (D, IC), as seen in [39] [40] [42].

4.2 Inconsistency-tolerant Repairs

According to their definition in [3], repairs are database states that are consistent with the integrity constraints imposed on them and that differ minimally from an inconsistent predecessor state. However, each database state can be mapped into each other database state by some update. Hence, repairs can also be defined as updates that eliminate the inconsistency of a given database state: all constraints in the integrity theory associated with the database schema that had been violated before the update, are satisfied in the repaired state. Additionally, the modifications comprised by updates that qualify as repairs are required to be minimal, in some sense, so that no superfluous changes are effected by a repair, in compliance with the well-known Occam's razor principle. We speak of such repairs as total repairs, since they are supposed to yield database states that are totally consistent with their integrity constraints.

As opposed to total repairs, inconsistency-tolerant repairs, as defined below, do not necessarily eliminate all violations, but only some, while not introducing new violations that would equal or exceed the previous violations. More precisely, an inconsistency-tolerant repair is an update U of a database state that is inconsistent with its constraints, such that the updated state becomes less inconsistent, and there is no subset of U that could achieve the same or a larger amount of inconsistency reduction. The amount of inconsistency before and after an update intended as a repair is compared by using some database inconsistency measure.

We are going to re-define total repairs in two steps: in Definition 7a, below, we define total inconsistency reductions, and in 7b, we characterize total repairs as minimal total inconsistency reductions. It is easy to see that this definition is equivalent to the original one in [3]. Analogously, we define inconsistency-tolerant repairs, based on some inconsistency measure μ, in 7c and d.

Definition 7 (*total and inconsistency-tolerant repair*)
Let D be a database, IC an integrity theory that is violated in D, U an update and μ a database inconsistency measure.
a) U is a *total inconsistency reduction* of (D, IC) if each constraint in IC is satisfied in D^U.
b) U is a *total repair* of (D, IC) if U is a total inconsistency reduction of (D, IC) and there is no total inconsistency reduction $U' \subsetneq U$ of (D, IC).
c) U is a *μ-based inconsistency reduction* of (D, IC) if $\mu(D^U, IC) < \mu(D, IC)$.
d) U is a *μ-based inconsistency-tolerant repair* of (D, IC) if U is a μ-based inconsistency reduction and there is no μ-based inconsistency reduction U' of (D, IC) such that $U' \subsetneq U$ and $\mu(D^{U'}, IC) \leq \mu(D^U, IC)$.

From now on, we may simply speak of an 'inconsistency-tolerant inconsistency reduction' or an 'inconsistency-tolerant repair' whenever μ is understood implicitly. We also may speak of a 'measure-based repair', or a 'μ-repair' when we want to emphasize the relevance of μ. Moreover, we may generically speak of a 'repair' U if U is a total or an inconsistency-tolerant repair.

In Subsection 4.2.1, we show that inconsistency-tolerant repairs (Def. 7d) are a proper generalization of total repairs (Def. 7b), and that the latter are border cases of the former. In Subsection 4.2.2, we have a closer look at the minimality condition imposed on repairs. In Subsection 4.2.3, we expose some useful properties of inconsistency reductions. In Subsection 4.2.4, we illustrate that repairs may depend significantly on the inconsistency measure on which they are based.

4.2.1 Measure-based Repairs Generalize Total Repairs

Clearly, Definition 7 excludes the case $U = \emptyset$, since the empty update does not repair anything. However, it includes the case $\mu(D^U, IC) = o$ (cf. Definition 5), i.e., total repairs are border cases of inconsistency-tolerant repairs. Thus, Definition 7d properly generalizes total repairs, as stated in Proposition 1 below.

As opposed to inconsistency-tolerant repairs (Def. 7d), the definition of total repairs (7b) does not recur on any database inconsistency measure. However, for each inconsistency measure μ whatsoever, each total repair is a μ-repair, as entailed by the following consequence of Definitions 4, 5 and 7. For Proposition 1b, recall that β-repairs are based on a measure with a binary range, as defined in Subsection 3.2.

Proposition 1
For each triple (D, IC, U) such that IC is violated in D, the following holds.
a) For each inconsistency measure μ, U is a total repair of (D, IC) if and only if U is a μ-repair of (D, IC) such that $\mu(D^U, IC) = o$.
b) U is a total repair of (D, IC) if and only if U is a β-repair of (D, IC).

Proof:
a) The 'if' part of Proposition 1a is entailed by Definitions 5 and 7a. For the 'only-if' part, let U be a total repair of (D, IC). Since (D, IC) is inconsistent and, by Definition 7a, (D^U, IC) is consistent, $o < \mu(D, IC)$ and $\mu(D^U, IC) = o$ hold, for each μ, by Definition 5. Hence, $\mu(D^U, IC) < \mu(D, IC)$ follows. It remains to verify the minimality of U, i.e., to show that there is no $U' \subsetneq U$ such that $\mu(D^{U'}, IC) \leq \mu(D^U, IC)$. Suppose there were such a U'. Since $\mu(D^U, IC) = o$, it would follow that $\mu(D^{U'}, IC) = o$. That, however, contradicts the premise that U is a total repair and hence minimal, according to Definition 7b, i.e., there is no $U' \subsetneq U$ such that $\mu(D^{U'}, IC) = o$. Hence, U is a μ-repair of (D, IC) such that $\mu(D^U, IC) = o$. □

b) For $\mu = \beta$, part *b*) follows from *a*) and the definition of β in Subsection 3.2. □

4.2.2 Minimality of Inconsistency-tolerant Repairs

Clearly, each total and each μ-repair is a total and, resp., μ-based inconsistency reduction, but not vice versa, due to the minimality conditions in Definition 7b and, resp., 7d. The following example features an update U that satisfies $\mu(D^U, IC) < \mu(D, IC)$ as in Definition 7c, but not the minimality condition of 7d. Typically, updates of that kind contain elements that do not contribute to the reduction of inconsistency. Example 5 also features an update that is an inconsistency-tolerant repair of (D, IC).

Example 5
Let $D = \{p, q, r\}$, $IC = \{\leftarrow q\}$, $U = \{delete\ q,\ insert\ s\}$. It is easy to verify that, for each $\mu \in \{\iota, |\iota|, \varsigma, |\varsigma|, \kappa, |\kappa|\}$, $\mu(D^U, IC) < \mu(D, IC)$ holds, i.e., U is a μ-based inconsistency reduction. However, U is not a μ-repair of (D, IC), since, for its proper subset $U' = \{delete\ q\}$, $\mu(D^{U'}, IC) = \mu(D^U, IC)$ holds, i.e., U' yields the same amount of inconsistency reduction as U. In fact, U' is a μ-repair of (D, IC), since the only proper subset of U' is $\emptyset$.

Note that the minimality condition of inconsistency-tolerant repairs in Definition 7d must not be weakened, so as to simply require that there is no proper subset U' of U such that $\mu(D^{U'}, IC) < \mu(D, IC)$, as illustrated by Example 6.

Example 6
Let $D = \{p, q, r, s\}$, $IC = \{\leftarrow q,\ \leftarrow r,\ \leftarrow s\}$, $U = \{delete\ r,\ delete\ s\}$, and $U' = \{delete\ r\}$. Thus, $D^U = \{p, q\}$ and $D^{U'} = \{p, q, s\}$. It is easy to verify that, for each $\mu \in \{\iota, |\iota|, \varsigma, |\varsigma|, \kappa, |\kappa|\}$, both U and U' are μ-repairs of (D, IC), i.e., inconsistency reductions that are minimal according to Definition 7d, although $U' \subsetneq U$. However, also $\mu(D^U, IC) < \mu(D^{U'}, IC)$ holds, i.e., U reduces inconsistency more than U', i.e., U' is not preferable to U.

Instead of the conditions in Definition 7b, 7d, several other, non-equivalent definitions of minimality are conceivable. For total repairs, that has been pointed out, e.g., in [6]. For inconsistency-tolerant repairs, one could, for instance, replace $U' \subsetneq U$ by $|U'| < |U|$, or, more generally, require that there is no U' that would be better than U according to some preference criteria. Such criteria could, e.g., be determined by assigning some weights to the cases of constraints, or to the causes of their violation, in order to differentiate between different degrees of tolerability associated with the respective inconsistencies. In this paper, we do not study such alternative minimality conditions, except to note that $U' \subsetneq U$ entails $|U'| < |U'|$, i.e., requiring minimality of the cardinality of U, is strictly more demanding than subset minimality, and that many preference criteria are application-dependent, as opposed to subset minimality.

4.2.3 Useful Properties of Inconsistency Reductions

The following result is useful for computing repairs.

Proposition 2
For each measure μ and each triple (D, IC, U) such that U is a μ-based inconsistency reduction of (D, IC), some subset of U is a μ-repair of (D, IC).

Proof:
If U is a repair, then we are done. So, suppose that U is not a repair. Then, by Definition 7, there is a proper subset U' of U such that $\mu(D^{U'}, IC) \leq \mu(D^U, IC)$.

Since U is an inconsistency reduction, i.e., $\mu(D^U, IC) < \mu(D, IC)$, it follows that U' is a μ-based inconsistency reduction of (D, IC). If U' is a μ-repair of (D, IC), then we are done. If not, we iterate the preceding argument inductively, until we arrive at a subset U^* of U' and hence of U that, by Definition 7d and the transitivity of $\leq$, is a μ-repair of (D, IC). □

Ad-hoc intents to reduce inconsistency by singleton updates often occur in practice. The following result states that each insertion or deletion of a single database item which is confirmed to be an inconsistency reduction does not have to be checked for minimality for qualifying as a repair.

Proposition 3
For each measure μ and each triple (D, IC, U), each singleton μ-based inconsistency reduction U of (D, IC) is a μ-repair of (D, IC).

Proof:
We only have to show that U is minimal, according to Definition 7d. The only proper subset U' of U is $\emptyset$, for which $\mu(D^{U'}, IC) \leq \mu(D^U, IC)$ never may hold, since $D^{U'} = D$. □

Inconsistency can often be reduced iteratively, by sequences of singleton updates so that constraint violations are repaired one by one while the overall inconsistency does not increase. However, to compose such a sequence into one atomic transaction does not necessarily yield a repair, as shown by Example 7. However, by the transitivity of the ordering of the range of measures, it follows that such a transaction always yields at least an inconsistency reduction.

Example 7
Let $D_f = \{s, t, t'\}$ be the fact base of a deductive database D, the rule base of which consists of the following five clauses: $p \leftarrow s, t$; $q \leftarrow s, t'$; $r \leftarrow \sim s, t$; $r \leftarrow s, \sim t'$; $r \leftarrow t, t'$. Further, let $IC = \{\leftarrow p, \leftarrow q, \leftarrow r\}$. For the sequence of updates $U_1 = \{delete\ t'\}$, $U_2 = \{delete\ s\}$, $U_3 = \{delete\ t\}$, we have $\iota(D, IC) = IC$, $\iota(D^U, IC) = \{\leftarrow p, \leftarrow r\}$, $\iota(D^{U_1 U_2}, IC) = \{\leftarrow r\}$ and $\iota(D^{U_1 U_2 U_3}, IC) = \emptyset$. By Proposition 3, U_1 is an ι-repair of (D, IC), U_2 is an ι-repair of (D^{U_1}, IC) and U_3 is an ι-repair of $(D^{U_1 U_2}, IC)$. However, neither the transactional update $U_1 \cup U_2$ nor $U^* = U_1 \cup U_2 \cup U_3$ is an ι-repair of D, since U_2 is not only an ι-repair of (D^{U_1}, IC) but also an ι-repair of (D, IC) such that $U_2 \subsetneq U_1 \cup U_2$ and $\iota(D^{U_2}, IC) \subseteq \iota(D^{U_1 U_2}, IC)$, and $U = \{delete\ s, delete\ t\}$ is a total ι-repair of (D, IC) such that $U \subsetneq U^*$.

4.2.4 Repairs Depend on Inconsistency Measures

By Definition 7c,d, repairs are parametrized by the inconsistency measure on which they are based. The choice of that measure may have a significant influence on repairs and repair checking. Examples 8a and b illustrate that, for two measures μ, μ', a μ-repair U is not necessarily a μ'-repair, since μ and μ' may measure the effect of U differently.

Example 8 Let $D = \{p, q, r\}$ and $IC = \{\leftarrow q, \leftarrow s\}$.

a) Let $U_a = \{delete\ q,\ insert\ s\}$. Clearly, U_a is a μ-repair of (D, IC) for each μ that assigns a higher weight of inconsistency to the violation of $\leftarrow q$ than to the violation of $\leftarrow s$. However, U_a is not a $|\iota|$-repair, nor even a $|\iota|$-based inconsistency reduction, of (D, IC) (recall: $|\iota|$ counts the violated constraints in IC), since $|\iota|(D^{U_a}, IC) = |\iota|(D, IC) = 1$.

b) Let $U_b = \{insert\ t\}$, and ξ be the measure that counts the facts in D that contribute to some integrity violation and then divides that count by the cardinality of D. (Similar inconsistency measures have been studied in [64] [52].) Since $\xi(D^{U_b}, IC) = 1/4$ and $\xi(D, IC) = 1/3$, U_b is a ξ-repair, by Proposition 3. However, for each $\mu \in \{|\iota|, |\varsigma|, |\kappa|\}$, U_b is clearly not a μ-repair of (D, IC), since $\mu(D, IC) = \mu(D^{U_b}, IC) = 1$.

4.3 Inconsistency-tolerant Repair Checking

Difficulties and complications of automated repairing of database inconsistency by event-condition-action rules are well-documented [14]. Unpredictability of what may happen tends to increase whenever a database is updated ad-hoc and 'by hand' for restoring consistency or for getting rid of some constraint violation. Thus, repairing may go wrong. Hence, repair checkers, i.e., methods for checking if an update actually is a repair, are needed.

In Subsection 4.3.1, we define repair checking in analogy to integrity checking. In 4.3.2, we modularize repair checking according to the definition of repairs as minimal inconsistency reductions into inconsistency reduction checking and minimality checking. In 4.3.3, we show that inconsistency reduction can be checked by integrity checking. In 4.3.4, we show that also minimality can be checked by integrity checking. This leads to the results in 4.3.5, that repair checking can be computed by integrity checking. In 4.3.6, the use of simplified integrity checking for computing repair checking is assessed.

The basic idea behind repair checking by integrity checking is that an update U reduces inconsistency if and only if its undo increases it. As we have seen in Subsection 4.1, it is easier to observe an increase than a decrease of inconsistency by measure-based integrity checking.

4.3.1 Repair Checking – Definition

The repair checking problem is to find out if an update U is a repair of (D, IC). Hence, each automated total repair checker can be described as an update checker. The output *yes* accepts U as a repair in compliance with Definition 7, and *no* means that U is not recognized as a repair.

Soundness and completeness of repair checkers are defined below. Definition 8 abstracts away from the distinction of repair checking of updates that are either meant to be total or inconsistency-tolerant repairs, in accordance with Proposition 1.

Definition 8 (*repair checker*)
Let rc be an update checker, and μ an inconsistency measure. rc is called a *sound*, resp., *complete μ-repair checker* if (*), resp., (**) holds, for each triple (D, IC, U).

(*) $rc(D, IC, U) = yes \Rightarrow U$ is a μ-repair
(**) U is a μ-repair $\Rightarrow rc(D, IC, U) = yes$

In words, rc is sound if its output $rc(D, IC, U) = yes$ identifies U as a repair of (D, IC), and complete if each repair U of (D, IC) is identified by rc.

Below, Proposition 4 highlights close relationships between repair checking and integrity checking.

Proposition 4
For each inconsistency measure μ, the following holds.
a) Each sound μ-repair checker is a sound μ-based integrity checker.
b) Each complete μ-based integrity checker is a complete μ-repair checker.

Proof:
By Definitions 6, 7 and 8. □

Neither the converse of Propositon 4*a* nor the converse of Proposition 4*b* hold, which also follows from Definitions 6, 7 and 8.

4.3.2 Repair Checking – Modularization

Corresponding to Definition 7, repair checking of an update U proceeds in two modular phases. First, check if U reduces inconsistency. We call this phase the *inconsistency reduction check*. If U has passed the inconsistency reduction check, the second phase of repair checking is to check if U is minimal. We call this phase the *minimality check*. For the first of the two phases of repair checking, the following definition characterizes sound and complete inconsistency reduction checking, according to Definition 7*c*.

Definition 9 (*inconsistency reduction checker*)
Let μ be an inconsistency measure, and ir an update checker. ir is called a *sound*, resp., *complete*, μ-based inconsistency reduction checker if (*), resp., (**) holds, for each triple (D, IC, U).

(*) $ir(D, IC, U) = yes \Rightarrow \mu(D^U, IC) < \mu(D, IC)$
(**) $\mu(D^U, IC) < \mu(D, IC) \Rightarrow ir(D, IC, U) = yes$

In words, ir is sound if $ir(D, IC, U) = yes$ correctly indicates that U reduces the inconsistency of (D, IC) measured by μ, and complete if each U that reduces inconsistency is checked correctly by rc.

Next, we define the soundness and completeness of the second phase, viz. measure-based minimality checking, according to Definition 7d.

Definition 10 (*minimality checker*)
Let μ be an inconsistency measure, and mc an update checker. mc is called a *sound*, resp., *complete*, μ-based minimality checker if (*), resp., (**) holds, for each triple (D, IC, U) such that U is a μ-based inconsistency reduction.

(*) $\quad mc(D, IC, U) = yes \;\Rightarrow\;$ for each $U' \subsetneq U$, $\mu(D^{U'}, IC) \not\leq \mu(D^U, IC)$

(**) for each $U' \subsetneq U$, $\mu(D^{U'}, IC) \not\leq \mu(D^U, IC) \;\Rightarrow\; mc(D, IC, U) = yes$

In words, mc is sound if $mc(D, IC, U) = yes$ correctly indicates that U is a minimal inconsistency reduction, and mc is complete if the minimality of each repair of (D, IC) is acknowledged by mc.

The following result is a straightforward consequence of Definitions 7–10. It entails that repair checking can be realized modularly by inconsistency reduction checking and, if the latter was successful, subsequent minimality checking.

Proposition 5 (*modularization of repair checking*)
Let μ be an inconsistency measure, ir a sound resp. complete μ-based inconsistency reduction checker, mc a sound resp. complete μ-based minimality checker, and rc an update checker. rc is a sound resp. complete μ-repair checker if and only if (*) resp. (**) holds, for each triple (D, IC, U).

(*) $\quad rc(D, IC, U) = yes \;\Rightarrow\; ir(D, IC, U) = yes$ and $mc(D, IC, U) = yes$

(**) $ir(D, IC, U) = yes$ and $mc(D, IC, U) = yes \;\Rightarrow\; rc(D, IC, U) = yes$

4.3.3 Inconsistency Reduction Checking by Integrity Checking

In this subsection, we show how inconsistency reduction can be implemented by measure-based integrity checking. Also, we specify conditions that guarantee the soundness and the completeness of inconsistency reduction checkers.

Part a of Lemma 1, below, shows that U is not a μ-based inconsistency reduction of (D, IC) if sound μ-based integrity checking of $(D^U, IC, \overline{U})$ accepts $\overline{U}$. Part b shows that complete inconsistency reduction checking is guaranteed by sound measure-based inconsistency checking.

Lemma 1
For each measure μ and each sound μ-based integrity checker ic, a) and b) hold.

a) For each triple (D, IC, U) such that $ic(D^U, IC, \overline{U}) = yes$, U is not a μ-based inconsistency reduction of (D, IC).

b) An update checker ir is a complete μ-based inconsistency reduction checker if $(*)$ holds, for each triple (D, IC, U).

$(*) \quad ir(D, IC, U) = no \Rightarrow ic(D^U, IC, \overline{U}) = yes$

Proof:
a) We have to show that $\mu(D^U, IC) \not< \mu(D, IC)$ if $ic(D^U, IC, \overline{U}) = yes$. If $ic(D^U, IC, \overline{U}) = yes$, then the soundness of ic entails $\mu(D^{U\overline{U}}, IC) \leq \mu(D^U, IC)$, i.e. $\mu(D, IC) \leq \mu(D^U, IC)$. Hence, $\mu(D^U, IC) \not< \mu(D, IC)$. □

b) We have to show that $\mu(D^U, IC) < \mu(D, IC) \Rightarrow ir(D, IC, U) = yes$ holds, according to Definition 9, under the premise of $(*)$. We show the contrapositive, i.e., $ir(D, IC, U) = no \Rightarrow \mu(D^U, IC) \not< \mu(D, IC)$. Let $ir(D, IC, U) = no$. That, by the contrapositive of $(*)$, entails $ic(D^U, IC, \overline{U}) = yes$. The soundness of ic entails $\mu(D^{U\overline{U}}, IC) \leq \mu(D^U, IC)$, i.e., $\mu(D, IC) \leq \mu(D^U, IC)$, hence $\mu(D^U, IC) \not< \mu(D, IC)$. □

Part a of Lemma 2, below, shows that U is a μ-based inconsistency reduction of (D, IC) if complete μ-based integrity checking of $(D^U, IC, \overline{U})$ rejects $\overline{U}$ and the range of μ is totally ordered. Part b shows that sound inconsistency reduction checking is guaranteed by complete measure-based inconsistency checking if the range of the measure is totally ordered.

Lemma 2
For each measure μ and complete μ-based integrity checker ic such that the range of μ is totally ordered, the following holds.

a) For each triple (D, IC, U) such that $ic(D^U, IC, \overline{U}) = no$, U is a μ-based inconsistency reduction of (D, IC).

b) An update checker ir is a sound μ-based inconsistency reduction checker if $(*)$ holds, for each triple (D, IC, U).

$(*) \quad ir(D, IC, U) = yes \Rightarrow ic(D^U, IC, \overline{U}) = no$

Proof:
a) Under the premise of Lemma 2, we have to show that $\mu(D^U, IC) < \mu(D, IC)$ if $ic(D^U, IC, \overline{U}) = no$. Since ic is complete, $ic(D^U, IC, \overline{U}) = no$ entails, according to Definition 6, that $\mu(D^{U\overline{U}}, IC) \not\leq \mu(D^U, IC)$, which is equivalent to $\mu(D, IC) \not\leq \mu(D^U, IC)$, since $D^{U\overline{U}} = D$. Since $\leq$ is a total order, it follows that $\mu(D^U, IC) < \mu(D, IC)$. □

b) We have to show that $ir(D, IC, U) = yes \Rightarrow \mu(D^U, IC) < \mu(D, IC)$ holds, according to Definition 9, under the premise of $(*)$. Let $ir(D, IC, U) = yes$. That, by $(*)$, entails $ic(D^U, IC, \overline{U}) = no$. From the completeness of ic, it follows that $\mu(D^{U\overline{U}}, IC) \not\leq \mu(D^U, IC)$, i.e., $\mu(D, IC) \not\leq \mu(D^U, IC)$, i.e., $\mu(D^U, IC) < \mu(D, IC)$, since $\leq$ is a total order. □

The condition in Lemma 2 that the range of μ is totally ordered cannot be waived, as shown by Example 9.

Example 9
Let $D = \{p, q\}$, $IC = \{\leftarrow q, \leftarrow r\}$, $U = \{insert\ r,\ delete\ q\}$ and ic_N be the integrity checker in [79]. For each triple (D, IC, U), ic_N focuses precisely on cases of constraints in IC that are violated in D^U by U. Thus, ic_N, based on ζ (as defined in Subsection 3.2), is complete for relational databases and constraints in clausal form, as already mentioned in Subsection 4.1. Clearly, we have $D^U = \{p, r\}$, $\overline{U} = \{delete\ r,\ insert\ q\}$, $\zeta(D^U, IC) = \{\leftarrow r\}$ and $\zeta(D, IC) = \{\leftarrow q\}$, thus $\zeta(D^U, IC) \not\subseteq \zeta(D, IC)$ and $ic_N(D^U, IC, \overline{U}) = $ no. However, U obviously is not a ζ-based inconsistency reduction of (D, IC).

Part a of Lemma 3, below, shows that U is a μ-based inconsistency reduction of (D, IC) if complete μ-based integrity checking of $(D^U, IC, \overline{U})$ rejects $\overline{U}$ and sound μ-based integrity checking of (D, IC, U) accepts U. Part b shows that sound and complete inconsistency reduction checking of U can be realized by sound integrity checking of U and complete integrity checking of $\overline{U}$.

Lemma 3
For each measure μ and each sound and complete μ-based integrity checker ic, the following holds.

a) For each triple (D, IC, U), U is a μ-based inconsistency reduction of (D, IC) if and only if $ic(D^U, IC, \overline{U}) = $ no and $ic(D, IC, U) = $ yes.

b) An update checker ir is a sound, resp., complete μ-based inconsistency reduction checker if and only if $(*)$, resp., $(**)$ holds, for each triple (D, IC, U).
$ir(D, IC, U) = $ yes $\Rightarrow$ $(ic(D^U, IC, \overline{U}) = $ no and $ic(D, IC, U) = $ yes$)$ $(*)$
$(ic(D^U, IC, \overline{U}) = $ no and $ic(D, IC, U) = $ yes$)$ $\Rightarrow$ $ir(D, IC, U) = $ yes $(**)$

Proof:
a) Under the premises of Lemma 3, we have to show that $\mu(D^U, IC) < \mu(D, IC)$ if and only if $ic(D^U, IC, \overline{U}) = $ no and $ic(D, IC, U) = $ yes. We first show the if-half. As in the proof of Lemma 2, we have that the completeness of ic entails $\mu(D, IC) \not\leq \mu(D^U, IC)$. From $ic(D, IC) = $ yes and the soundness of ic, it follows that $\mu(D^U, IC) \leq \mu(D, IC)$. From $\mu(D, IC) \not\leq \mu(D^U, IC)$, it follows that $\mu(D^U, IC) < \mu(D, IC)$.

For showing the only-if half, let U be a μ-based inconsistency reduction of (D, IC), i.e., $\mu(D^U, IC) < \mu(D, IC)$. That entails $\mu(D^U, IC) \leq \mu(D, IC)$. From that and the completeness of ic, it follows that $ic(D, IC, U) = $ yes. It remains to show that $ic(D^U, IC, \overline{U}) = $ no, which follows from the soundness of ic. □

b) To prove that ir is sound under the premise of $(*)$, we have to show that $ir(D, IC, U) = $ yes $\Rightarrow$ $\mu(D^U, IC) < \mu(D, IC)$ holds, according to Definition 8.

Suppose that $ir(D, IC, U) = yes$. By $(*)$, that entails $ic(D^U, IC, \overline{U}) = no$ and $ic(D, IC, U) = yes$. Completeness of ic entails that $\mu(D^{U\overline{U}}, IC) \not\leq \mu(D^U, IC)$, and soundness of ic entails that $\mu(D^U, IC) \leq \mu(D, IC)$. That is equivalent to $\mu(D, IC) \not\leq \mu(D^U, IC)$ and $\mu(D^U, IC) \leq \mu(D, IC)$. That entails $\mu(D, IC) \neq \mu(D^U, IC)$ and $\mu(D^U, IC) \leq \mu(D, IC)$, hence $\mu(D^U, IC) < \mu(D, IC)$.

To prove that the soundness of ir entails $(*)$, let $ic(D, IC, U) = yes$. We have to show $ic(D^U, IC, \overline{U}) = no$ and $ic(D, IC, U) = yes$. The soundness of ir entails $\mu(D^U, IC) < \mu(D, IC)$, hence $\mu(D, IC) \not\leq \mu(D^U, IC)$ and $\mu(D^U, IC) \leq \mu(D, IC)$. Clearly, $\mu(D, IC) \not\leq \mu(D^U, IC)$ is the same as $\mu(D^{U\overline{U}}, IC) \not\leq \mu(D^U, IC)$, from which $ic(D^U, IC, \overline{U}) = no$ follows by the soundness of ic. From $\mu(D^U, IC) \leq \mu(D, IC)$, the completeness of ic entails $ic(D, IC, U) = yes$.

To prove that ir is complete under the premise of $(**)$, we show the contrapositive $ir(D, IC, U) = no \Rightarrow \mu(D^U, IC) \not< \mu(D, IC)$ of the implication $\mu(D^U, IC) < \mu(D, IC) \Rightarrow ir(D, IC, U) = yes$. Suppose that $ir(D, IC, U) = no$. By $(**)$, it follows that $ic(D^U, IC, \overline{U}) = yes$ or $ic(D, IC, U) = no$. If $ic(D^U, IC, \overline{U}) = yes$, then the same argument as in the proof of Lemma 1a applies. So, assume that $ic(D, IC, U) = no$. By the completeness of ic, that entails $\mu(D^U, IC) \not\leq \mu(D, IC)$, hence $\mu(D^U, IC) \not< \mu(D, IC)$.

To prove that the completeness of ir entails $(**)$, let $ic(D^U, IC, \overline{U}) = no$ and $ic(D, IC, U) = yes$. Thus, we have to show that $ir(D, IC, U) = yes$. From $ic(D, IC, U) = yes$, the soundness of ic entails $\mu(D^U, IC) \leq \mu(D, IC)$. From $ic(D^U, IC, \overline{U}) = no$, the completeness of ic entails $\mu(D^{U\overline{U}}, IC) \not\leq \mu(D^U, IC)$, i.e., $\mu(D, IC) \not\leq \mu(D^U, IC)$. From $\mu(D^U, IC) \leq \mu(D, IC)$ and $\mu(D, IC) \not\leq \mu(D^U, IC)$, it follows that $\mu(D^U, IC) < \mu(D, IC)$. Hence, the completeness of ir entails $ir(D, IC, U) = yes$. □

Note that, for U to be recognized as a μ-based inconsistency reduction, both Lemmata 2 and 3 require running a complete μ-based integrity checker ic on $\overline{U}$. However, Lemma 2 additionally requires a totally ordered range of μ, while Lemma 3 additionally requires also the soundness of ic and running it also on U.

4.3.4 Minimality Checking by Integrity Checking

In Subsection 4.3.3, we have seen how the inconsistency reduction of updates can be checked by measure-based integrity checking. In Lemma 4 and Proposition 6, below, we are going to see that also the minimality check of inconsistency reductions can be accomplished by measure-based integrity checking.

Lemma 4
For each measure μ, each sound resp. complete μ-based integrity checker ic and each triple (D, IC, U) such that U is a μ-based inconsistency reduction U of (D, IC), $(*)$ resp. $(**)$ holds.

(∗) U is not minimal if there is a proper non-empty subset U' of U such that $ic(D^U, IC, \overline{U'}) = yes$

(∗∗) U is not minimal only if there is a proper non-empty subset U' of U such that $ic(D^U, IC, \overline{U'}) = yes$

Proof:
(∗) Let U' be a proper subset of U such that $ic(D^U, IC, \overline{U'}) = yes$. Thus, we have to show that U is not minimal. Let $U'' = U \setminus U'$. Since $U' \neq U$ and $U' \neq \emptyset$, it follows that $U'' \subsetneq U$. Hence, it suffices to show that $\mu(D^{U''}, IC) \leq \mu(D^U, IC)$. It follows from the soundness of ic and $ic(D^U, IC, \overline{U'}) = yes$ that $\mu(D^{U\overline{U'}}, IC) \leq \mu(D^U, IC)$, i.e., $\mu(D^{U''}, IC) \leq \mu(D^U, IC)$, since $U\overline{U'} = U''$. □

(∗∗) Assume that U is not minimal. Thus, by Definition 7d, there is a proper non-empty subset U''' (say) of U such that $\mu(D^{U'''}, IC) \leq \mu(D^U, IC)$. For $U' = U \setminus U''$, this is equivalent to $\mu(D^{U\overline{U'}}, IC) \leq \mu(D^U, IC)$, since $U'' = U\overline{U'}$. Since ic is complete, $ic(D^U, IC, \overline{U'}) = yes$ follows, by Definition 6. □

The contrapositive of Lemma 4, together with Definition 7d, yields Proposition 6, below. We state it explicitly, for facilitating the evidence of theorems in Subsection 4.3.5. Informally, it describes how to obtain sound and complete minimality checkers by complete and, resp., sound integrity checkers.

Proposition 6
For each measure μ, each complete resp. sound μ-based integrity checker ic, and each triple (D, IC, U) such that U is a μ-based inconsistency reduction U of (D, IC), an update checker mc is a sound resp. complete μ-based minimality checker if (∗) resp. (∗∗) holds.

(∗) $mc(D, IC, U) = yes \Rightarrow$ for each non-empty $U' \subsetneq U$, $ic(D^U, IC, \overline{U'}) = no$
(∗∗) for each non-empty $U' \subsetneq U$, $ic(D^U, IC, \overline{U'}) = no \Rightarrow mc(D, IC, U) = yes$

Proof:
From the premise that ic is complete and (∗) in Proposition 6, the soundness of mc follows by (∗∗) in Lemma 4. From the premise that ic is sound and (∗∗) in Proposition 6, the completeness of mc follows by (∗) in Lemma 4. □

4.3.5 Repair Checking by Integrity Checking

Lemmata 1–4 and Proposition 6 provide handles for unfolding Proposition 5 into the main results of this paper, as presented in Theorems 1–3, below. They state how inconsistency-tolerant repairs can be verified or falsified by measure-based integrity checking.

Theorem 1 devises a way to see that a given update is not a repair. By requiring that the used integrity checker be not just sound but also complete, Theorem 3 devises a sound and complete way to check if U is or is not a repair. While only catering for the soundness of repair checking, Theorem 2 additionally requires that the range of the used inconsistency measure be totally ordered, but in turn only needs the completeness of integrity checking, and one run of integrity checking less than repair checking according to Theorem 3.

Theorem 1
For each measure μ and each sound μ-based integrity checker ic, a) and b) hold.

a) For each triple (D, IC, U) such that $ic(D^U, IC, \overline{U}') = yes$ for some $U' \subseteq U$, U is not a μ-repair of (D, IC).

b) An update checker rc is a complete μ-repair checker if $(*)$ holds, for each triple (D, IC, U).

$$(*) \quad rc(D, IC, U) = no \;\Rightarrow\; ic(D^U, IC, \overline{U}') = yes, \text{ for some } U' \subseteq U,\, U \neq \emptyset.$$

Proof:
By Lemma 1 and $(*)$ in Lemma 4, we have the following.

a) For each triple (D, IC, U) such that $ic(D^U, IC, \overline{U}) = yes$ or $ic(D^U, IC, \overline{U}') = yes$ for some non-empty $U' \subsetneq U$, U is not a μ-repair of (D, IC).

b) An update checker rc is a complete μ-repair checker if the following holds, for each triple (D, IC, U).

$$rc(D, IC, U) = no \;\Rightarrow\; ic(D^U, IC, \overline{U}) = yes, \text{ or there is a } U' \subsetneq U,\, U \neq \emptyset,$$
$$\text{such that } ic(D^U, IC, \overline{U}') = yes$$

From that, Theorem 1 follows. □

Theorem 2
For each inconsistency measure μ with a totally ordered range, and each complete μ-based integrity checker ic, the following holds.

a) For each triple (D, IC, U), U is a μ-repair of (D, IC) if, for each $U' \subseteq U$, $ic(D^U, IC, \overline{U}') = no$.

b) An update checker rc is a sound μ-repair checker if, for each triple (D, IC, U) and each $U' \subseteq U$, $(*)$ holds.

$$rc(D, IC, U) = yes \;\Rightarrow\; ic(D^U, IC, \overline{U}') = no,\; ic(D^U, IC, \overline{U}') = no \quad (*)$$

Proof:
By Lemma 2 and $(*)$ in Proposition 6, we have the following.

a) For each triple (D, IC, U), U is a μ-repair of (D, IC) if $ic(D^U, IC, \overline{U}) = no$ and $ic(D^U, IC, \overline{U}') = no$, for each $U' \subsetneq U$.

b) An update checker rc is a sound μ-repair checker if $(*)$ holds, for each triple (D, IC, U).

$$rc(D, IC, U) = yes \;\Rightarrow\; ic(D^U, IC, \overline{U}) = no \text{ and} \qquad (*)$$
$$\text{for each } U' \subsetneq U,\; ic(D^U, IC, \overline{U}') = no.$$

From that, Theorem 2 follows □

Theorem 3
For each measure μ and each sound and complete μ-based integrity checker ic, the following holds.
a) For each triple (D, IC, U), U is a μ-repair of (D, IC) if and only if, for each non-empty $U' \subseteq U$, $ic(D^U, IC, \overline{U'}) = no$, and $ic(D, IC, U) = yes$.
b) An update checker rc is a sound, resp. complete μ-repair checker if and only if, for each triple (D, IC, U), $(*)$, resp. $(**)$ holds.

$rc(D,IC,U) = yes \Rightarrow$ for each non-empty $U' \subseteq U$, $ic(D^U,IC,\overline{U'}) = no$
and $ic(D, IC, U) = yes$ $\qquad(*)$

For each non-empty $U' \subseteq U$, $ic(D^U, IC, \overline{U'}) = no$
and $ic(D, IC, U) = yes$ $\qquad\Rightarrow rc(D,IC,U) = yes$ $(**)$

Proof:
By Lemma 3 and Proposition 6, we have the following.
a) For each triple (D, IC, U), U is a μ-repair of (D, IC) if and only if $ic(D^U, IC, \overline{U}) = no$ and $ic(D, IC, U) = yes$ and, for each $U' \subsetneq U$, $ic(D^U, IC, \overline{U'}) = no$,
b) An update checker rc is a sound, resp. complete μ-repair checker if and only if, for each triple (D, IC, U), $(*)$, resp. $(**)$ holds.

$rc(D, IC, U) = yes \Rightarrow ic(D^U, IC, \overline{U})) = no$ and $ic(D, IC, U) = yes$,
and for each $U' \subsetneq U$, $ic(D^U, IC, \overline{U'}) = no$ $\qquad(*)$

$ic(D^U, IC, \overline{U}) = no$ and $ic(D, IC, U) = yes$
and for each $U' \subsetneq U$, $ic(D^U, IC, \overline{U'}) = no \quad \Rightarrow rc(D, IC, U) = yes$ $(**)$

From that, Theorem 3 follows. □

Theorems 1–3 suggest a computation of repair checking by measure-based integrity checking. The output $ic(D^U, IC, \overline{U}) = yes$ means that U is not a μ-repair, by Theorem 1. For U to be a μ-repair, the output $ic(D, IC, U) = yes$ is only necessary, but not sufficient. Sufficient conditions to identify updates as μ-repairs are given by Theorems 2 and 3, which both require the use of a complete μ-based integrity checker.

In any case, however, for computing repair checking, a measure-based integrity checker should be used that simplifies the evaluation of constraints, such as the method in [79], or one of the many methods that have been developed as modifications, refinements or extensions of that approach [67] [73].

We are going to assess simplified repair checking according to Theorem 3, by the two phases of inconsistency reduction and minimality checking. Note, however, that, if the range of μ is totally ordered, then the cost of μ-based repair checking according to Theorem 2 is lower than that of a computation according to Theorem 3, since the totally ordered range of μ enables a less costly inconsistency reduction check.

Phase 1:
Check if U is a μ-based inconsistency reduction of (D, IC) by computing $ic(D^U, IC, \overline{U})$ and $ic(D, IC, U)$, according to Lemma 3.

Phase 2:
Check if U is minimal in the sense of Definition 7d by checking whether $ic(D^U, IC, \overline{U'}) = no$ holds, for each non-empty $U' \subsetneq U$, according to Proposition 6.

For Phase 1, at most two runs of *ic* are needed. For Phase 2, at most $2^n - 2$ runs of *ic* are needed, where n is the cardinality of U. Hence, maximally 2^n runs of *ic* are needed for deciding if U is a μ-repair of (D, IC) or not. Thus, the actual cost of repair checking by integrity checking depends on *ic*. If a sound and complete simplified integrity checker is available, then running such a method tends to be much less costly than brute-force integrity checking, as employed for total repair checking.

In [38], it is shown that the cost of total repair checking involves $k \times (m-1)$ unsimplified constraint evaluations, where $m = 2^n$ and k is the cardinality of IC. For ease of comparison, suppose that all constraints in IC are expressed by a single constraint formula I (the conjunction of all constraints). Then, for total repair checking, we'd have in the order of m-1 evaluations of I against D^{U_i} ($1 \leq i \leq m-1$), where the U_i are the non-empty subsets of U. Compared to that, m evaluations of a simplification of I for simplified repair checking (one more against D than as for total repair checking) obviously tends to be much less costly.

The significant cost savings obtainable by simplified integrity checking have been noted in many studies in the literature, among them, e.g., [79] [19] [13] [15]. They essentially amount to the difference between having to evaluate universally quantified formulas, as for brute-force repair checking, and evaluating their simplified instances obtained from ground substitutions of their global variables, as for most methods of simplified integrity checking. Often, that corresponds to the difference between evaluating possibly huge joins of relations and simple lookups of ground instances of such joins.

A last remark: If, in Phase 1, U turns out to be an inconsistency reduction then, by Proposition 2, at least one total repair of (D, IC) exists. From each μ-based inconsistency reduction of (D, IC), a μ-repair of (D, IC) can obviously be obtained by iterating Phase 2, as in the proof of Proposition 2.

5 Mathematical and Classical Measures versus Database Inconsistency Measures

In this section, we compare mathematical measures and classical inconsistency measures to database inconsistency measures, pointing out similarities and essential differences.

In Subsection 5.1, we compare the formal definitions. In Subsection 5.2, we compare syntactic and semantic similarities and differences, except those related to consistency and inconsistency, which are dealt with in Subsection 5.3. In Subsection 5.4, we compare the validity and significance of various axiomatic postulates for different kinds of inconsistency measures. In Subsection 5.5, differences with regard to typical applications are pointed out.

5.1 Definitions

In mathematical measure theory [4], a measure μ maps elements (typically, sets) of a *measure space* $\mathbb{S}$ (sets of sets) to some *metric space*, which most often consists of a set of real numbers, i.e., a totally ordered range of values, that sometimes is augmented with an additional unique greatest element ∞. Intuitively, for $S \in \mathbb{S}$, $\mu(S)$ tells how 'big' S is. Well-known examples of mathematical measures are probability measures, whose range is the real number interval [0,1], for indicating the likelihood of certain "events", formalized as elements of the measure space [56].

Likewise, classical inconsistency measures are usually described as mappings from sets of logical formulae to a numerical range of values, possibly augmented with an ∞ element. Classical inconsistency measures are conceptually simpler than mathematical measures, inasmuch as the measure space of the former consists of discrete elements, while the elements of mathematical measure spaces typically are sets of real number vectors that usually are not discrete.

As opposed to mathematical measures and classical inconsistency measures, the range of database inconsistency measures is not necessarily numerical. Moreover, the ordering of non-numerical ranges of database inconsistency measures is not necessarily total, i.e., there may be incomparable amounts of database inconsistency. Examples of such measures are ι, ζ and κ. In fact, ι and ζ arguably are more useful than their numerical counterparts $|\iota|$ and $|\zeta|$, since computing the latter involves the computation of ι and ζ, as well as a count of violated constraints or cases, or of causes of such violations, respectively, all of which is unnecessary when ι or ζ are used directly, by running well-known integrity checkers such as those in [79], [21], [69], [88], and others.

On the other hand, there is no obvious reason why classical inconsistency measures should be limited to totally ordered ranges. For instance, it is reasonable to conceive of an inconsistency measure σ (say) that maps each set P

of propositional formulas to the union of all minimally inconsistent subsets of P. A set P then can be taken to be equally or less inconsistent than a set P' if $\sigma(P) \subseteq \sigma(P')$. Thus, σ considers the inconsistency of two sets P, P' such that neither $\sigma(P)$ is a subset of $\sigma(P')$ nor vice versa, as incomparable. That is plausible, at least as long as there is no quantifiable knowledge about the 'meaning' or 'significance' of the formulas in the symmetric difference of $\sigma(P)$ and $\sigma(P')$, which could be used to refine σ. Similarly to the database inconsistency measures ι, ζ and κ, the inconsistency measure σ seems useful for comparing input arguments P and P' such that P' can be obtained by some change of P which leaves a significant part of P intact.

In general, not only the range but also the domain of classical inconsistency measures and database inconsistency measures is different. The latter are defined for pairs (D, IC). To bring that definition closer to the definition of the domain of classical inconsistency measures, one could simply replace (D, IC) by $D \cup IC$. However, such unions still are not just sets of predicate logic formulas. Rather, databases and integrity constraints conform to a particular syntax and sematics, which sets them apart from the usual input of classical inconsistency measures. These distinctions are looked at in more detail in Subsections 5.2 and 5.3.

5.2 Syntax and Semantics

Let us recall some pertinent syntactic and semantic similarities and differences between classical logic and database logic, and their repercussions on the respective forms of inconsistency measures.

Database logic and classical logic have been closely intertwined since the inception of relational databases [17]. This connection has been boosted and further tightened by the rise of deductive databases [48]. Indeed, each relational and each deductive database essentially consists of logic sentences in some specific syntactic form.

However, essential differences between classical inconsistency measures and database inconsistency measures have to do with the classical logic syntax and semantics underlying classical inconsistency measures, on the one hand, and the syntax and semantics of database logic underlying database inconsistency measures, on the other. More precisely, the interpretation of the syntax of database clauses and integrity constraints may deviate from both the proof- and the model-theoretic semantics of classical logic (for example, the Clark completion [16] or some variant of it for proof theory, and stable or well-founded semantics [50] [94] for model theory).

A fundamental difference consists of the monotonicity of classical logic and the non-monotonicity of database negation, which impedes a straightforward generalization of classical inconsistency measures to databases. That difference is going to be illustrated in Subsection 5.4.

Even more intricate differences may manifest themselves by some specifics of interpreting integrity constraints (cf. [65], Summary and Comments to Chapters 6, 9, 10 and 13). Constraints may involve two kinds of negation: the non-monotonic negation of negative literals, as well as classical logic negation, e.g., if constraints are represented by denial clauses. Moreover, there is a deontic flavor adherent to the semantics of constraints: they are meant to be actively satisfied by the database. In Subsection 5.3, we are going to come back to the syntax and semantics of integrity constraints in connection with inconsistency measuring.

Another significant difference between the syntax and semantics of classical inconsistency measures and database inconsistency measures: the logic used for processing applications of the latter involves resolution-based reasoning with database clauses, i.e., quantified predicate logic sentences, while the logic deployed for classical inconsistency measures is essentially propositional. Even the classical measures that assess the inconsistency of sets of predicate logic sentences do not process quantified formulas, but only ground literals of atoms from the Herbrand base, i.e., formulas which are not essentially different from propositional ones.

5.3 Inconsistency

Yet another significant difference becomes evident by scrutinizing the distinct concepts of inconsistency in classical logic and database logic. But let us begin with a commonality of both. It is well-known that each sentence whatsoever is derivable in inconsistent classical logic theories, according to the *ex falso quodlibet* property (EFQ). Due to that, inconsistency may amount to a problem of monstrous proportions. Not so, however, for inconsistency measures, neither for classical ones nor for database inconsistency measures, although for different reasons. Below, we are going to see why.

From the point of view of classical inconsistency measures, inconsistency is a semantic, i.e., model-theoretic issue. For avoiding the devastating effects of EFQ, the usual model-theoretic semantics is abandoned by giving way to three- or four-valued or quasi-classical or probabilistic or possibilistic semantics of inconsistent pairs (D, IC). As opposed to that, database logic does not need to resort to multi-valued or paraconsistent semantics for coping with inconsistency, since database inconsistency is a mere syntactic issue. Let us see why.

A set D of database clauses can never be inconsistent, but the union of D with a set IC of integrity constraints can. Yet, inconsistency in databases can be hedged syntactically, simply by representing each constraint in denial form (which may involve the inclusion of additional clauses in D) and replacing each denial constraint of the form $\leftarrow B$ in IC by the database clause $violated \leftarrow B$, where *violated* is a distinguished 0-ary predicate that does not occur in the database, thus obtaining a set IC' of database clauses about the predicate *violated*. Then, the inconsistency of $D \cup IC$ is equivalent to the answer '*yes*' to the query $\leftarrow$ *violated* against the database $D' \cup IC'$, where D' is the union of D with the additional clauses needed to represent the constraints in IC in denial form. Clearly, for each constraint I, and particularly for each I that is violated in D (i.e., $D \cup \{I\}$ is inconsistent), $D' \cup \{violated \leftarrow \sim I\}$ is not inconsistent. Thus, by a mechanical rewriting, the inconsistency of (D, IC) boils down to the derivability of the predicate *violated*. Such a rewriting is described in more detail, e.g., in [22].

Even if an inconsistent (D, IC) is not rewritten to $(D' \cup IC', \{\leftarrow violated\})$, the usual inference rules employed in database query answering (no matter if logic-programming-style backward reasoning or fixpoint-generating forward reasoning procedures are used) do not at all compute arbitrary answers, contrary to what is predicted by EFQ. In fact, database logic is paraconsistent, i.e., inconsistency-tolerant, since the disjunctive thinning inference rule of classical logic is not used in database logic, as observed in [65] [23] [24] [26]. The inconsistency tolerance of our generalizations of integrity checking as well as of repairs and repair checking is evidenced in Section 4.

5.4 Postulates

Standard properties of mathematical measures μ are that they are *positive-definite*, i.e., $\mu(\mathcal{S}) \geq 0$ for each $\mathcal{S} \in \mathbb{S}$ and $\mu(\emptyset) = 0$, *monotone*, i.e., for each $S, S' \in \mathbb{S}$ such that $S \subset S'$, $\mu(S) \leq \mu(S')$ holds, and *additive*, i.e., $\mu(S \cup S') = \mu(S) + \mu(S')$, for disjoint sets $S, S' \in \mathbb{S}$. Similarly, classical inconsistency measures are often postulated to be positive-definite mappings.

As already mentioned in Subsection 3.1, there is a vivid discussion in the literature about several properties that can or should be imposed on classical inconsistency measures. Often, those properties are sculpted to be similar to standard postulates for mathematical measures, such as monotony or additivity. However, for classical inconsistency measures, there is no consensus on their stringency. As observed in [90], "there is no common agreement on the variety of rationality postulates".

In Subsection 3.1, we have postulated positive definiteness also for database inconsistency measures. In principle, however, database inconsistency measures that are not positive-definite are conceivable. For instance, one could imagine assigning different degrees of consistency to different pairs (D, IC) such that IC is satisfied in D, for distinguishing different levels of the propensity of D to become violated, or for distinguishing, among consistent pairs (D, IC), between the satisfiability and the theoremhood of constraints, if the two do not coincide. Also, as argued in [36], it is possible to imagine database inconsistency measures that map tuples (D, IC) with only negligible inconsistency to o. In [40] [41] [29] [31] [32], a property is postulated that is weaker than positive-definite but still guarantees that for each pair of pairs $(D, IC), (D', IC'), \mu(D, IC) < \mu(D', IC')$

holds if (D, IC) is consistent and (D', IC') is not. Anyway, for simplicity, we have chosen to stick exclusively to positive-definite database inconsistency measures in this paper.

Despite a lack of unanimity, frequently encountered postulates in the literature about classical inconsistency measures are monotony and, to a lesser extent, additivity, which is stronger than monotony: for $S \subset S'$ and $S'' = S' \backslash S$, additivity of μ entails $\mu(S) \leq (\mu(S) + \mu(S'')) = \mu(S \cup S'') = \mu(S')$. For database inconsistency measures μ, the monotony property is that $(*)$ holds

$(*)$ $D \subseteq D'$, $IC \subseteq IC'$ $\Rightarrow$ $\mu(D, IC) \leq \mu(D', IC')$

for each pair of pairs (D, IC), (D', IC').

For relational and definite databases with integrity theories consisting of conjunctive denials only, the database inconsistency measures β, ι, $|\iota|$, ς, $|\varsigma|$, κ, $|\kappa|$ are easily seen to be monotone, but they no longer comply with $(*)$ if database negation is allowed in the body of clauses, as shown by Example 10.

Example 10
Let D consist of the rule $r(x) \leftarrow s(x, z)$ and the two facts $q(1, 2)$ and $s(2, 1)$. Further, let $IC = \{\leftarrow q(x, y), \sim r(x)\}$. Clearly, (D, IC) is inconsistent and, for $D' = D \cup \{s(1, 1)\}$ and $IC' = IC$, $(*)$ does not hold for any positive-definite database inconsistency measure, since (D', IC') is consistent. This example also shows that database consistency is not compact, in the sense that each subset pair of a consistent pair of a database state and an integrity theory would also be consistent: (D', IC') is consistent but its subset pair (D, IC) is not. Hence, non-monotonic database negation impedes the use of minimally inconsistent or maximally consistent subsets in a way that classical inconsistency measures are defined, e.g., in [60] [54].

See also the discussion of several other postulates (e.g., a weaker form of additivity) for database inconsistency measures in [36].

5.5 Applications

The applications of mathematical measures are often of probabilistic or statistical kind. The main application of conventional inconsistency measures is to quantify logical inconsistency in sets of logic sentences. As opposed to that, the main purpose of database inconsistency measures is not to quantify the amount of inconsistency in a given database state (although they can be used for that too), but rather to compare the amounts of integrity violation in consecutive database states, i.e., before and after updates, in order to support decisions of whether to accept, reject or amend the changes.

Related to their respective applications just mentioned, there is yet another essential difference between conventional inconsistency measures and database inconsistency measures. Database inconsistency measures do not have to be computed, or at least not for the applications as featured in Section 4. Rather, it suffices to run an integrity checker for gaining evidence of the difference (i.e., decrease or increase or invariance or incomparability) between the measured amounts of inconsistency in consecutive database states, in order to support decisions about accepting or dismissing updates.

To illustrate this again, let us recall Definition 6. It entails that, for determining if an update U of a database D changes the amount of constraint violation in IC, as gauged by some measure μ, it suffices to run a μ-based integrity checker ic. For instance, each ι- or ζ-based ic rejects U if U violates a constraint, or an instance thereof, that had been satisfied before updating D with U. In that case, the obvious conclusion that U does not decrease the set of extant integrity violations, nor does it leave it invariant, does not need a computation of the value of ι or, resp., ζ.

However, there are applications where the values of μ for given input arguments do not have to be computed, also for mathematical measures μ. For instance, to compare and decide who is the taller one of two persons, their measured heights needs not be known; rather, it suffices to put the two side by side. As opposed to that, classical inconsistency measures are meant to be computed, which can be notoriously complex, as shown in [91].

For fairness, it remains to be mentioned that there is at least one application, viz. query answering with integrity [25] [28] [35], for which a database inconsistency measure, viz. κ (described in Subsection 3.2), has to be computed. Roughly, the idea of that application is that an answer to a query in a database that is inconsistent with its constraints has integrity if there is no overlap between the causes that explain the answer and the causes of integrity violation, where the causes are computed by κ. Computing the causes of integrity violation can be done in quiescent states, and hence does not have to cope with real-time requirements that are more likely to occur at update time, i.e., the time when causes of answers are computed. The computation of κ can be done while constructing proof trees of computing answers to queries and constraints processed as queries. Details of that are described in the cited references.

6 Conclusion

We have recapitulated the background, the definition and several applications of database inconsistency measures. Also, their originality and significance has been pointed out by comparing them to conventional mathematical measures and classical inconsistency measures of sets of logical formulas.

In fact, a comparative study of the commonalities, similarities and differences of classical measures and database inconsistency measures had been the original main objective of this article. However, database inconsistency measures have so far been investigated only by a single author, viz. the writer of these lines, while classical measures are dealt with, or at least touched upon, by quite a number of authors, as witnessed by virtually each reference list of any paper on the subject, except the inaugural one by John Grant. Thus, an abstract comparison between classical inconsistency measures and another kind of inconsistency measures which still is a maverick and a pet subject of the author, is not likely to attract much interest. Hopefully, a more promising way to stir up some curiosity in database inconsistency measures is what constitutes the actual structure of the paper: we have first illustrated the use and utility of database inconsistency measures for inconsistency-tolerant generalizations of several well-established applications of database change management, and have postponed the comparison with classical measures toward the end.

Classical inconsistency measures and database inconsistency measures have in common an innovative feature: their constructive, so-to-speak positivistic handling of inconsistency, which is uninhibited by the no-go connotations of inconsistency and its potentially devastating effects in classical logic theories [96]. That confident posture is not original of inconsistency measuring, but the latter has helped much to pave the way for the acceptance of inconsistency tolerance and paraconsistency in database theory and artificial intelligence [7] [24]. It is hoped that it may have a similar effect in the development of (as of yet still depauperate) formal foundations of data quality theory. In that field, a lot of metrics and measures are discussed in the literature, although most of them do not even come close to the formal solidity of inconsistency measures as addressed in the papers of this collection.

Apart from a foundation for measures of data quality, future work is due also for a formal study of the complexity of repair checking by integrity checking, as well as for a study of how to obtain repairs from the output of cause-based measures as addressed in Subsection 3.2. Moreover, the following hitherto neglected technical issue waits to be addressed. So far, the modification of an attribute value for eliminating faulty information in a stored database fact A is dealt with by a transaction of the form $\{delete\ A,\ insert\ A'\}$, where A' is the modification of A obtained by replacing a bad attribute value by a good one. That is unsatisfactory, as shown in the following example.

Example 11 *Let D be a relational database that contains the fact $p(a, a)$, and let $\leftarrow p(x, x)$ be the only constraint concerning p in IC. Then, the update $U = \{delete\ p(a,a),\ insert\ p(a,b)\}$ is an inconsistency reduction but not a repair of (D, IC), since the subset $U' = \{delete\ p(a,a)\}$ of U is a repair that eliminates the same integrity violation.*

Given that, in Example 11, the intention to repair (D, IC) by fixing its wrong attribute value is served by U but not by U', updates such as U should be captured as legitimate repairs. Policies of repairing by modifying attribute updates have been studied in [95] [75] [74]. We are taking these papers as an orientation for upcoming research on extending repair checking by integrity checking, to cover also attribute-based repairs.

Appendix: Index of Definitions

The list below is an index of definitions in this paper, with brief descriptions of the respective subjects. Each definition is identified by its number and, in brackets, the number of the section or subsection where it appear in the text.

Definition 1 (2.1.2): *undo*.
The inverse map of update U, denoted by $\overline{U}$.

Definition 2a (2.1): *global variable*.
Universally quantified logic variable, not in the scope of any existential quantifier.

Definition 2b (2.1): *case*.
Instantiation of some integrity constraint, obtained by a substitution of its global variables.

Definition 2c (2.1): *basic case*.
Instantiation of some integrity constraint, obtained by a ground substitution of its global variables.

Definition 3 (2.2): *update checker*
Function that checks some properties of updates.

Definition 4 (3.1): *database inconsistency measure*.
Function that quantifies the amount of integrity violation in databases.

Definition 5 (3.1): *positive definite measure*.
Normal database inconsistency measure. Examples ι, $|\iota|$, ζ, $|\zeta|$, β, κ, $|\kappa|$, υ of such measures are defined in Subsection 3.2.

Definition 6 (4.1): *inconsistency-tolerant measure-based integrity checker*.
Function that checks if updates preserve integrity.

Definition 7 (4.2): *total* and *inconsistency-tolerant repair*.
Update that reduces inconsistency completely or, resp., partially.

Definition 8 (4.3.1): *repair checker*.
Function that checks if updates are repairs.

Definition 9 (4.3.2): *inconsistency reduction checkers*.
Function that checks if updates reduce inconsistency.

Definition 10 (4.3.2): *minimality checker*
Function that checks if updates that reduce inconsistency are minimal.

Acknowledgements
A preliminary extended abstract of this paper has been published in [43]. With regard to measure-based repair checking, this paper also draws from [38]. John Grant had provided valuable comment on early drafts, and also the reviewers of the predecessor version of this paper have helped a lot to improve the presentation of its main ideas. Maria Vanina Martinez has provided very much appreciated feedback for polishing the camera-ready version.

References

[1] S. Abiteboul, R. Hull, and V. Vianu. *Foundations of Databases*. Addison-Wesley, 1995.

[2] C. Alchourron, A. Gardenfors, and D. Makinson. On the Logic of Theory Change: Partial Meet Contraction and Revision Functions. *J. Symbolic Logic*, 50(2):510–521, 1985.

[3] M. Arenas, L. Bertossi, and J. Chomicki. Consistent Query Answers in Inconsistent Databases. In *Proceedings of PODS*, pages 68–79. ACM Press, 1999.

[4] H. Bauer. *Maß- und Integrationstheorie*. De Gruyter, 2nd edition, 1992.

[5] M. Ben-Ari. *Mathematical Logic for Computer Science*. Springer, 3rd edition, 2012.

[6] L. Bertossi. Consistent Query Answering in Databases. *SIGMOD Record*, 35(2):6876, 2006.

[7] L. Bertossi, A. Hunter, and T. Schaub. *Inconsistency Tolerance*, volume 3300 of *LNCS*. Springer, 2005.

[8] P. Besnard. Revisiting Postulates for Inconsistency Measures. In *Proc. 14th JELIA*, volume 8761 of *LNAI*, pages 383–396. Springer, 2014.

[9] P. Besnard. Forgetting-Based Inconsistency Measure. In *Proc. 10th Scalable Uncertainty Management (SUM 2016)*, volume 9858 of *LNAI*, pages 331–337. Springer, 2016.

[10] P. Besnard. Basic Postulates for Inconsistency Measures. *TLDKS-XXXIV*, 10620:to appear, 2017.

[11] F. Bry, H. Decker, and R. Manthey. A Uniform Approach to Constraint Satisfaction and Constraint Satisfiability in Deductive Databases. In *Advances in Database Technology - EDBT'88*, volume 303 of *LNCS*, pages 488–505. Springer, 1988.

[12] L. Cavedon. Acyclic Logic Programs and the Completeness of SLDNF-Resolution. *Theor. Comput. Sci.*, 86(1):81–92, 1991.

[13] M. Celma, C. Garcia, L. Mota, and H. Decker. Comparing and Synthesizing Integrity Checking Methods for Deductive Databases. In *Proc. 10th ICDE*, pages 214–222. IEEE Computer Society, 1994.

[14] S. Ceri, R. Cochrane, and J. Widom. Practical Applications of Triggers and Constraints: Success and Lingering Issues (10-Year Award). In *Proc. 26th VLDB*, pages 254–262. Morgan Kaufmann, 2000.

[15] H. Christiansen and D. Martinenghi. Incremental Integrity Checking: Limitations and Possibilities. In *Proceedings of LPAR 2005*, volume 3835 of *LNCS*, pages 712–727. Springer, 2005.

[16] K. Clark. Negation as Failure. In *Logic and Data Bases*, pages 293–322. Plenum Press, 1978.

[17] E. Codd. A Relational Model of Data for Large Shared Data Banks. *CACM*, 13(6):377–387, 1970.

[18] N. da Costa. On the Theory of Inconsistent Formal Systems. *Notre Dame J. Formal Logic*, 15(4):497–510, 1974.

[19] S. K. Das and H. Williams. Integrity Checking Methods in Deductive Databases: A Comparative Evaluation. In *Proc. 7th BNCOD*, British National Conference on Databases, pages 85–116. CUP, 1989.

[20] G. de Bona and M. Finger. Notes on Measuring Inconsistency in Probabilistic Logic. Technical report, University of Sao Paulo, 2014.

[21] H. Decker. Integrity Enforcement on Deductive Databases. In *Expert Database Systems*, pages 381–395. Benjamin Cummings, 1987.

[22] H. Decker. The Range Form of Databases and Queries or: How to Avoid Floundering. In *Proc. 5th ÖGAI*, volume 208 of *Informatik-Fachberichte*, pages 114–123. Springer, 1989.

[23] H. Decker. Historical and Computational Aspects of Paraconsistency in View of the Logic Foundation of Databases. In *Semantics in Databases*, volume 2582 of *LNCS*, pages 63–81. Springer, 2003.

[24] H. Decker. A Case for Paraconsistent Logic as a Foundation of Future Information Systems. In *17th CAiSE Workshops*, volume 2, pages 451–461. FEUP Ediçoes, 2005.

[25] H. Decker. Basic Causes for the Inconsistency Tolerance of Query Answering and Integrity Checking. In *Proc. 21st DEXA Workshops*, pages 318–322. IEEE CSP, 2010.

[26] H. Decker. How to Confine Inconsistency or, Wittgenstein only Scratched the Surface. In *Proceedings 8th ECAP*, pages 70–75. Tech. Univ. Munich, 2010.

[27] H. Decker. Toward a Uniform Cause-based Approach to Inconsistency-tolerant Database Semantics. In *Proc. 9th ODBASE, OTM Conferences, Part II*, volume 6427 of *LNCS*, pages 983–998. Springer, 2010.

[28] H. Decker. Answers that Have Integrity. In *Proc. 4th SDKB*, volume 6834 of *LNCS*, pages 54–72. Springer, 2011.

[29] H. Decker. Causes of the Violation of Integrity Constraints for Supporting the Quality of Databases. In *Proc. 12th ICCSA, Part V*, volume 6786 of *LNCS*, pages 283–292. Springer, 2011.

[30] H. Decker. Consistent Explanations of Answers to Queries in Inconsistent Knowledge Bases. In *Explanation-aware Computing, Proc. 22nd IJCAI Workshop ExaCt*, pages 71–80, 2011.

[31] H. Decker. Inconsistency-Tolerant Integrity Checking Based on Inconsistency Metrics. In *Proc. 16th KES, Part II*, volume 6882 of *LNAI*, pages 548–558. Springer, 2011.

[32] H. Decker. Partial Repairs that Tolerate Inconsistency. In *Proc. 15th ADBIS*, volume 6909 of *LNCS*, pages 389–400. Springer, 2011.

[33] H. Decker. Axiomatizing Inconsistency Metrics for Integrity Maintenance. In *Proc. 16th KES*, pages 1243–1252. IOS Press, 2012.

[34] H. Decker. New Measures for Maintaining the Quality of Databases. In *Proc. 13th ICCSA, Part IV*, volume 7336 of *LNCS*, pages 170–185. Springer, 2012.

[35] H. Decker. Answers that Have Quality. In *Proc. 14th ICCSA, Part II*, volume 7972 of *LNCS*, pages 543–558. Springer, 2013.

[36] H. Decker. Measure-Based Inconsistency-Tolerant Maintenance of Database Integrity. In *Proc. 5th SDKB*, volume 7693 of *LNCS*, pages 149–173. Springer, 2013.

[37] H. Decker. Database Inconsistency Measuring. submitted, 2017.

[38] H. Decker. Inconsistency-tolerant Database Repairs and Simplified Repair Checking by Measure-based Integrity Checking. In *Consistency and Inconsistency in Data-centric Applications*, volume 10620 of *LNCS*, pages 153–183. Springer, 2017.

[39] H. Decker and D. Martinenghi. A Relaxed Approach to Integrity and Inconsistency in Databases. In *Proc. 13th LPAR*, volume 4246 of *LNCS*, pages 287–301. Springer, 2006.

[40] H. Decker and D. Martinenghi. Classifying Integrity Checking Methods with regard to Inconsistency Tolerance. In *Proceedings of the 10th international ACM SIGPLAN conference on Principles and Practice of Declarative Programming*, pages 195–204. ACM Press, 2008.

[41] H. Decker and D. Martinenghi. Modeling, Measuring and Monitoring the Quality of Information. In *Proc. 28th ER Workshops*, volume 5833 of *LNCS*, pages 212–221. Springer, 2009.

[42] H. Decker and D. Martinenghi. Inconsistency-tolerant Integrity Checking. *IEEE Transactions of Knowledge and Data Engineering*, 23(2):218–234, 2011.

[43] H. Decker and S. Misra. Database Inconsistency Measures and their Applications. In *Proceedings ICIST'17*, pages 101–112, 2017.

[44] H. Decker, F. Muñoz, and S. Misra. Data Consistency: Toward a Terminological Clarification. In *Proc. 16th ICCSA, Part V*, volume 9159 of *LNCS*, pages 206–220. Springer, 2015.

[45] H. Decker, E. Teniente, and T. Urpí. How to Tackle Schema Validation by View Updating. In *Proceedings of 5th EDBT*, volume 1057 of *LNCS*, pages 535–549. Springer, 1996.

[46] J. H. Doorn and L. C. Rivero. *Database Integrity: Challenges and Solutions*. Idea Group Publishing, 2002.

[47] D. Dubois and H. Prade. Possibilistic Logic: A Retrospective and Prospective View. *Fuzzy Sets and Systems*, 144(1):3–23, 2004.

[48] H. Gallaire, J. Minker, and J.-M. Nicolas. Logic and Databases: A Deductive Approach. *ACM Comput. Surv.*, 16(2):153–185, 1984.

[49] M. Gaye, O. Sall, M. Bousso, and M. Lo. Measuring Inconsistencies Propagation from Change Operation Based on Ontology Partitioning. In *Proceedings 11th SITIS*, pages 178–184, 2015.

[50] M. Gelfond and V. Lifschitz. The Stable Model Semantics for Logic Programming. In *Proceedings 5th ICLP*, pages 1070–1080. MIT Press, 1988.

[51] J. Grant. Classifications for Inconsistent Theories. *Notre Dame Journal of Formal Logic*, 19(3):435–444, 1978.

[52] J. Grant and A. Hunter. Measuring Inconsistency in Knowledgebases. *Journal of Intelligent Information Systems*, 27(2):159–184, 2006.

[53] J. Grant and A. Hunter. Analysing Inconsistent First-order Knowledgebases. Technical report, University College London, 2007.

[54] J. Grant and A. Hunter. Measuring the Good and the Bad in Inconsistent Information. In *Proc. 22nd IJCAI*, pages 2632–2637. IJCAI-AAAI, 2011.

[55] A. Gupta, Y. Sagiv, J. D. Ullman, and J. Widom. Constraint Checking with Partial Information. In *Proceedings of PODS 1994*, pages 45–55. ACM Press, 1994.

[56] D. J. Hand. Statistics and the Theory of Measurement. *J. R. Statist. Soc.*, 159(3):445–492, 1996.

[57] A. Hunter. Measuring Inconsistency in Knowledge via Quasi-Classical Models. In *Proc. 18th AAAI & 14th IAAI*, pages 68–73. MIT Press, 2002.

[58] A. Hunter. Evaluating Significance of Inconsistencies. In *Proceedings of the Workshop on Formal Bases for Data Bases, 18th IJCAI*, pages 468–478, 2003.

[59] A. Hunter and S. Konieczny. Approaches to Measuring Inconsistent Information. In *Inconsistency Tolerance*, volume 3300 of *LNCS*, pages 191–236. Springer, 2005.

[60] A. Hunter and S. Konieczny. Measuring Inconsistency through Minimal Inconsistent Sets. In *Proceedings 11th Conf. on Principles of Knowledge Representation and Reasoning*, pages 358–366. AAAI, 2008.

[61] A. Hunter and S. Konieczny. On the Measure of Conflicts: Shapley Inconsistency Values. *Artificial Intelligence*, 174(14):1007–1026, 2010.

[62] S. Jabbour, Y. Ma, B. Raddaoui, and L. Sais. Quantifying Conflicts in Propositional Logic through Prime Implicates. *J. Approximate Reasoning*, 2017.

[63] K. Knight. Measuring Inconsistency. *J. Philosophical Logic*, 31:77–98, 2002.

[64] S. Konieczny, J. Lang, and P. Marquis. Quantifying Information and Contradiction in Propositional Logic through Epistemic Tests. In *Proc. 18th IJCAI*, pages 106–111. Morgan Kaufmann, 2003.

[65] R. A. Kowalski. *Logic for Problem Solving, Revisited*. Computer Science Essentials. Books on Demand, 2014. Edited by Thom Frühwirth, originally published in 1979.

[66] J. Lang and P. Marquis. Reasoning under Inconsistency: A Forgetting-based Approach. *Artificial Intelligence*, 174(12-13):799–823, 2010.

[67] T. W. Ling and S. Y. Lee. A Survey of Integrity Constraint Checking Methods in Relational Databases. In *Database Systems for Next-Generation Applications*, volume 1 of *Advanced Database Research and Development Series*, pages 68–78. World Scientific, 1993.

[68] J. W. LLoyd. *Foundations of Logic Programming, 2nd edition*. Springer, 1987.

[69] J. W. Lloyd, L. Sonenberg, and R. W. Topor. Integrity Constraint Checking in Stratified Databases. *J. Logic Programming*, 4(4):331–343, 1987.

[70] E. Lozinskii. Information and Evidence in Logic Systems. *JETAI*, 6(2):163–193, 1994.

[71] Y. Ma and P. Hitzler. Distance-based Measures of Inconsistency and Incoherency for Description Logics. In *Proc. 23rd DL*, volume 573 of *CEUR Workshop*, pages 475–485, 2010.

[72] Y. Ma, G. Qi, and P. Hitzler. Computing Inconsistency Measure Based on Paraconsistent Semantics. *J. Logic and Computation*, 21(6):1257–1281, 2011.

[73] D. Martinenghi, H. Christiansen, and H. Decker. Integrity Checking and Maintenance in Relational and Deductive Databases and Beyond. In *Intelligent Databases: Technologies and Applications*, pages 238–285. IGI Global, 2007.

[74] M. V. Martinez, F. Parisi, A. Pugliese, G. I. Simari, and V. S. Subrahmanian. Policy-based Inconsistency Management in Relational Databases. *International Journal of Approximate Reasoning*, 55(2):501–528, 2014.

[75] M. V. Martinez, A. Pugliese, G. I. Simari, V. S. Subrahmanian, and H. Prade. How Dirty is Your Relational Database? An Axiomatic Approach. In *Proc. 9th ESQARU*, volume 4724 of *Lecture Notes in Computer Science*, pages 103–114. Springer, 2007.

[76] K. McAreavey, W. Liu, and P. Miller. Computational Approaches to Finding and Measuring Inconsistency in Arbitrary Knowledge Bases. *J. Approximate Reasoning*, 55:16591693, 2014.

[77] A. R. N. Brisaboa, M. Luaces. An Inconsistency Measure of Spatial Data Sets with respect to Topological Constraints. *International Journal of Geographical Information Science*, 28(1):56–82, 2014.

[78] J.-M. Nicolas. A Property of Logical Formulas Corresponding to Integrity Constraints on Data Base Relations. In *Proceedings of the Workshop on Formal Bases for Data Bases*, 1979.

[79] J.-M. Nicolas. Logic for Improving Integrity Checking in Relational Data Bases. *Acta Informatica*, 18:227–253, 1982.

[80] C. Ordonez and J. García-García. Referential Integrity Quality Metrics. *Decision Support Systems*, 44(2):495–508, 2008.

[81] K. Popper. *Objective Knowledge: an Evolutionary Approach*. Oxford University Press, revised edition, 1979.

[82] G. Priest. Paraconsistent Logic. In *Handbook of Philosophical Logic*, volume 6. Kluwer, 2002.

[83] G. Qi, P. Haase, S. Schenk, S. Stadtmller, and P. Hitzler. Inconsistency-tolerant Reasoning with Networked Ontologies. Technical report, University of Karlsruhe, 2008. Deliverable D1.2.4 of NeOn project.

[84] G. Qi and A. Hunter. Measuring Incoherence in Description Logic-based Ontologies. In *Proceedings 6th ISWC*, pages 381–394, 2007.

[85] R. Ramakrishnan and J. Gehrke. *Database Management Systems*. McGraw-Hill, 2003.

[86] K. Ramamritham and P. Chrysanthis. A Taxonomy of Correctness Criteria in Database Applicationsss. *VLDB Journal*, 5:85–97, 1996.

[87] R. Reiter. What Should a Database Know? *J. Logic Programming*, 14(1&2):127–153, 1992.

[88] F. Sadri and R. Kowalski. A Theorem-Proving Approach to Database Integrity. In *Foundations of Deductive Databases and Logic Programming*, pages 313–362. Morgan Kaufmann, 1988.

[89] C. Shannon. A Mathematical Theory of Communication. *Bell System Tech. Journal*, 27(3):379–423, 1948.

[90] M. Thimm. On the Compliance of Rationality Postulates for Inconsistency Measures: A more or less Complete Picture. *KI*, 31(1):31–39, 2017.

[91] M. Thimm and J. Wallner. Some Complexity Results on Inconsistency Measurement. In *Proc. 15th Knowledge Representation*, pages 114–123. AAAI, 2016.

[92] R. Topor. Safety and Domain Independence. In *Encyclopedia of Database Systems*, pages 2463–2466. Springer, 2009.

[93] M. Ulbricht, M. Thimm, and G. Brewka. Measuring Inconsistency in Answer Set Programs. In *Proceedings 15th JELIA*, pages 577–583, 2016.

[94] A. Van Gelder, K. Ross, and J. Schlipf. The Well-Founded Semantics for General Logic Programs. *J. ACM*, 38(3):620–650, 1991.

[95] J. Wijsen. Database Repairing Using Updates. *Transaction on Database Systems*, 30(3):722–768, 2005.

[96] Wikipedia. Principle of Explosion. https://en.wikipedia.org/wiki/Principle_of_explosion. Downloaded January 4, 2018.

On Measuring Inconsistency in Spatio-Temporal Databases

John Grant[1], Maria Vanina Martinez[2],
Cristian Molinaro[3], Francesco Parisi[3]

[1]University of Maryland, College Park, USA
grant@cs.umd.edu,
[2]Universidad Nacional del Sur in Bahia Blanca, Argentina
mvm@cs.uns.edu.ar,
[3]DIMES Department, Università della Calabria, Italy
{cmolinaro, fparisi}@dimes.unical.it

Abstract

The problem of managing spatio-temporal data arises in many applications, such as location-based services, environmental monitoring, and geographic information systems. Often spatio-temporal data arising from such applications turn out to be inconsistent, i.e., representing an impossible situation in the real world. Several inconsistency measures have been proposed to quantify in a principled way inconsistency in classical knowledge bases. We start by showing how some of these measures carry over to spatio-temporal databases. Then, we define and investigate new measures that are particularly suitable for dealing with inconsistent spatio-temporal information because they explicitly take into account the spatial and temporal dimensions as well as the dimension concerning the identifiers of the monitored objects.

1 Introduction

Recent years have seen a great deal of interest in tracking moving objects. For this reason, researchers have investigated in detail the representation and processing of spatio-temporal databases [22, 27, 17, 2, 24, 14].

In this paper, we focus on databases representing atomic statements of the form "object id is/was/will be inside region r at time t", denoted as atoms of the form (id, r, t). This allows the representation of information concerning moving objects in several application domains. For instance, a cell phone provider is interested in knowing which cell phones will be in the range of some towers at a given time [5]. A transportation company is interested in predicting the vehicles that will be on a given road at a given time in order to avoid congestion [16]. A retailer is interested in knowing the positions of the shoppers moving in a shopping mall in order to offer suitable customized coupons on discounts [18]. These are just a few examples of a common issue.

Most previous work on spatial-temporal databases assumes that a *consistent* version of the database is available before processing queries or performing updates. However, in real life scenarios, this assumption is rather simplistic in knowledge bases in general, but especially for those where the information changes dynamically and in a distributed manner. Inconsistency in those settings can appear for various reasons, such as sensing errors that can affect the estimation and generation of spatio-temporal information. For instance, in smart video surveillance [25], an identification system may fail to assign the same identifier to the same object monitored by cameras having different views [4, 31], thus generating wrong object identifiers that do not correctly model the monitored situation. Likewise, different objects may be assigned the same identifier after exiting and re-entering the monitored space [6], yielding an inconsistent representation of the situation being monitored. Other errors can be due to imprecise estimation of regions occupied by objects, and inaccurate time-stamps associated with sensor readings, because of the inherent imprecision of the systems originating spatio-temporal data [23].

Moreover, as data are often collected from heterogeneous sources, such as on-board GPS devices and roadside sensors [30], the same monitored situation (an object being in a given place at a given time) can result in different sets of spatio-temporal data corresponding to the different sources. Thus, data integrated from different sources are likely to generate inconsistent spatio-temporal databases, where an inconsistency is due to the presence of data representing conflicting information on the monitored scenario. For instance, an on-board GPS device could detect the car in which it is installed quite far away from the place where it actually is at the given time (*e.g.*, on a route parallel to that being used), while a roadside sensor could detect (the licence plate number of) that car on a different street at the same time. The information provided by these two sources, which are independent of each other, may be inconsistent since they entail the presence of the same car in different places at the same time. For that matter, each of the two sources may provide wrong information: for instance, the detection region of the GPS device may be inaccurate and/or the licence plate number detected by the roadside sensor may be wrong.

Considering the specific aspects and dimensions that spatio-temporal information contains, as indicated in the examples, we observe that defining, identifying, measuring, and deciding how to deal with such data appear to be more complex than dealing with theories given in propositional logic. In this chapter we study the use of measures of inconsistency in spatio-temporal databases. Inconsistency measures are mathematical functions that provide a way to determine the quantity of inconsistency in a knowledge base. A measure of inconsistency equal to zero implies that no inconsistency is present, while the larger this number, the "dirtier" the information is.

The plan of this chapter is as follows. In Section 2 we introduce the syntax and semantics of spatio-temporal (ST) databases. We also give examples to illustrate these concepts. In Section 3 we show how to transform an ST database into an equivalent propositional knowledge base. We use this transformation in Section 4 to show how classical inconsistency measures apply to ST databases. Then, in Section 5 we propose several inconsistency measures that take into account special features of spatio-temporal databases. We conclude the paper in Section 6.

2 Spatio-Temporal Databases

This section introduces the syntax and semantics of Spatio-Temporal (ST) databases.

2.1 Syntax

Assume the existence of 3 finite sets: *ID* is the set of object ids, *T* is the set of integer time values, and *Space* is the set of point locations; below we explain in further detail what these sets represent. We assume that an object can be in only one location at a time, but a single location may contain more than one object.

Definition 1 (ST atom/database). *An* ST *atom is a tuple* (id, r, t)*, where* $id \in ID$ *is an object id,* $\emptyset \subset r \subseteq Space$ *is a nonempty region, and* $t \in T$ *is a time value. An* ST *database is a finite set of* ST *atoms.*

Intuitively, the ST atom (id, r, t) says that object id is/was/will be inside region r at time t. Hence, ST atoms can represent information about the past and the present (such as from techniques for interpreting RFID readings [9, 10]), but also information about the future, such as that derived from methods for predicting the destination of moving objects [19, 15, 26], or from querying predictive databases [3, 1, 21].

Although in general a region may be any non-empty set of points in *Space*, in our examples for convenience we use a square grid for *Space* within which a region is a rectangle. In the grid, each location can be written as (x, y), where x and y are integers and $0 \leq x, y \leq N$ for some integer N. Thus *Space* contains $(N+1)^2$ points. A rectangular region can be specified by 2 points, (x_1, y_1) and (x_2, y_2), where each value $x_1, y_1, x_2,$ and y_2 is an integer, $0 \leq x_1 \leq x_2 \leq N$, and $0 \leq y_1 \leq y_2 \leq N$. Then (x_1, y_1) is the lower left endpoint and (x_2, y_2) is the upper right endpoint of the region. Sometimes it will be useful to refer

Id	Area	Time
id_1	d	1
id_1	b	3
id_1	c	3
id_2	b	1
id_2	e	2
id_3	e	1

(a)

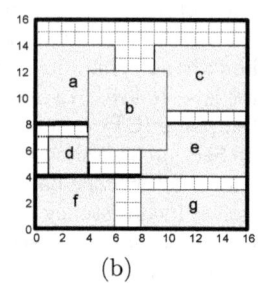

(b)

$I(id_1, 1) = (3, 6)$
$I(id_1, 2) = (7, 5)$
$I(id_1, 3) = (10, 10)$
$I(id_2, 1) = (5, 7)$
$I(id_2, 2) = (12, 6)$
$I(id_2, 3) = (8, 7)$
$I(id_3, 1) = (9, 6)$
$I(id_3, 2) = (5, 5)$
$I(id_3, 3) = (5, 6)$

(c)

Figure 1: (a) ST database $\mathcal{S}_{lab}$; (b) Areas of the lab; (c) Interpretation I.

to the size of a region by which we mean the number of points with integer coordinates in the region. We write $|r|$ for the size of region r. Other (more complex) representations of spatial regions and time are clearly possible, though in this work we focus on this simpler model for the clarity of the presentation. Clearly, depending on the application domain there might be a requirement for more complex representations: this may include the modeling of vectors for spatial-temporal information. These knowledge representation issues are outside the scope of this chapter and left to future investigation.

Example 1. *Consider a laboratory where data coming from biometric sensors are collected and analyzed. Biometric data such as faces, voices, and fingerprints recognized by sensors are matched against given profiles (such as those of people having access to the lab) and tuples like the ones shown in Figure 1(a) are obtained. Every tuple consists of the profile* id *resulting from the matching phase, the area of the lab where the sensor recognizing the profile is operating, and the time at which the profile has been recognized. For instance, the tuple in the first row of the table in Figure 1(a), representing the* ST *atom* $(id_1, d, 1)$, *says that the profile having id* id_1 *was in region d at time 1. We use* $\mathcal{S}_{lab}$ *to denote the* ST *database consisting of the* ST *atoms represented by the tuples in Figure 1(a). In Figure 1(b), the plan of the lab and the areas covered by sensors are shown. In area d a fingerprint sensor is located. After fingerprint authentication,* id_1 *was recognized at time 3 in areas b and c. In some cases, such as for* id_1 *at time 2,* $\mathcal{S}_{lab}$ *contains no information. In this example* $ID = \{id_1, id_2, id_3\}$ *and* $T = \{1, 2, 3\}$.

Given an ST database $\mathcal{S}$, an object id, and a time t, we use the notation $\mathcal{S}_{id,t}$ to refer to the set $\mathcal{S}_{id,t} = \{(id', r', t') \in \mathcal{S} \mid id' = id \wedge t' = t\}$, that is, the set of ST atoms in $\mathcal{S}$ that refer to the specific object identifier id and time t.

2.2 Semantics

The meaning of an ST database is given by the interpretations that satisfy it.

Definition 2 (ST interpretation). *An* ST *interpretation I is a function, $I : ID \times T \to Space$.*

An interpretation specifies a trajectory for each $id \in ID$. That is, for each $id \in ID$, I says where in *Space* object id was/is/will be at each time $t \in T$. In particular, this means that an object can be in only one location at a time.

Example 2. *An interpretation I for the* ST *database S_{lab} of Example 1 is shown in Figure 1(c).*

We now define satisfaction and ST models.

Definition 3 (Satisfaction and ST model). *Let $a = (id, r, t)$ be an* ST *atom and let I be an* ST *interpretation. We say that I satisfies a (denoted $I \models a$) iff $I(id, t) \in r$. I satisfies an* ST *database S (denoted $I \models S$) iff $\forall a \in S$, $I \models a$. If I satisfies* ST *atom a (resp.* ST *database S), we say that I is a model for a (resp. S).*

Given two ST atoms a and a', we write $a \models a'$ if every model of a is a model of a'.

Example 3. *In our running example, interpretation I is a model for the* ST *atom $(id_1, d, 1)$ as id_1 at time value 1 is assigned to the spatial point $(3, 6)$, which is a point in area d. Reasoning analogously, it is easy to see that I is a model for all of the atoms in Figure 1(a). Hence, I is a model for S_{lab}.*

Example 4. *Let I_1 be the interpretation that is equal to I except that $I_1(id_1, 1) = (12, 12)$. It is easy to check that I_1 is not a model for the* ST *atom $(id_1, d, 1)$, because object id_1 is not assigned to a point in region d at time 1. Since $(id_1, d, 1)$ is in S_{lab} (see the first row of the table in Figure 1(a)), then I_1 is not a model for S_{lab}.*

We use $\mathbf{M}(S)$ to denote the set of all models for an ST database S. In the following, we will use the symbol M to refer to interpretations that are models, that is, elements in $\mathbf{M}(S)$.

Definition 4 (Consistency). *An* ST *database S is consistent iff $\mathbf{M}(S) \neq \emptyset$.*

Example 5. *The model of Example 3 shows that S_{lab} is consistent. Consider now, the* ST *database S' in Figure 2 where a simpler scenario of the lab fingerprint authentication is shown. In this case, $ID = \{id_1, id_2\}$ and $T = \{1, 3\}$. Note that there is no interpretation for S' since it is impossible to define a function that assigns a spatial point to id_1 at time values 1 and 3, and can satisfy at the same time all the* ST *atoms in which it is involved (there is no overlap between areas a and c, and b and c, respectively).*

ID	Area	Time
id_1	a	1
id_1	b	3
id_1	c	1
id_2	b	1
id_1	c	3
id_2	a	1

(a)

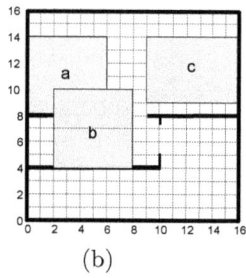

(b)

Figure 2: (a) ST database $\mathcal{S}'$; (b) Areas of the lab.

There is an important connection between the consistency of $\mathcal{S}$ and its subsets $\mathcal{S}_{id,t}$.

Proposition 1. *An ST database $\mathcal{S}$ is consistent iff for each $id \in ID$ and $t \in T$, $\mathcal{S}_{id,t}$ is consistent.*

Proof. If $\mathcal{S}$ is consistent then it has a model which must also be a model of each $\mathcal{S}_{id,t}$ for $id \in ID$ and $t \in T$. Hence each $\mathcal{S}_{id,t}$ is consistent. Conversely, suppose that each $\mathcal{S}_{id,t}$ for $id \in ID$ and $t \in T$ is consistent. In this case there must be a model $M^{id,t}$ for each $\mathcal{S}_{id,t}$. Let M be such that $M(id,t) = M^{id,t}(id,t)$. Then M is a model for $\mathcal{S}$ and hence $\mathcal{S}$ is consistent. □

We call each minimal inconsistent subset of $\mathcal{S}$ an *inconsistency*. Hence, the ST database $\mathcal{S}'$ of Example 5 has 2 inconsistencies: $\{(id_1, a, 1), (id_1, c, 1)\}$ and $\{(id_1, b, 3), (id_1, c, 3)\}$.

3 Transforming an ST Database to a Propositional Knowledge Base

Several inconsistency measures have been proposed in the literature for measuring inconsistency in propositional knowledge bases. Therefore, the first natural approach to measuring inconsistency in ST databases is to translate them into semantically equivalent classical propositional knowledge bases and then apply known measures. We assume the reader is familiar with propositional logic.

Given an ST database $\mathcal{S}$, we define an equivalent propositional knowledge base $K_\mathcal{S}$ as follows. Let U be the set of all propositions $x_{id,p,t}$ such that $id \in ID$, $p \in Space$, and $t \in T$. What we call "proposition" is often called "atom" in propositional logic. We avoid this terminology to avoid confusion with ST atoms. Each $x_{id,p,t}$ corresponds to the tuple $(id, \{p\}, t)$. The intuitive meaning of $x_{id,p,t}$ is that it is assigned true if object id is at point p at time t.

Definition 5 (Transformation). *For a given* **ST** *database* $\mathcal{S}$, $K_\mathcal{S}$ *consists of the following (3 groups of) formulas:*

1) *For each* **ST** *atom* $(id, r, t) \in \mathcal{S}$, *the formula* $f_{id,r,t} = \bigvee\limits_{p \in r} x_{id,p,t}$ *belongs to* $K_\mathcal{S}$. *Thus,* $f_{id,r,t}$ *states that at least one variable among those corresponding to the pair id, t, and to the points in the region r must be true. That is, id must be in region r at time t.*

2) *For each* $id \in ID$ *and* $t \in T$, *the formula* $f_{id,t}^{Space} = \bigvee\limits_{p \in Space} x_{id,p,t}$ *belongs to* $K_\mathcal{S}$. *Thus,* $f_{id,t}^{Space}$ *states that, for the fixed pair id, t, at least one of the variables* $x_{id,p,t}$ *must be true where p ranges over Space. So the set* $\{f_{id,t}^{Space} \mid id \in ID \text{ and } t \in T\}$ *states that each object must be in Space for each time value.*

3) *For each* $id \in ID$, $t \in T$, *the formula*
$f_{id,t} = \bigwedge\limits_{p_i, p_j \in Space \text{ and } p_i \neq p_j} (\neg x_{id,p_i,t} \vee \neg x_{id,p_j,t})$
belongs to $K_\mathcal{S}$. *Thus for fixed id and t,* $f_{id,t}$ *states that at most one of the variables in the set* $\{x_{id,p,t} \mid p \in Space\}$ *can be true. So the set* $\{f_{id,t} \mid id \in ID \text{ and } t \in T\}$ *states that each object can be at no more than one point in Space for each time value.*

Example 6. *Consider* $\mathcal{S}'$ *from Example 5. The formula encoding the atom* $(id_1, c, 1)$ *is* $\bigvee\limits_{p \in c} x_{id_1, p, 1}$ *which is a disjunction of* $|c| = 8 \times 6 = 48$ *propositions. Recalling that* $|ID| = 2$ *and* $|T| = 2$, *and there are 6* **ST** *atoms in* $\mathcal{S}'$, $K_{\mathcal{S}'}$ *has 6 formulas of type 1),* $2 \times 2 = 4$ *formulas of type 2), and* $2 \times 2 = 4$ *formulas of type 3).*

In the usual way, an interpretation for the set of propositions U is a truth assignment for each proposition. We identify an interpretation with the set of propositions assigned true. A model for a propositional knowledge base K is an interpretation for which all the formulas of K are true.

Let $\mathbf{M}(K_\mathcal{S})$ denote the set of all models of $K_\mathcal{S}$. The following theorem states that there is a one-to-one correspondence between the set of models of $K_\mathcal{S}$ and that of $\mathcal{S}$.

Theorem 1. *For each* **ST** *database* $\mathcal{S}$, *there is a bijection* $b : \mathbf{M}(\mathcal{S}) \to \mathbf{M}(K_\mathcal{S})$.

Proof. Given a model M for $\mathcal{S}$, we define the interpretation τ for the propositions in U such that $\tau(x_{id,p,t})$ is *true* iff $M(id, t) = p$. We show that $\{x_{id,p,t} \mid \tau(x_{id,p,t}) = true\}$ is a model for $K_\mathcal{S}$ that corresponds in a one-to-one way to M.

($\Rightarrow$) Since M is a function assigning to each id, t pair exactly one point $p \in Space$, it follows that, for each id, t, exactly one proposition of the form $x_{id,p,t}$

with $p \in \textit{Space}$ is assigned *true* by τ and thus the formulas of the forms $f_{id,t}^{Space}$ and $f_{id,t}$ are satisfied by τ. Moreover, M being a model means that $M(id,t) = p \in r$ for each ST atom $a = (id, r, t) \in \mathcal{S}$, which entails that the proposition $x_{id,p,t}$ assigned true by τ is that corresponding to a point p in region r of the ST atom of the form $a = (id, r, t)$. Thus, all formulas of the form $f_{id,r,t}$ are also satisfied by τ, which then defines a model for $K_{\mathcal{S}}$.

($\Leftarrow$) Since τ satisfies all the formulas in $K_{\mathcal{S}}$, i) for each id, t pair, exactly one proposition of the form $x_{id,p,t}$ with $p \in \textit{Space}$ is assigned *true* by τ, and ii) for each $f_{id,r,t}$, a proposition $x_{id,p,t}$ corresponding to a point $p \in r$ is assigned true. Thus, M is a model for $\mathcal{S}$. $\square$

The following corollary follows.

Corollary 1. *For each* ST *database* $\mathcal{S}$, $\mathcal{S}$ *is consistent iff* $K_{\mathcal{S}}$ *is consistent.*

As our interest is in measuring inconsistency, consider how $K_{\mathcal{S}}$ can be inconsistent. No type 2) formulas can conflict with any other formulas; hence they make no contribution to an inconsistency. Also, there cannot be an inconsistency of a type 1) formula $f_{id,r,t}$ with any type 1) formula of the form $f_{id',r',t'}$ where either $id' \neq id$ or $t' \neq t$. Similarly, $f_{id,r,t}$ cannot have an inconsistency with any type 3) formula of the form $f_{id',t'}$ where either $id' \neq id$ or $t' \neq t$. This shows that an inconsistency must be caused by a type 3) formula $f_{id,t}$ with some type 1) formulas $f_{id,r_1,t}, \ldots, f_{id,r_j,t}$ with the same id, t pair. These type 1) formulas in $K_{\mathcal{S}}$ stand exactly for the ST atoms $(id, r_1, t), \ldots, (id, r_j, t)$ of $\mathcal{S}$ for which there is an inconsistency in case $\cap_{k=1}^{j} r_k = \emptyset$. The reason is that in this case $\mathcal{S}$ states that the object id at time t must be at a point, in the intersection of the r_k regions, that does not exist. An inconsistent set of formulas is *minimal* if no proper subset is inconsistent. Minimal inconsistent subsets contain only formulas of types 1) and 3).

When we perform a transformation from $\mathcal{S}$ to $K_{\mathcal{S}}$ we create a propositional knowledge base that is much larger than the ST database. It contains $|ID| \times |\textit{Space}| \times |T|$ propositions in $|\mathcal{S}| + 2 \times |ID| \times |T|$ formulas. So whenever possible we will deal with $\mathcal{S}$ directly, recognizing that a set of j ST atoms of $\mathcal{S}$ that we are interested in, $(id, r_1, t), \ldots, (id, r_j, t)$ (all with the same id, t pair) corresponds in $K_{\mathcal{S}}$ to the set of $j + 1$ formulas $f_{id,r_1,t}, \ldots, f_{id,r_j,t}, f_{id,t}$ for the purpose of inconsistency.

The following definition defines a particularly useful class of ST databases for which the calculation of inconsistency measures is much easier than for the general case.

Definition 6. *An* ST *database is* singular *if for each id, t pair there is at most one inconsistency (minimal inconsistent subset).*

Example 7. *The* ST *databases of Example 1 and Example 5 are both singular. The former has no inconsistencies. The latter has 2 inconsistencies, one for the pair $id_1, 1$ and one for the pair $id_1, 3$.*

4 Classical Inconsistency Measures for ST Databases

In this section we define the concept of an inconsistency measure and show how six such measures can be applied to ST databases using the transformation from Section 3.

4.1 The Definition of Inconsistency Measure

An inconsistency measure is a function that assigns a nonnegative real value to every knowledge base. Inconsistency measures can be classified in various ways and may satisfy certain properties. One distinction is between absolute measures that measure the total amount of inconsistency and relative measures that use a ratio to determine how big a portion of the knowledgebase is inconsistent. We will deal in this paper only with absolute inconsistency measures for which the following definition is appropriate.

Definition 7 (Inconsistency measure). *Let $\mathcal{L}$ be a propositional language. An inconsistency measure $\mathcal{I} : 2^{\mathcal{L}} \to \mathcal{R}^{\geq 0}$ is a function such that the following two properties hold for any $K \in 2^{\mathcal{L}}$,*

1. *(Consistency) $\mathcal{I}(K) = 0$ iff K is consistent.*
2. *(Monotony) If $K \subseteq K'$, then $\mathcal{I}(K) \leq \mathcal{I}(K')$.*

These two properties ensure that all and only consistent knowledge bases get a measure of 0 and that the measure is monotonic for subsets. Our method in this section is to consider various classical inconsistency measures $\mathcal{I}_x$ for propositional knowledge bases and define the inconsistency measures $\mathcal{I}_x$ for spatio-temporal databases as $\mathcal{I}_x(\mathcal{S}) = \mathcal{I}_x(K_\mathcal{S})$. However, because of the equivalences between sets of atoms of $\mathcal{S}$ and sets of formulas of $K_\mathcal{S}$ for inconsistency, in many cases we can deal directly with $\mathcal{S}$.

4.2 Counting Minimal Inconsistent Subsets

Given a propositional language $\mathcal{L}$ and $K \subseteq \mathcal{L}$ a propositional knowledge base, a minimal inconsistent subset of K is any set $MIS \subseteq K$ such that MIS is inconsistent and there is no $MIS' \subsetneq MIS$ that is inconsistent. We write $\mathcal{MI}(K)$ for the set of all minimal inconsistent subsets of K. Analogously, we can define minimal inconsistent subsets for an ST database $\mathcal{S}$ and use the notation $\mathcal{MI}(\mathcal{S})$ to denote the set of all minimal inconsistent subsets of $\mathcal{S}$.

Example 8. *Consider the database $\mathcal{S}'$ of Example 5. In this case there are 2 minimal inconsistent subsets. Depending on whether we are considering $\mathcal{S}'$ or*

$K_{\mathcal{S}'}$ we obtain:

$$\mathcal{MI}(\mathcal{S}') = \{\{(id_1, a, 1), (id_1, c, 1)\}, \{(id_1, b, 3), (id_1, c, 3)\}\} \text{ and}$$
$$\mathcal{MI}(K_{\mathcal{S}'}) = \{\{f_{id_1,a,1}, f_{id_1,c,1}, f_{id_1,1}\}, \{f_{id_1,b,3}, f_{id_1,c,3}, f_{id_1,3}\}\}.$$

The inconsistency measure

$$\mathcal{I}_{MIS}(K) = |\mathcal{MI}(K)|$$

counts the number of minimal inconsistent subsets. Note how a minimal inconsistent subset of $\mathcal{S}$ of size n corresponds to a minimal inconsistent subset of $K_{\mathcal{S}}$ of size $n+1$ because of the additional formula of type 3), namely $f_{id,t}$.

4.3 Using a Hitting Set

For the next measure we form in a minimal way a set of formulas that contains at least one formula from each minimal inconsistent subset. Formally, we write

$$\mathcal{I}_H(K) = min\{|X| \text{ s.t. } X \subseteq K \text{ and } \forall MIS \in \mathcal{MI}(K), X \cap MIS \neq \emptyset\}.$$

For a singular spatio-temporal database there is an important property of minimal inconsistent subsets that we will use next.

Proposition 2. *For a singular ST database $\mathcal{S}$, no formula can be a member of two different minimal inconsistent subsets of $\mathcal{S}$ (or $K_{\mathcal{S}}$).*

Proof. Singularity means that for every id, t pair there is at most one minimal inconsistent subset. So the id, t pair must be different for each minimal inconsistent subset. □

We next show how that $\mathcal{I}_{MIS}$ and $\mathcal{I}_H$ are equivalent for singular ST databases.

Proposition 3. *For a singular ST database $\mathcal{S}$, $\mathcal{I}_H(\mathcal{S}) = \mathcal{I}_{MIS}(\mathcal{S})$.*

Proof. By Proposition 2 we must choose one formula from each minimal inconsistent subset of $\mathcal{S}$ to obtain a hitting set. □

Example 9. *Continuing with our example we obtain $\mathcal{I}_H(\mathcal{S}') = \mathcal{I}_{MIS}(\mathcal{S}') = 2$.*

4.4 Using the Sizes of Inconsistent Subsets

Consider the following 4 sets of formulas in propositional logic: $F_1 = \{p \wedge \neg p\}$, $F_2 = \{p_1 \vee p_2, \neg p_1 \wedge \neg p_2\}$, $F_3 = \{p_1, p_2, \neg p_1 \vee \neg p_2\}$, and $F_4 = \{p_1, p_2, p_3, \neg p_1 \vee \neg p_2 \vee \neg p_3\}$. We chose the sets so that $|F_i| = i$ for $1 \leq i \leq 4$. Each set is minimally inconsistent. Intuition suggests that F_4 is somewhat less inconsistent than F_3, F_3 is somewhat less inconsistent than F_2, and F_2 is somewhat less inconsistent than F_1. That is, when a minimal inconsistent set contains more formulas than another minimal inconsistent set, the former is intuitively less inconsistent than the latter. This concept can be captured by assigning a fraction to the inconsistency of a minimal inconsistent set using its size in the denominator.

$$\mathcal{I}_Q(K) = \begin{cases} 0, & \text{if } K \text{ is consistent} \\ \sum_{X \in \mathcal{MI}(K)} \frac{1}{|X|}, & \text{otherwise} \end{cases}$$

It is easy to see that for each positive integer n there are minimal inconsistent sets of propositional logic formulas of size n. We next show that this result, except for $n = 1$, also holds for **ST** databases where a region is any set of points.

Proposition 4. *For every integer $n \geq 2$ there is an **ST** database such that $\mathcal{S}$ has a minimal inconsistent subset of size n.*

Proof. For a given integer $n \geq 2$ we show how to construct n **ST** atoms that are minimally inconsistent. Let $|Space| \geq n$. Choose a region R such that $|R| = n$. Let $\mathcal{R}^{n-1} = \{r \subset R \text{ s.t. } |r| = n-1\}$ and $\mathcal{S} = \{(id, r, t) \mid r \in \mathcal{R}^{n-1}\}$ for a fixed id, t pair. Then $|\mathcal{S}| = |\mathcal{R}^{n-1}| = n$ and $\bigcap_{r \in \mathcal{R}^{n-1}} r = \emptyset$. Also, for every point $p \in R$, p must belong to $n-1$ (all except one) regions in $\mathcal{R}^{n-1}$. Hence the intersection of any $n-1$ regions is not empty. Therefore $\{(id, r, t) \mid r \in \mathcal{R}^{n-1}\}$ is a minimally inconsistent set of size n. □

So a minimal inconsistent subset of $\mathcal{S}$ can have any size at least 2 and correspondingly a minimal inconsistent subset of $K_\mathcal{S}$ can have any size at least 3 because of the additional formula $f_{id,t}$ (assuming that *Space* is large enough). However, this is not the case when the regions are rectangles, as we show next.

Proposition 5. *When all the regions are rectangles, every minimal inconsistent subset of every **ST** database $\mathcal{S}$ has size 2.*

Proof. We have previously explained that the size of a minimal inconsistent subset of $\mathcal{S}$ must be at least 2. Hence we need to show that it cannot be greater than 2. It suffices to show that it cannot be 3. This would require 3 rectangles R_1, R_2, and R_3 such that each pair has a nonempty intersection, but $R_1 \cap R_2 \cap R_3 = \emptyset$. We now show why this is impossible. We write $R_i = ((x_i^\ell, y_i^\ell), (x_i^u, y_i^u))$ for $i = 1, 2, 3$. We start with R_1 and R_2 as given and $R_1 \cap R_2 \neq \emptyset$. By symmetry the number of distinct cases can be reduced to 3.

1. $R_2 \subseteq R_1$.
   In this case if $p \in R_2 \cap R_3$ then $p \in R_1 \cap R_2 \cap R_3$.

2. $R_2 \not\subseteq R_1$, $x_1^\ell \leq x_2^\ell \leq x_1^u \leq x_2^u$, and $y_1^\ell \leq y_2^\ell \leq y_2^u \leq y_1^u$.
   As $R_2 \cap R_3 \neq \emptyset$, w.l.o.g. we can consider the case where $y_3^\ell \geq y_2^\ell$. Then $R_3 \cap R_1 \neq \emptyset$ implies that $x_3^\ell \leq x_1^u$. Hence $(x_1^u, y_3^\ell) \in R_1 \cap R_2 \cap R_3$.

3. $R_2 \not\subseteq R_1$, $x_1^\ell \leq x_2^\ell \leq x_1^u \leq x_2^u$, and $y_1^\ell \leq y_2^\ell \leq y_1^u \leq y_2^u$.
As $R_3 \cap R_1 \neq \emptyset$, we must have $x_3^\ell \leq x_1^u$ and $y_3^\ell \leq y_1^u$. Now $R_3 \cap R_2 \neq \emptyset$ implies $y_3^u \geq y_2^\ell$. Hence $(x_3^\ell, y_3^u) \in R_1 \cap R_2 \cap R_3$.

□

Corollary 2. *For every* ST *database* $\mathcal{S}$ *where every region is a rectangle,* $I_Q(\mathcal{S}) = \frac{1}{3} \times I_{MIS}(\mathcal{S})$.

Proof. The minimal inconsistent set of 2 ST atoms comprises 3 formulas in the corresponding translation into propositional logic. □

Example 10. *Continuing with our example,* $\mathcal{I}_Q(\mathcal{S}') = \mathcal{I}_Q(K_{\mathcal{S}'}) = \frac{1}{3} + \frac{1}{3} = \frac{2}{3}$.

4.5 Counting the Problematic Formulas

In work on inconsistency measures it is a common practice to divide the formulas of a knowledge base into 2 groups. The formulas that appear in no minimal inconsistent set are called *free* formulas; the formulas that are not free are called *problematic*. Formally, we write

$$Problematic(K) = \bigcup_{MIS \in \mathcal{MI}(K)} MIS$$

and

$$Free(K) = K \setminus Problematic(K).$$

The next inconsistency measure counts the number of problematic formulas:

$$\mathcal{I}_P(K) = |Problematic(K)|.$$

Again we find a simple calculation for singular spatio-temporal databases.

Proposition 6. *Suppose that* $\mathcal{S}$ *is a singular* ST *database,* $\mathcal{I}_{MIS}(\mathcal{S}) = j$, *and the sizes of the minimal inconsistent subsets are* $n_1, n_2, \ldots, n_j$. *Then,* $\mathcal{I}_P(\mathcal{S}) = \sum_{i=1}^j n_i$.

Proof. As in a singular spatio-temporal database no formula appears in more than one minimal inconsistent subset, we can simply add the sizes of all the minimal inconsistent subsets to obtain the number of problematic formulas. □

Example 11. *In our running example there are 2 minimal inconsistent subsets of size 3 each. Hence* $\mathcal{I}_P(\mathcal{S}') = 3 + 3 = 6$.

4.6 Counting the Maximal Consistent Subsets

Another inconsistency measure counts the maximal consistent subsets of the knowledge base. We write $MC(K)$ for this set. In order for this to be a consistency measure for knowledge bases it is defined as

$$\mathcal{I}_A(K) = |MC(K)| + |Selfcontradictions(K)| - 1$$

where a self-contradiction is a single formula that is a contradiction, such as $p \wedge \neg p$. The reason we must add the contradictory formulas is that they don't appear in any way in a maximal consistent set. Then the reason that 1 must be subtracted is to obtain $\mathcal{I}_A(K) = 0$ for a consistent K because every consistent knowledge base has a maximal consistent subset, namely K itself.

Let us now consider what $\mathcal{I}_A(\mathcal{S}) = \mathcal{I}_A(K_\mathcal{S})$ might be for a spatio-temporal database $\mathcal{S}$. Recall that each inconsistency is caused by n ST atoms of $\mathcal{S}$ or equivalently n formulas of type 1) and the appropriate formula $f_{id,t}$ of type 3) for $K_\mathcal{S}$. Looking at $\mathcal{S}$, this gives rise to n consistent subsets where each such set contains $n-1$ of the n formulas of the minimal inconsistent subset under consideration, and similarly for $K_\mathcal{S}$ with $n+1$ substituted for n. But these subsets are usually not maximal consistent because they do not include for instance atoms for some other id, t pair that do not cause an inconsistency and hence are in every maximal consistent set. The analogous situation occurs for $K_\mathcal{S}$. The calculation can be complicated in general; however, we can do it for singular spatio-temporal databases.

Proposition 7. *Suppose that $\mathcal{S}$ is a singular spatio-temporal database, $\mathcal{I}_{MIS}(\mathcal{S}) = j$, and the sizes of the j minimal inconsistent subsets are $n_1, n_2, \ldots, n_j$. Then $\mathcal{I}_A(\mathcal{S}) = (\prod_{i=1}^{j} n_i) - 1$.*

Proof. The hypothesis assures that no formula appears in more than one minimal inconsistent subset. Hence for each minimal inconsistent set of size n_i, all n_i subsets of size $n_i - 1$ are consistent. Therefore, each of those can be extended to a maximal consistent subset by combining it with the $n_k - 1$ subsets of all other minimal inconsistent subsets of size n_k as well as all the free formulas. Then $|MC(K_\mathcal{S})| = \prod_{i=1}^{j} n_i$ and hence $\mathcal{I}_A(\mathcal{S}) = (\prod_{i=1}^{j} n_i) - 1$. □

Example 12. *In our running example there are 2 minimal inconsistent subsets of size 3 each (in $K_{\mathcal{S}'}$), hence $\mathcal{I}_A(\mathcal{S}') = 3 \times 3 - 1 = 8$.*

4.7 Using a 3-valued Logic

A different approach to measuring the inconsistency of a knowledge base uses a 3-valued logic with the three values T, F, and B, where B stands for "Both" and signifies an inconsistency. The truth table is given in Figure 3.

α	T	T	T	B	B	B	F	F	F
β	T	B	F	T	B	F	T	B	F
$\alpha \vee \beta$	T	T	T	T	B	B	T	B	F
$\alpha \wedge \beta$	T	B	F	B	B	F	F	F	F
$\neg \alpha$	F	F	F	B	B	B	T	T	T

Figure 3: Truth table for three valued logic (3VL)

A 3-valued *interpretation* for a knowledge base K is a function $i :$ Propositions$(K) \to \{F, B, T\}$, where Propositions(K) denotes the set of all propositions appearing in K. An interpretation i is called a *model* of K if for all formulas $\phi \in K$, $i(\phi) = T$ or $i(\phi) = B$, where $i(\phi)$ is the value assigned to ϕ by i (using the truth table in Figure 3). We write $Conflictbase(i) = \{\alpha \in $ Propositions$(K) \mid i(\alpha) = B\}$. The *contention measure* counts the minimal number of propositions that must be in the $Conflictbase$ in order to get a *model* of K as follows:

$$\mathcal{I}_C(K) = Min\{|Conflictbase(i)| \mid i \text{ is a model of } K\}.$$

Next we show how to compute $\mathcal{I}_C(\mathcal{S})$ for singular spatio-temporal databases.

Theorem 2. *If $\mathcal{S}$ is a singular spatio-temporal database then $\mathcal{I}_C(\mathcal{S}) = \mathcal{I}_{MIS}(\mathcal{S})$.*

Proof. The transformation of $\mathcal{S}$ to $K_\mathcal{S}$ in Section 3 creates $|ID| \times |Space| \times |T|$ propositions. Assume now that the id, t pair is fixed. We show how to construct the restriction of a model i to the propositions of the form $x_{id,p',t}$ for all $p' \in Space$. There are 3 cases.

- There are no formulas of type 1) for the id, t pair.
  This means that there is no inconsistency for that id, t pair. Choose any point $p \in Space$. Let $i(x_{id,p,t}) = T$ and $i(x_{id,p',t}) = F$ for all $p' \in Space \setminus \{p\}$. For this restriction of a model i to the id, t pair there are 0 atoms in $Conflictbase(i)$.

- All formulas of type 1) for the id, t pair are in $Free(K_\mathcal{S})$.
  Again, this means that there is no inconsistency for that id, t pair. Suppose there are j such formulas of type 1) of the form $f_{id,r_k,t}$ for $r_k, 1 \leq k \leq j$. As there is no inconsistency, $\bigcap_{k=1}^{j} r_k \neq \emptyset$. Pick any point $p \in \bigcap_{k=1}^{j} r_k$. Let $i(x_{id,p,t}) = T$ and $i(x_{id,p',t}) = F$ for all $p' \in Space \setminus \{p\}$. As in the previous case we get the restriction of a model i to the id, t pair with 0 atoms in $Conflictbase(i)$.

- A minimal inconsistent set for the id, t pair contains a set of formulas of type 1) and a formula of type 3).

By singularity there can be only one such set for the id, t pair. Suppose that this set has j formulas of type 1): $\{f_{id,r_k,t} \mid 1 \leq k \leq j\}$. As the intersection of every $j-1$ of these j regions is not empty, there must be a point $p_1 \in \bigcap_{k=1}^{j-1} r_k$. Also, choose a point $p_2 \in r_j$. Then $p_2 \notin \bigcap_{k=1}^{j-1} r_k$. For the restriction of a model i for the id, t pair, let $i(x_{id,p_1,t}) = B$, $i(x_{id,p_2,t}) = T$, and $i(x_{id,p',t}) = F$ for all $p' \in \mathit{Space} \setminus \{p_1, p_2\}$. In this case there is 1 atom in $\mathit{Conflictbase}(i)$.

When we combine the interpretations i for each id, t pair we obtain a model that has exactly one atom in $\mathit{Conflictbase}(i)$ for each minimal inconsistent set. This shows that $\mathcal{I}_C(\mathcal{S}) = \mathcal{I}_C(K_\mathcal{S}) = \mathcal{I}_{MIS}(K_\mathcal{S}) = \mathcal{I}_{MIS}(\mathcal{S})$. □

Example 13. *Continuing with our example, we obtain* $\mathcal{I}_C(\mathcal{S}') = \mathcal{I}_{MIS}(\mathcal{S}') = 2$.

5 New Inconsistency Measures for ST Databases

In this section we propose several inconsistency measures that are not possible to define for propositional knowledge bases but are relevant for ST databases. We use the fact that ST databases can be considered along three dimensions: objects, time, and space. This allows us to measure the inconsistency along one or a combination of dimensions. We start by defining several possible properties, usually called postulates, for inconsistency measures in order to determine which of these hold for the new measures.

Then we define the new measures. Separating the dimensions of ST databases requires looking inside the formulas. Consider what such a step means for propositional knowledge bases. The formulas there contain propositions and logical connectives (as well as parentheses). As absolute inconsistency measures typically use only the problematic formulas, a natural way of measuring inconsistency is to count the number of distinct propositions in the problematic formulas. Let us call a proposition p *problematic* if p appears in a problematic formula. Then we can define $\mathcal{I}_P(K) = |\{p \mid p \text{ is a problematic proposition}\}|$. Actually, we did not find this definition in the literature on inconsistency measures. However, it is the absolute version of a relative inconsistency measure studied in [29]. So $\mathcal{I}_P$ is our inspiration for measuring inconsistency along the three dimensions.

5.1 Some Rationality Postulates

In order to make the chapter self-contained we write several properties that some inconsistency measures possess. These are just a few of the properties that have been discussed as desirable for inconsistency measures. See [28] for a thorough study of these properties, usually called rationality postulates.

Each property has some intuitive rationale, even though some properties are not compatible with one another.

In Definition 7 we have already given two such properties that we believe all inconsistency measures must have. In Definition 8, we present a list of six additional properties that have counterparts for classical knowledge bases. As in this section we are considering these properties directly for ST databases, we write the postulates as they apply to ST databases rather than propositional knowledge bases.

Before giving the definitions of the postulates, we introduce some notation and terminology analogous to that of propositional knowledge bases. For an ST database $\mathcal{S}$, the ST atoms that appear in no minimal inconsistent set are called *free* ST atoms; the ST atoms that are not free are called *problematic*. Formally, we write $Problematic(\mathcal{S}) = \bigcup_{MIS \in \mathcal{MI}(\mathcal{S})} MIS$ and $Free(\mathcal{S}) = \mathcal{S} \setminus Problematic(\mathcal{S})$.

Definition 8. *Let $\mathcal{I}$ be an inconsistency measure, and S, S' be two ST databases. We have the following postulates:*

1. (Free-Formula Independence) *If $(id, r, t) \in Free(\mathcal{S})$ then $\mathcal{I}(\mathcal{S}) = \mathcal{I}(\mathcal{S} \setminus \{(id, r, t)\})$.*

2. (Penalty) *If $(id, r, t) \in Problematic(\mathcal{S})$ then $\mathcal{I}(\mathcal{S}) > \mathcal{I}(\mathcal{S} \setminus \{(id, r, t)\})$.*

3. (Dominance) *If (id, r, t) and (id, r', t) are ST atoms such that $r \subseteq r'$ then $\mathcal{I}(\mathcal{S} \cup \{(id, r, t)\}) \geq \mathcal{I}(\mathcal{S} \cup \{(id, r', t)\})$.*

4. (Super-Additivity) *If $\mathcal{S} \cap \mathcal{S}' = \emptyset$ then $\mathcal{I}(\mathcal{S} \cup \mathcal{S}') \geq \mathcal{I}(\mathcal{S}) + \mathcal{I}(\mathcal{S}')$.*

5. (Attenuation) *If $MIS, MIS' \in \mathcal{MI}(\mathcal{S})$ and $|MIS| < |MIS|'$ then $\mathcal{I}(MIS) > \mathcal{I}(MIS')$.*

6. (Equal Conflict) *If $MIS, MIS' \in \mathcal{MI}(\mathcal{S})$ and $|MIS| = |MIS'|$ then $\mathcal{I}(MIS) = \mathcal{I}(MIS')$.*

Free-Formula Independence states that free formulas do not contribute anything to the inconsistency measure. As in this chapter we deal only with absolute inconsistency measures, all of which satisfy Free-Formula Independence, *we will not deal with this property further.* Penalty states that problematic formulas increase the inconsistency.

In the formulation for propositional knowledge bases Dominance states that for two formulas α and β where $\alpha \models \beta$ and α is not a contradiction, adding α to a knowledge base increases its inconsistency at least as much as adding β. For ST databases the only way that an ST atom α can logically imply an ST atom β is if the *id* and *t* values are the same and the region of α is a subset of the region of β. We do not need to check for a contradiction as no ST atom by itself can be contradictory.

Super-additivity deals with two ST databases that do not intersect, in which case the inconsistency of the union is at least as great as the inconsistency of the sum. Attenuation and Equal Conflict refer to minimal inconsistent subsets. The former states that inconsistency is inversely related to size; while the latter requires minimal inconsistent subsets of the same size to have the same inconsistency.

As we investigate these properties for the new inconsistency measures, we will find that in most cases these properties are not satisfied. We do not consider this to mean that the postulates are not good ones or that our inconsistency measures are not good; it simply means that these specific postulates are not appropriate for the new measures. The counterexamples we give provide some insight into the meaning of the postulates and what our inconsistency measures do or do not measure.

5.2 Measuring Inconsistency along the Object and Time Dimensions Individually

In some cases we may just be interested in how many objects or how many time values are involved in an inconsistency. Those are the two cases we deal with below.

5.2.1 Measuring Inconsistency along the Object Dimension

We now define an inconsistency measure based strictly on objects:

$$\mathcal{I}_O(\mathcal{S}) = |\{id \in ID \mid \exists (id, r, t) \in MIS \text{ where } MIS \in \mathcal{MI}(\mathcal{S})\}|.$$

It is clear that $\mathcal{I}_O$ is an inconsistency measure. Let us now consider the other properties that we have defined in the previous subsection.

Theorem 3. *The $\mathcal{I}_O$ inconsistency measure satisfies Dominance and Equal Conflict, but does not satisfy Penalty, Super-Additivity, and Attenuation.*

Proof. We consider the properties individually.

1. Penalty. Let $\mathcal{S}_1 = \{(id, r_1, t_1), (id, r_2, t_1), (id, r_1, t_2), (id, r_2, t_2)\}$ where $t_1 \neq t_2$ and $r_1 \cap r_2 = \emptyset$. All the formulas are problematic because of the two inconsistencies but only one object is involved. Hence $\mathcal{I}_O(\mathcal{S}_1) = 1 = \mathcal{I}_O(\mathcal{S}_1 \setminus \{(id, r_2, t_2)\})$.

2. Dominance. Let $\mathcal{S}$ be an ST database. Notice that either $\mathcal{I}_O(\mathcal{S} \cup \{(id, r', t)\}) = \mathcal{I}_O(\mathcal{S})$ or $\mathcal{I}_O(\mathcal{S} \cup \{(id, r', t)\}) = \mathcal{I}_O(\mathcal{S}) + 1$. In the first case, the result is immediate. In the second case, there must be a set of ST atoms $(id, r_1, t), \ldots, (id, r_j, t)$ in $\mathcal{S}$ such that $(\cap_{i=1}^{j} r_i) \cap r' = \emptyset$. But then $r \subseteq r'$ implies that $(\cap_{i=1}^{j} r_i) \cap r = \emptyset$, and therefore $\mathcal{I}_O(\mathcal{S} \cup \{(id, r, t)\}) = \mathcal{I}_O(\mathcal{S}) + 1$.

3. Super-Additivity. Suppose that both $\mathcal{S}$ and $\mathcal{S}'$ involve inconsistencies for a single common object. Then, $1 = \mathcal{I}_O(\mathcal{S}) = \mathcal{I}_O(\mathcal{S}') = \mathcal{I}_O(\mathcal{S} \cup \mathcal{S}') < \mathcal{I}_O(\mathcal{S}) + \mathcal{I}_O(\mathcal{S}') = 2$.

4. Attenuation. Let
$\mathcal{S}_2 = \{(id, r_1, t_1), (id, r_2, t_1), (id, r_3, t_2), (id, r_4, t_2), (id, r_5, t_2)\}$ where $r_1 \cap r_2 = \emptyset$, $r_3 \cap r_4 \neq \emptyset$, $r_3 \cap r_5 \neq \emptyset$, $r_4 \cap r_5 \neq \emptyset$, and $r_3 \cap r_4 \cap r_5 = \emptyset$.
Then, $\mathcal{MI}(\mathcal{S}_2) = \{MIS, MIS'\}$ where $MIS = \{(id, r_1, t_1), (id, r_2, t_1)\}$ and $MIS' = \{(id, r_3, t_2), (id, r_4, t_2), (id, r_5, t_2)\}$.
Here, $2 = |MIS| < |MIS'| = 3$, but $\mathcal{I}_O(MIS) = 1 = \mathcal{I}_O(MIS')$.

5. Equal Conflict. This follows because any minimal inconsistent subset can deal with only one object, hence if MIS is a minimal inconsistent subset then $\mathcal{I}_O(MIS) = 1$.

$\square$

Example 14. *Consider again our running example. It is easy to see that id_1 is involved in inconsistencies and id_2 is not. Thus, $\mathcal{I}_O(\mathcal{S}') = 1$.*

5.2.2 Measuring Inconsistency along the Time Dimension

Similar to the inconsistency measure along the object dimension, a natural inconsistency measure along the time dimension counts how many time values are involved in an inconsistency.

$$\mathcal{I}_T(\mathcal{S}) = |\{t \in T \mid \exists (id, r, t) \in MIS \text{ where } MIS \in \mathcal{MI}(\mathcal{S})\}|.$$

Theorem 4. *The $\mathcal{I}_T$ inconsistency measure satisfies Dominance and Equal Conflict, but does not satisfy Penalty, Super-Additivity, and Attenuation.*

Proof. We deal with the properties individually.

1. Penalty. Let $\mathcal{S}_3 = \{(id_1, r_1, t), (id_1, r_2, t), (id_2, r_1, t), (id_2, r_2, t)\}$ where $id_1 \neq id_2$ and $r_1 \cap r_2 = \emptyset$. All the formulas are problematic because of the two inconsistencies but only one time point is involved. Hence $\mathcal{I}_T(\mathcal{S}_3) = 1 = \mathcal{I}_T(\mathcal{S}_3 \setminus \{(id_1, r_1, t)\})$.

2. Dominance. The proof for the Dominance of $\mathcal{I}_O$ works here also.

3. Super-Additivity. The proof for the Super-Additivity of $\mathcal{I}_O$ works here after substituting "time" for "object".

4. Attenuation. The proof for the Attenuation of $\mathcal{I}_O$ works here also.

5. Equal Conflict. The proof for the Equal Conflict of I_O works here after substituting "time value" for "object".

Example 15. *Consider again our running example. Here id_1 is involved in inconsistencies at time values 1 and 3, while and id_2 is not involved in any inconsistency. Thus, $\mathcal{I}_T(\mathcal{S}') = 2$.*

5.3 Measuring Inconsistency along the Object and Time Dimensions Together

ST databases treat objects and time in a similar manner and differently from space. That is why $\mathcal{I}_O$ and $\mathcal{I}_T$ satisfy the same properties with similar proofs as we have just seen. So it is natural to combine the object and time dimensions before dealing with the spatial dimension. This can be done by combining the two dimensions individually, that is, computing $\mathcal{I}_O$ and $\mathcal{I}_T$ and then applying some operation(s) to the two numbers to get a result. Instead of doing so we observe that in many cases we are dealing with id, t pairs. For instance, a single inconsistency (minimal inconsistent set) must have the same id and t values. Hence, in this subsection we define an inconsistency measure on id, t pairs.

We start with the appropriate definition.

$$\mathcal{I}_{OT}(\mathcal{S}) = |\{(id, t) \mid \exists (id, r, t) \in MIS \text{ where } MIS \in \mathcal{MI}(\mathcal{S})\}|.$$

It should not be surprising that $\mathcal{I}_{OT}$ satisfies exactly the same properties as $\mathcal{I}_O$ and $\mathcal{I}_T$.

Theorem 5. *The $\mathcal{I}_{OT}$ inconsistency measure satisfies Dominance, and Equal Conflict, but does not satisfy Penalty, Super-Additivity, and Attenuation.*

Proof. We deal with each property individually.

1. Penalty. Let $\mathcal{S}_3 = \{(id, r_1, t), (id, r_2, t), (id, r_3, t)\}$ where $r_1 \cap r_2 = \emptyset$ and $r_2 \cap r_3 = \emptyset$. All the formulas are problematic. Hence $\mathcal{I}_{OT}(\mathcal{S}_3) = 1 = \mathcal{I}_{OT}(\mathcal{S}_3 \setminus \{(id, r_1, t)\})$.

2. Dominance. The proof for the Dominance of $\mathcal{I}_O$ works here also.

3. Super-Additivity. Let $\mathcal{S}_4 = \{(id, r_1, t), (id, r_2, t)\}$ and $\mathcal{S}_5 = \{(id, r_3, t), (id, r_4, t)\}$, where $r_1 \cap r_2 = \emptyset$ and $r_3 \cap r_4 = \emptyset$. Then $\mathcal{I}_{OT}(\mathcal{S}_4) = \mathcal{I}_{OT}(\mathcal{S}_5) = \mathcal{I}_{OT}(\mathcal{S}_4 \cup \mathcal{S}_5) = 1$. Hence, $\mathcal{I}_{OT}(\mathcal{S} \cup \mathcal{S}') < \mathcal{I}_{OT}(\mathcal{S}) + \mathcal{I}_{OT}(\mathcal{S}')$.

4. Attenuation. The proof for the Attenuation of $\mathcal{I}_O$ works here also.

5. Equal Conflict. The proof for the Equal Conflict of $\mathcal{I}_O$ works here also after changing "object value" to "id, t pair".

Example 16. *Consider again the* **ST** *database* $\mathcal{S}'$ *of Example 5.* $\mathcal{I}_{OT}(\mathcal{S}') = 2$ *as there are two object id, time pairs involved in inconsistencies, namely* $(id_1, 1)$ *and* $(id_1, 3)$.

5.4 Measuring Inconsistency along the Space Dimension

In this subsection we measure inconsistency by the amount of space that the inconsistencies involve. We give two different measures. The first one is similar to what we did for objects and time. The second involves distance.

5.4.1 The $\mathcal{I}_S$ Inconsistency Measure

For a given **ST** database $\mathcal{S}$ we define a region $R_\mathcal{S}$ as follows:

$$R_\mathcal{S} = \{p \in Space \mid \exists (id, r, t) \in MIS \text{ where } p \in r \text{ and } MIS \in \mathcal{MI}(\mathcal{S})\}.$$

Then, we define $\mathcal{I}_S(\mathcal{S}) = |R_\mathcal{S}|$.
So $\mathcal{I}_S$ counts the number of points that are in regions that are involved in an inconsistency. Again, $\mathcal{I}_S$ is an inconsistency measure and we now consider the other properties we have defined for it.

Theorem 6. *The $\mathcal{I}_S$ inconsistency measure does not satisfy any of the five other properties (i.e., properties 2-6 of Definition 8) we have considered.*

Proof. We deal with each property individually.

1. Penalty. Let $\mathcal{S}_6 = \{(id_1, \{p_1\}, t), (id_1, \{p_2\}, t), (id_2, \{p_1\}, t), (id_2, \{p_2\}, t)\}$ where $id_1 \neq id_2$ and $p_1 \neq p_2$. All the formulas are problematic because of the two inconsistencies. However, $\mathcal{I}_S(\mathcal{S}_6) = 2 = \mathcal{I}_S(\mathcal{S}_6 \setminus \{(id_1, \{p_1\}, t)\})$.

2. Dominance. Let $\mathcal{S}_7 = \{(id, \{p_1\}, t)\}$, $r = \{p_2\}$, and $r' = \{p_2, p_3\}$, where $p_1, p_2,$ and p_3 are distinct points. Here $2 = \mathcal{I}_S(\mathcal{S}_7 \cup \{(id, r, t)\}) < \mathcal{I}_S(\mathcal{S}_7 \cup \{(id, r', t)\}) = 3$.

3. Super-Additivity. Consider the **ST** databases $\mathcal{S}_8 = \{(id_1, \{p_1\}, t), (id_1, \{p_2\}, t)\}$ and $\mathcal{S}_9 = \{(id_2, \{p_1\}, t), (id_2, \{p_2\}, t)\}$, where $id_1 \neq id_2$ and $p_1 \neq p_2$. Then, $\mathcal{I}_S(\mathcal{S}_8) = \mathcal{I}_S(\mathcal{S}_9) = \mathcal{I}_S(\mathcal{S}_8 \cup \mathcal{S}_9) = 2$. Hence, $\mathcal{I}_S(\mathcal{S}_8 \cup \mathcal{S}_9) < \mathcal{I}_S(\mathcal{S}_8) + \mathcal{I}_S(\mathcal{S}_9)$.

4. Attenuation. Consider the **ST** database $\mathcal{S}_{10} = \{(id, \{p_1\}, t_1), (id, \{p_2\}, t_1), (id, \{p_1, p_2\}, t_2), (id, \{p_2, p_3\}, t_2), (id, \{p_1, p_3\}, t_2)\}$, where $t_1 \neq t_2$, and $p_1, p_2,$ and p_3 are distinct points. Clearly,
$MIS_4 = \{(id, \{p_1\}, t_1), (id, \{p_2\}, t_1)\}$ and
$MIS'_4 = \{(id, \{p_1, p_2\}, t_2), (id, \{p_2, p_3\}, t_2), (id, \{p_1, p_3\}, t_2)\}$ are in $\mathcal{MI}(\mathcal{S}_{10})$, $2 = |MIS_4| < |MIS'_4| = 3$, but $2 = \mathcal{I}_S(MIS_4) < \mathcal{I}_S(MIS'_4) = 3$.

5. Equal Conflict. Let
$S_{11} = \{(id, \{p_1\}, t_1), (id, \{p_2\}, t_1), (id, \{p_1\}, t_2), (id, \{p_2, p_3\}, t_2)\}$ where
$t_1 \neq t_2$ and p_1, p_2, and p_3 are distinct points. Then
$MIS_5 = \{(id, \{p_1\}, t_1), (id, \{p_2\}, t_1)\}$ and
$MIS'_5 = \{(id, \{p_1\}, t_2), (id, \{p_2, p_3\}, t_2)\}$ are in $\mathcal{MI}(S_{11})$, $|MIS_5| = 2 = |MIS'_5|$, but $\mathcal{I}_S(MIS_5) = 2$ and $\mathcal{I}_S(MIS'_5) = 3$.

□

Example 17. *Continuing with our running example,* $\mathcal{I}_S(S') = 49 + 49 + 48 - 15 = 131$, *which is the overall number of points of regions a, b, c, that are the regions appearing in atoms involved in inconsistencies. 15 is subtracted in order not to count the points in* $a \cap b$ *twice.*

5.4.2 The $\mathcal{I}_D$ Inconsistency Measure

For this measure we require a distance function on the points of *Space*: $d(p, p')$ satisfying the usual distance axioms. This allows us to define a function on regions as follows:

$$d(r, r') = min\{d(p, p') \mid p \in r, p' \in r'\}.$$

Next we define distance for minimal inconsistent subsets. Let *MIS* be a minimal inconsistent subset of S: $MIS = \{(id, r_1, t), \ldots, (id, r_j, t)\}$. We start by defining j new regions, one for each i, $1 \leq i \leq j$,

$$R_i = r_1 \cap \ldots \cap r_{i-1} \cap r_{i+1} \cap \ldots \cap r_j$$

(with the appropriate adjustment so that for $i = 1$ there is no r_{i-1} and for $i = j$ there is no r_{i+1}). That is, R_i is the intersection of all the r sets except for r_i. We know from our analysis of minimal inconsistent sets in our framework that for each i, $1 \leq i \leq j$, $R_i \neq \emptyset$, but $\cap_{i=1}^{j} r_i = \emptyset$. We define a distance for each minimal inconsistent set *MIS* as follows:

$$d(MIS) = min\{d(R_i, r_i) \mid 1 \leq i \leq j\}.$$

We can think of $d(MIS)$ as the minimal distance in *Space* that would resolve the inconsistency. Then we define

$$\mathcal{I}_D(S) = \sum_{MIS \in \mathcal{MI}(S)} d(MIS).$$

Thus $\mathcal{I}_D$ sums the minimal distances that resolve each inconsistency. It is an inconsistency measure.

Theorem 7. *The* $\mathcal{I}_D$ *inconsistency measure satisfies Penalty, Dominance, and Super-Additivity, but not Attenuation and Equal Conflict.*

Proof. We deal with each property individually.

1. Penalty. First of all, notice that if $\mathcal{S} \subseteq \mathcal{S}'$, then $\mathcal{MI}(\mathcal{S}) \subseteq \mathcal{MI}(\mathcal{S}')$. Let $\mathcal{S}$ be an **ST** database. If $(id, r, t) \in Problematic(\mathcal{S})$ then there exists $MIS \in \mathcal{MI}(\mathcal{S})$ s.t. $MIS \notin \mathcal{MI}(\mathcal{S} \setminus \{(id, r, t)\})$. Hence, $\mathcal{I}_D(\mathcal{S}) > \mathcal{I}_D(\mathcal{S} \setminus \{(id, r, t)\})$.

2. Dominance. First, suppose that $(id, r', t) \notin Problematic(\mathcal{S} \cup \{(id, r', t)\})$. Then $\mathcal{I}_D(\mathcal{S} \cup \{(id, r', t)\}) = \mathcal{I}_D(\mathcal{S}) \leq \mathcal{I}_D(\mathcal{S} \cup \{(id, r, t)\})$. The other case is where $(id, r', t) \in MIS' \in \mathcal{MI}(\mathcal{S} \cup \{(id, r', t)\})$. This means that there is a set of **ST** atoms $(id, r'_1, t), \ldots, (id, r'_j, t)$ in $\mathcal{S}$ such that writing R'_j for $\cap_{i=1}^{j} r'_i$, $R'_j \neq \emptyset$, but $R'_j \cap r' = \emptyset$. But then $r \subseteq r'$ implies that $R'_j \cap r = \emptyset$ also. Hence $(id, r, t) \in MIS \in \mathcal{MI}(\mathcal{S} \cup \{(id, r, t)\})$. Consider how $d(MIS)$ is calculated as the distance between 2 points, say p and p'. One of those points must be in r, say p. Let $p' \in R'_j$ such that $d(MIS) = d(p, p')$. By $r \subseteq r'$, $p \in r'$, and there may be points in r' closer to R'_j than p. Hence $d(MIS') \leq d(MIS)$. This is true for every MIS' such that $(id, r', t) \in MIS'$. Therefore $\mathcal{I}_D(\mathcal{S} \cup \{(id, r, t)\}) \geq \mathcal{I}_D(\mathcal{S} \cup \{(id, r', t)\})$.

3. Super-Additivity. $\mathcal{S} \cap \mathcal{S}' = \emptyset$ implies that $\mathcal{MI}(\mathcal{S}) \cap \mathcal{MI}(\mathcal{S}') = \emptyset$. Then $\mathcal{MI}(\mathcal{S}) \cup \mathcal{MI}(\mathcal{S}') \subseteq \mathcal{MI}(\mathcal{S} \cup \mathcal{S}')$. So $\mathcal{I}_D(\mathcal{S} \cup \mathcal{S}') = \sum_{MIS \in \mathcal{MI}(\mathcal{S} \cup \mathcal{S}')} d(MIS) \geq \mathcal{I}_D(\mathcal{S}) + \mathcal{I}_D(\mathcal{S}') = \sum_{MIS \in \mathcal{MI}(\mathcal{S})} d(MIS) + \sum_{MIS \in \mathcal{MI}(\mathcal{S}')} d(MIS)$.

4. Attenuation. Consider the **ST** database
$\mathcal{S}_{12} = \{(id, \{p_1\}, t_1), (id, \{p_2\}, t_1), (id, \{p_3, p_4\}, t_2), (id, \{p_4, p_5\}, t_2), (id, \{p_3, p_5\}, t_2)\}$, where $t_1 \neq t_2$, all the points p_1, p_2, p_3, p_4, p_5 are distinct, $d(p_1, p_2) = 1$, $d(p_3, p_4) = 2$, $d(p_4, p_5) = 2$, and $d(p_3, p_5) = 2$. Then $MIS = \{(id, \{p_1\}, t_1), (id, \{p_2\}, t_1)\}$ and $MIS' = \{(id, \{p_3, p_4\}, t_2), (id, \{p_4, p_5\}, t_2), (id, \{p_3, p_5\}, t_2)\}$ are in $\mathcal{MI}(\mathcal{S})$, $2 = |MIS| < |MIS'| = 3$, but $1 = \mathcal{I}_D(MIS) < \mathcal{I}_D(MIS') = 2$.

5. Equal Conflict. Let $\mathcal{S}_{13} = \{(id, \{p_1\}, t), (id, \{p_2\}, t), (id, \{p_3\}, t)\}$, where p_1, p_2, p_3 are all distinct, $d(p_1, p_2) = 1$, and $d(p_1, p_3) = 2$. Then $MIS = \{(id, \{p_1\}, t), (id, \{p_2\}, t)\}$ and $MIS' = \{(id, \{p_1\}, t), (id, \{p_3\}, t)\}$ are in $\mathcal{MI}(\mathcal{S})$, $|MIS| = 2 = |MIS'|$, but $\mathcal{I}_D(MIS) = 1$ and $\mathcal{I}_D(MIS') = 2$.

□

Example 18. *Consider again the* **ST** *database* $\mathcal{S}'$ *of Example 5. There are two minimal inconsistent subsets, namely* $MIS = \{(id_1, a, 1), (id_1, c, 1)\}$ *and* $MIS' = \{(id_1, b, 3), (id_1, c, 3)\}$. *As* $d(MIS) = 3$ *and* $d(MIS') = 1$, *we get* $\mathcal{I}_D(\mathcal{S}') = 4$.

5.5 Measuring Inconsistency by Combining the Three Dimensions

When we started measuring inconsistency along the three dimensions individually we took our inspiration from counting the number of propositions in the problematic formulas of a propositional knowledge base. But for combining dimensions we must start from scratch as there is no analogy from propositional knowledge bases.

We now present three inconsistency measures that combine the three dimensions. In each case we use the $\mathcal{I}_S$ inconsistency measure for space. The first inconsistency measure, $\mathcal{I}_{OST1}$, deals separately with each dimension; the second, $\mathcal{I}_{OST2}$, combines objects and time with space; while the third, $\mathcal{I}_{OST3}$, combines the three dimensions as a unit.

5.5.1 Measuring Inconsistency by Combining the Three Dimensions Separately

We start by defining $\mathcal{I}_{OST1}$:

$$\mathcal{I}_{OST1}(\mathcal{S}) = a \times \mathcal{I}_O(\mathcal{S}) + b \times \mathcal{I}_T(\mathcal{S}) + c \times \mathcal{I}_S(\mathcal{S})$$

where a, b, and c are positive real numbers. Thus $\mathcal{I}_{OST1}$ is not a single inconsistency measure but depends on the three parameters a, b, and c, that allow for the assignment of weights to the amount of inconsistency along the object, time, and space dimensions. Our results hold for all assignments.

Theorem 8. *The $\mathcal{I}_{OST1}$ inconsistency measure does not satisfy any of the properties 2–6 of Definition 8.*

Proof. We deal with each property individually.

1. Penalty. Let
   $\mathcal{S}_{14} = \{(id_1, \{p_1\}, t_1), (id_1, \{p_2\}, t_1), (id_1, \{p_1\}, t_2), (id_1, \{p_2\}, t_2),$
   $(id_2, \{p_1\}, t_1), (id_2, \{p_2\}, t_1), (id_2, \{p_1\}, t_2), (id_2, \{p_2\}, t_2)\}.$
   Then $\mathcal{I}_{OST1}(\mathcal{S}_{14}) = 2a + 2b + 2c = \mathcal{I}_{OST1}(\mathcal{S}_{14} \setminus \{(id_1, \{p_1\}, t_1)\}).$

2. Dominance. We use $\mathcal{S}_7$, r, and r' from the proof of Theorem 6. Then
   $\mathcal{I}_{OST1}(\mathcal{S}_7 \cup \{(id, r, t)\}) = a + b + 2c < \mathcal{I}_{OST1}(\mathcal{S}_7 \cup \{(id, r', t)\}) = a + b + 3c.$

3. Super-Additivity. Consider $\mathcal{S}_8$ and $\mathcal{S}_9$ from the proof of Theorem 6. Then
   $\mathcal{I}_{OST1}(\mathcal{S}_8) = a + b + 2c = \mathcal{I}_{OST1}(\mathcal{S}_9)$ and $\mathcal{I}_{OST1}(\mathcal{S}_8 \cup \mathcal{S}_9) = 2a + b + 2c < \mathcal{I}_{OST1}(\mathcal{S}_8) + \mathcal{I}_{OST1}(\mathcal{S}_9) = 2a + 2b + 4c.$

4. Attenuation. Consider $\mathcal{S}_{10}$, MIS_4, and MIS'_4 from the proof of Theorem 6. Then $2 = |MIS_4| < |MIS'_4| = 3$, but $\mathcal{I}_{OST1}(MIS_4) = a + b + 2c < \mathcal{I}_{OST1}(MIS'_4) = a + b + 3c.$

5. Equal Conflict. Consider S_{11}, MIS_5, and MIS'_5 from the proof of Theorem 6. Then $2 = |MIS_5| < |MIS'_5| = 3$, but $\mathcal{I}_{OST1}(MIS_5) = a+b+2c < \mathcal{I}_{OST1}(MIS'_5) = a+b+3c$.

$\square$

Example 19. *Consider again the* **ST** *database S' of Example 5. Suppose we choose $a = 100$, $b = 100$, and $c = 1$. Then $\mathcal{I}_{OST1}(S') = 100 + 200 + 131 = 431$.*

5.5.2 Measuring Inconsistency by Combining Objects and Time with Space

We start by defining $\mathcal{I}_{OST2}$:

$$\mathcal{I}_{OST2}(S) = c \times \mathcal{I}_S(S) + d \times \mathcal{I}_{OT}(S)$$

where c and d are positive real numbers. Again, c and d are parameters used to assign weights to $\mathcal{I}_S$ and $\mathcal{I}_{OT}$.

Theorem 9. *The $\mathcal{I}_{OST2}$ inconsistency measure does not satisfy any of the five properties.*

Proof. We deal with each property individually.

1. Penalty. Let
$S_{15} = \{(id_1, \{p_1\}, t), (id_1, \{p_2\}, t), (id_1, \{p_3\}, t)$
$(id_2, \{p_1\}, t), (id_2, \{p_2\}, t), (id_2, \{p_3\}, t)\}$.
Then $\mathcal{I}_{OST2}(S_{15}) = 3c + 2d = \mathcal{I}_{OST1}(S_{15} \setminus \{(id_1, \{p_1\}, t)\})$.

2. Dominance. Again, we use S_7, r, and r' from the proof of Theorem 6. Then $\mathcal{I}_{OST2}(S_7 \cup \{(id, r, t)\}) = 2c + d < \mathcal{I}_{OST2}(S_7 \cup \{(id, r', t)\}) = 3c + d$.

3. Super-Additivity. Consider S_8 and S_9 from the proof of Theorem 6. Then $\mathcal{I}_{OST2}(S_8) = 2c + d = \mathcal{I}_{OST2}(S_9)$ and $\mathcal{I}_{OST2}(S_8 \cup S_9) = 2c + 2d < \mathcal{I}_{OST2}(S_8) + \mathcal{I}_{OST2}(S_9) = 4c + 2d$.

4. Attenuation. Consider S_{10}, MIS_4, and MIS'_4 from the proof of Theorem 6. Then $2 = |MIS_4| < |MIS'_4| = 3$, but $\mathcal{I}_{OST2}(MIS_4) = 2c + d < \mathcal{I}_{OST2}(MIS'_4) = 3c + d$.

5. Equal Conflict. Consider S_{11}, MIS_5, and MIS'_5 from the proof of Theorem 6. Then $2 = |MIS_5| < |MIS'_5| = 3$, but $\mathcal{I}_{OST2}(MIS_5) = 2c + d \neq \mathcal{I}_{OST2}(MIS'_5) = 3c + d$.

$\square$

Example 20. *Consider again the* **ST** *database S' of Example 5. Suppose we choose $c = 1$ and $d = 100$. Then $\mathcal{I}_{OST2}(S') = 131 + 200 = 331$.*

5.5.3 Measuring Inconsistency by Combining the Three Dimensions as a Unit

We start with the definition.

$\mathcal{I}_{OST3}(\mathcal{S}) = |\{(id, \{p\}, t) \mid \exists (id, r, t) \in \mathit{MIS} \text{ where } p \in r \text{ and } \mathit{MIS} \in \mathcal{MI}(\mathcal{S})\}|$.

So $\mathcal{I}_{OST3}$ counts the number of **ST** atoms with point regions for which there is an **ST** atom (id, r, t) with the point in r that is in a minimal inconsistent subset.

Theorem 10. *The $\mathcal{I}_{OST3}$ inconsistency measure does not satisfy any of the five properties we have considered.*

Proof. We deal with each property individually.

1. Penalty. Let $\mathcal{S}_{16} = \{(id, \{p_1\}, t), (id, \{p_2\}, t), (id, \{p_1, p_2\}, t), (id, \{p_3\}, t)\}$ where p_1, p_2, p_3 are distinct points. It is easy to see that all the formulas are problematic. However, $\mathcal{I}_{OST3}(\mathcal{S}_{16}) = 3 = \mathcal{I}_{OST3}(\mathcal{S}_{16} \backslash \{(id_1, \{p_1\}, t)\})$.

2. Dominance. Again, we use $\mathcal{S}_7$, r, and r' from the proof of Theorem 6. Then $2 = \mathcal{I}_{OST3}(\mathcal{S}_7 \cup \{(id, r, t)\}) < \mathcal{I}_{OST3}(\mathcal{S}_7 \cup \{(id, r', t)\}) = 3$.

3. Super-Additivity. Consider the **ST** databases
   $\mathcal{S}_{17} = \{(id, \{p_1\}, t), (id, \{p_2\}, t)\}$ and $\mathcal{S}_{18} = \{(id, \{p_1, p_2\}, t), (id, \{p_3\}, t)\}$, where p_1, p_2, p_3 are distinct points. Then, $\mathcal{I}_{OST3}(\mathcal{S}_{17}) = 2$ and $\mathcal{I}_{OST3}(\mathcal{S}_{18}) = \mathcal{I}_{OST3}(\mathcal{S}_{17} \cup \mathcal{S}_{18}) = 3$. Hence, $\mathcal{I}_{OST3}(\mathcal{S}_{17} \cup \mathcal{S}_{18}) = 3 < \mathcal{I}_{OST3}(\mathcal{S}_{17}) + \mathcal{I}_{OST3}(\mathcal{S}_{18}) = 5$.

4. Attenuation. Again we use $\mathcal{S}_{10}, \mathit{MIS}_4,$ and MIS'_4 from the proof of Theorem 6. So $2 = |\mathit{MIS}_4| < |\mathit{MIS}'_4| = 3$, but $2 = \mathcal{I}_{OST3}(\mathit{MIS}_4) < \mathcal{I}_{OST3}(\mathit{MIS}'_4) = 3$.

5. Equal Conflict. Again, we use $\mathcal{S}_{11}, \mathit{MIS}_5,$ and MIS'_5 from the proof of Theorem 6. $|\mathit{MIS}_5| = 2 = |\mathit{MIS}'_5|$, but $\mathcal{I}_{OST3}(\mathit{MIS}_5) = 2 \neq \mathcal{I}_{OST3}(\mathit{MIS}'_5) = 3$.

□

Example 21. *Continuing with our running example,* $\mathcal{I}_{OST3}(\mathcal{S}') = 49 + 48 + 49 + 48 = 194$.

5.6 A Tractable Algorithm to Compute $\mathcal{I}_{OT}$

As shown in [28] calculating inconsistency measures for propositional knowledge bases is intractable. Here we show that calculating the $\mathcal{I}_{OT}$ measure is much easier.

Algorithm $Compute \mathcal{I}_{OT}$ *measure*
Input: $\mathcal{S} = \{(id_{a_i}, r_{b_i}, t_{c_i}) \mid 1 \leq i \leq n\}$, where $|\mathcal{S}| = n$.
Output: $\mathcal{I}_{OT}(\mathcal{S})$

1. **let** $\mathcal{I}_{OT}(\mathcal{S}) = 0$
2. **for** $k = 1$ **to** n
3.   **let** $\mathcal{P}_{a_k, c_k} = \emptyset$
4. **end for**
5. **for** $k = 1$ **to** n
6.   **if** $\mathcal{P}_{a_k, c_k} \neq Space$ **then**
7.   **let** $\mathcal{P}_{a_k, c_k} = \mathcal{P}_{a_k, c_k} \cup (Space \setminus r_{b_k})$
8.   **if** $\mathcal{P}_{a_k, c_k} = Space$ **then**
9.   **let** $\mathcal{I}_{OT}(\mathcal{S}) = \mathcal{I}_{OT}(\mathcal{S}) + 1$
10.   **end if**
11.   **end if**
12. **end for**
13. **return** $\mathcal{I}_{OT}(\mathcal{S})$.

Figure 4: Algorithm to calculate the $\mathcal{I}_{OT}$ measure.

The calculation of $\mathcal{I}_{OT}(\mathcal{S})$ is accomplished by the algorithm in Figure 4. It proceeds as follows. We initialize the inconsistency measure counter to 0 (line 1). For each id_i, t_j pair we use $\mathcal{P}_{i,j}$ to indicate the amount of space *not in the intersection* of all the regions for the id_i, t_j pair. If $\mathcal{P}_{i,j}$ becomes equal to *Space* then the intersection of all the regions for that pair is empty, signifying an inconsistent id, t pair. The first loop (lines 2–4) traverses all the ST atoms in $\mathcal{S}$ and initializes the amount of space not in the intersections of all the regions for all id, t pairs to the empty set. The next loop (lines 5–12) again traverses all the ST atoms and for each such atom it first (line 6) checks to see if the amount of space already not in the intersection of all the regions for the id, t pair of that atom is not all of *Space* — we assume this check can be accomplished in constant time. If it is all of *Space* then we have already counted that id, t pair as inconsistent. Otherwise, let r_{b_k} be the region corresponding to k-th atom in $\mathcal{S}$. We then add (line 7) the complement of the region of that atom to the amount of space not in the intersection of all the regions for that id, t pair. If the result is all of *Space* (line 8) then we have a new inconsistency for that id, t pair and add 1 to our inconsistency measure counter (line 9). At the end (line 13), the inconsistency measure counter, $\mathcal{I}_{OT}(\mathcal{S})$, is set correctly. Assuming that the basic union operation in line 7 can be done in constant time, we have that the running time of the algorithm is $O(|\mathcal{S}|)$; however, in the worst case this operation could take time $O(|Space|)$. Therefore, the general result is that the running time of the algorithm is $O(|\mathcal{S}| * C_\cup)$, where $C_\cup$ is the cost of computing the union of two regions.

We can use this algorithm with a slight modification to obtain $\mathcal{I}_O$ (resp. $\mathcal{I}_T$). The modification is that we also keep a list of objects (resp. time values) that we initialize to the empty set. Then, as we obtain a new inconsistency in the above algorithm, that is, $\mathcal{P}_{a_k,c_k} = Space$, we check to see if the object (resp. time value) is already in the list. We increment the inconsistency measure counter, $\mathcal{I}_O(\mathcal{S})$ (resp. $\mathcal{I}_T(\mathcal{S})$), and insert the object id_{a_k} (resp. the time value t_{c_k}) into the list only if the object (resp. time value) is not already in the list.

6 Conclusion and Future Work

We are aware of a number of works involving inconsistency in spatial databases as well as temporal databases. In a few cases the authors deal with measuring inconsistency, such as [8] where inconsistency measures are introduced for qualitative constraint networks, and [7] where inconsistency measures are proposed to evaluate how dirty a spatial dataset is with respect to a set of topological constraints. However, as far as we know this is the first work on inconsistency measures in spatio-temporal databases. We showed how to transform a spatio-temporal database, given in a special format, as an ST database, to a propositional logic knowledge base. Using this transformation we applied six known inconsistency measures to spatio-temporal databases. We obtained various results for a special class of spatio-temporal databases that have at most one inconsistency for any specific pair of object and time values. Then we used objects, time, and space as dimensions to obtain new inconsistency measures and showed some properties that they have. We illustrated all these measures on an example.

Many interesting issues concerning inconsistency measures in spatio-temporal databases remain to be investigated. We introduced eight entirely new measures using the dimension concept, but additional measures based on ideas other than counting things, such as objects, are possible. Many of our results in Section 4 apply only to a special class, the singular ST databases, and we would like to obtain results that hold in general. We have also not developed any relative inconsistency measures. We would also like to extend our results to the case where an ST database is augmented with probabilities [12, 13] and integrity constraints [11, 20]. Furthermore, we would like to consider practical applications involving GIS databases and data stored using RDF/ontologies.

Acknowledgement We thank the reviewers for their helpful comments.

References

[1] D. Agarwal, D. Chen, L. ji Lin, J. Shanmugasundaram, and E. Vee. Forecasting High-dimensional Data. In *SIGMOD Conference*, pages 1003–1012, 2010.

[2] P. K. Agarwal, L. Arge, and J. Erickson. Indexing Moving Points. *J. Comput. Syst. Sci.*, 66(1):207–243, 2003.

[3] M. Akdere, U. Çetintemel, M. Riondato, E. Upfal, and S. B. Zdonik. The Case for Predictive Database Systems: Opportunities and Challenges. In *CIDR*, pages 167–174, 2011.

[4] L. An, X. Chen, M. Kafai, S. Yang, and B. Bhanu. Improving Person Re-identification by Soft Biometrics based Reranking, booktitle = Proc. of International Conference on Distributed Smart Cameras (ICDSC), pages = 1–6, year = 2013.

[5] M. A. Bayir, M. Demirbas, and N. Eagle. Mobility Profiler: A Framework for Discovering Mobility Profiles of Cell Phone Users. *Pervasive and Mobile Computing*, 6(4):435 – 454, 2010.

[6] A. Bedagkar-Gala and S. K. Shah. A Survey of Approaches and Trends in Person Re-identification. *Image Vision Comput.*, 32(4):270–286, 2014.

[7] N. R. Brisaboa, M. R. Luaces, M. A. Rodríguez, and D. Seco. An Inconsistency Measure of Spatial Data Sets with respect to Topological Constraints. *International Journal of Geographical Information Science*, 28(1):56–82, 2014.

[8] J.-F. Condotta, B. Raddaoui, and Y. Salhi. Quantifying Conflicts for Spatial and Temporal Information. In *Proceedings of KR 2016*, pages 443–452, 2016.

[9] B. Fazzinga, S. Flesca, F. Furfaro, and F. Parisi. Offline Cleaning of RFID Trajectory Data. In *Int. Conf. on Scientific and Statistical Database Management (SSDBM)*, page 5, 2014.

[10] B. Fazzinga, S. Flesca, F. Furfaro, and F. Parisi. Exploiting Integrity Constraints for Cleaning Trajectories of RFID-Monitored Objects. *ACM Trans. Database Syst. (TODS)*, 41(4), 2016.

[11] J. Grant, C. Molinaro, and F. Parisi. Count Queries in Probabilistic Spatio-Temporal Knowledge Bases with Capacity Constraints. In *Proc. of European Conference on Symbolic and Quantitative Approaches to Reasoning with Uncertainty (ECSQARU)*, pages 459–469, 2017.

[12] J. Grant, F. Parisi, A. Parker, and V. S. Subrahmanian. An AGM-style Belief Revision Mechanism for Probabilistic Spatio-temporal Logics. *Artif. Intell.*, 174(1):72–104, 2010.

[13] J. Grant, F. Parisi, and V. S. Subrahmanian. Research in probabilistic spatiotemporal databases: The spot framework. In *Advances in Probabilistic Databases for Uncertain Information Management*, volume 304 of *Studies in Fuzziness and Soft Computing*, pages 1–22. Springer, 2013.

[14] M. Hadjieleftheriou, G. Kollios, V. J. Tsotras, and D. Gunopulos. Efficient Indexing of Spatiotemporal Objects. In *EDBT*, pages 251–268, 2002.

[15] T. Hammel, T. J. Rogers, and B. Yetso. Fusing Live Sensor Data into Situational Multimedia Views. In *Multimedia Information Systems*, pages 145–156, 2003.

[16] A. Karbassi and M. Barth. Vehicle route prediction and time of arrival estimation techniques for improved transportation system management. In *Proceedings of the 2013 IEEE Intelligent Vehicles Symposium*, pages 511–516, 2003.

[17] G. Kollios, D. Gunopulos, and V. J. Tsotras. On Indexing Mobile Objects. In *Int. Symposium on Principles of Database Systems (PODS)*, pages 261–272, 1999.

[18] S. Kurkovsky and K. Harihar. Using Ubiquitous Computing in Interactive Mobile Marketing. *Personal Ubiquitous Comput.*, 10(4):227–240, Mar. 2006.

[19] R. Mittu and R. Ross. Building upon the Coalitions Agent Experiment (COAX) - Integration of Multimedia Information in GCCS-M using IMPACT. In *Multimedia Inf. Syst.*, pages 35–44, 2003.

[20] F. Parisi and J. Grant. Knowledge Representation in Probabilistic Spatio-Temporal Knowledge Bases. *J. Artif. Intell. Res. (JAIR)*, 55:743–798, 2016.

[21] F. Parisi, A. Sliva, and V. S. Subrahmanian. A Temporal Database Forecasting Algebra. *Int. J. of Approximate Reasoning*, 54(7):827–860, 2013.

[22] M. Pelanis, S. Saltenis, and C. S. Jensen. Indexing the Past, Present, and Anticipated Future Positions of Moving Objects. *ACM Trans. Database Syst.*, 31(1):255–298, 2006.

[23] D. Pfoser and C. S. Jensen. Capturing the Uncertainty of Moving-Object Representations. In *Proc. of International Symposium on Advances in Spatial Databases (SSD)*, pages 111–132, 1999.

[24] D. Pfoser, C. S. Jensen, and Y. Theodoridis. Novel Approaches in Query Processing for Moving Object Trajectories. In *VLDB*, pages 395–406, 2000.

[25] M. H. Sedky, M. Moniri, and C. C. Chibelushi. Classification of Smart Video Surveillance Systems for Commercial Applications. In *Proc. of IEEE International Conference on Advanced Video and Signal Based Surveillance (AVSS)*, pages 638–643, 2005.

[26] F. Southey, W. Loh, and D. F. Wilkinson. Inferring Complex Agent Motions from Partial Trajectory Observations. In *IJCAI*, pages 2631–2637, 2007.

[27] Y. Tao, D. Papadias, and J. Sun. The TPR*-tree: An Optimized Spatio-Temporal Access Method for Predictive Queries. In *VLDB*, pages 790–801, 2003.

[28] M. Thimm. On the Evaluation of Inconsistency Measures. In *Measuring Inconsistency in Information*. College Publications, 2018.

[29] G. Xiao and Y. Ma. Inconsistency Measurement Based on Variables in Minimal Unsatisfiable Subsets. In *ECAI'12*, pages 864–869, 2012.

[30] B. Zhang and G. Trajcevski. The Tale of (fusing) Two Uncertainties. In *Proceedings of ACM SIGSPATIAL International Conference on Advances in Geographic Information Systems, 2014*, pages 521–524, 2014.

[31] L. Zhang, D. V. Kalashnikov, S. Mehrotra, and R. Vaisenberg. Context-based Person Identification Framework for Smart Video Surveillance. *Mach. Vis. Appl.*, 25(7):1711–1725, 2014.

www.ingramcontent.com/pod-product-compliance
Lightning Source LLC
Chambersburg PA
CBHW050122170426
43197CB00011B/1683